# Preface

This book contains the solutions of the exercises of my book: Introduction to Differential Geometry of Space Curves and Surfaces. These solutions are sufficiently simplified and detailed for the benefit of readers of all levels particularly those at introductory level.
Taha Sochi
London, December 2018

# Contents

# Nomenclature

In the following table, we define most symbols, notations and abbreviations that we used in the book to provide easy access to the reader.

| | |
|---|---|
| $\nabla$ | nabla differential operator |
| $\nabla^2$ | Laplacian operator |
| $\sim$ | isometric to |
| , subscript | partial derivative with respect to the following index(es) |
| ; subscript | covariant derivative with respect to the following index(es) |
| 1D, 2D, 3D, $n$D | one-dimensional, two-dimensional, three-dimensional, $n$-dimensional |
| overdot (e.g. $\dot{\mathbf{r}}$) | derivative with respect to general parameter $t$ |
| prime (e.g. $\mathbf{r}'$) | derivative with respect to natural parameter $s$ |
| $\delta/\delta t$ | absolute derivative with respect to $t$ |
| $\partial_\alpha$, $\partial_i$ | partial derivative with respect to $\alpha^{th}$ and $i^{th}$ variables |
| $a$ | determinant of surface covariant metric tensor |
| $\mathbf{a}$ | surface covariant metric tensor |
| $a_{11}, a_{12}, a_{21}, a_{22}$ | coefficients of surface covariant metric tensor |
| $a^{11}, a^{12}, a^{21}, a^{22}$ | coefficients of surface contravariant metric tensor |
| $a_{\alpha\beta}$, $a^{\alpha\beta}$, $a_\alpha^\beta$ | surface metric tensor or its components |
| $b$ | determinant of surface covariant curvature tensor |
| $\mathbf{b}$ | surface covariant curvature tensor |
| $\mathbf{B}$ | binormal unit vector of space curve |
| $b_{11}, b_{12}, b_{21}, b_{22}$ | coefficients of surface covariant curvature tensor |
| $b_{\alpha\beta}$, $b^{\alpha\beta}$, $b_\alpha^\beta$ | surface curvature tensor or its components |
| $C$ | curve |
| $\bar{C}_\mathbf{B}$, $\bar{C}_\mathbf{N}$, $\bar{C}_\mathbf{T}$ | spherical indicatrices of curve $C$ |
| $C_e$, $C_i$ | evolute and involute curves |
| $C^n$ | of class $n$ |
| $c_{\alpha\beta}$, $c^{\alpha\beta}$, $c_\alpha^\beta$ | tensor of third fundamental form or its components |
| $\mathbf{d}$ | Darboux vector |
| $\mathbf{d}_1$, $\mathbf{d}_2$ | unit vectors in Darboux frame |
| det | determinant of matrix |
| $ds$ | length of infinitesimal element of curve |
| $ds_\mathbf{B}, ds_\mathbf{N}, ds_\mathbf{T}$ | length of line element in binormal, normal and tangent directions |
| $d\sigma$ | area of infinitesimal element of surface |
| $e, f, g$ | coefficients of second fundamental form |
| $E, F, G$ | coefficients of first fundamental form |
| $\mathcal{E}$, $\mathcal{F}$, $\mathcal{V}$ | number of edges, faces and vertices of polyhedron |
| $\mathbf{E}_i$, $\mathbf{E}^j$ | covariant and contravariant space basis vectors |
| $\mathbf{E}_\alpha$, $\mathbf{E}^\beta$ | covariant and contravariant surface basis vectors |
| Eq./Eqs. | Equation/Equations |

| | |
|---|---|
| $f$ | function |
| $\mathfrak{g}$ | topological genus of closed surface |
| $g_{ij}$, $g^{ij}$ | space metric tensor or its components |
| $H$ | mean curvature |
| $I_S$, $II_S$, $III_S$ | first, second and third fundamental forms |
| $\mathbf{I}_S$, $\mathbf{II}_S$ | tensors of first and second fundamental forms |
| *iff* | if and only if |
| $J$ | Jacobian of transformation between two coordinate systems |
| $\mathbf{J}$ | Jacobian matrix |
| $K$ | Gaussian curvature |
| $K_t$ | surface total curvature |
| $L$ | length of curve |
| $\mathbf{n}$ | normal unit vector to surface |
| $\mathbf{N}$ | principal normal unit vector to curve |
| $P$ | point |
| $r$, $R$ | radius |
| $\mathcal{R}$ | Ricci curvature scalar |
| $\mathbf{r}$ | position vector |
| $\mathbf{r}_\alpha$, $\mathbf{r}_{\alpha\beta}$ | $1^{st}$ and $2^{nd}$ partial derivative of $\mathbf{r}$ with respect to subscripted variables |
| $R_1$, $R_2$ | principal radii of curvature |
| $\mathbb{R}^n$ | $n$-dimensional space (usually Euclidean) |
| $R_{ij}$, $R^i_j$ | Ricci curvature tensor of $1^{st}$ and $2^{nd}$ kind for space |
| $R_{\alpha\beta}$, $R^\alpha_\beta$ | Ricci curvature tensor of $1^{st}$ and $2^{nd}$ kind for surface |
| $R_{ijkl}$ | Riemann-Christoffel curvature tensor of $1^{st}$ kind for space |
| $R_{\alpha\beta\gamma\delta}$ | Riemann-Christoffel curvature tensor of $1^{st}$ kind for surface |
| $R^i_{jkl}$ | Riemann-Christoffel curvature tensor of $2^{nd}$ kind for space |
| $R^\alpha_{\beta\gamma\delta}$ | Riemann-Christoffel curvature tensor of $2^{nd}$ kind for surface |
| $R_\kappa$ | radius of curvature |
| $R_\tau$ | radius of torsion |
| $r, \theta, \phi$ | spherical coordinates of 3D space |
| $s$ | natural parameter of curve representing arc length |
| $S$ | surface |
| $S_T$ | tangent surface of space curve |
| $t$ | general parameter of curve |
| $T$ | function period |
| $\mathbf{T}$ | tangent unit vector of space curve |
| $T_PS$ | tangent space of surface $S$ at point $P$ |
| tr | trace of matrix |
| $\mathbf{u}$ | geodesic normal vector |
| $u^1, u^2$ | surface coordinates |
| $u^\alpha$ | surface coordinate |
| $u, v$ | surface coordinates |
| $x^i$ | space coordinate |

| | |
|---|---|
| $x^i_\alpha$ | surface basis vector in full tensor notation |
| $x, y, z$ | coordinates in 3D space (usually Cartesian) |
| $[ij, k]$ | Christoffel symbol of $1^{st}$ kind for space |
| $[\alpha\beta, \gamma]$ | Christoffel symbol of $1^{st}$ kind for surface |
| $\Gamma^k_{ij}$ | Christoffel symbol of $2^{nd}$ kind for space |
| $\Gamma^\gamma_{\alpha\beta}$ | Christoffel symbol of $2^{nd}$ kind for surface |
| $\delta_{ij}$, $\delta^{ij}$, $\delta^j_i$ | covariant, contravariant and mixed Kronecker delta |
| $\delta^{ij}_{kl}$ | generalized Kronecker delta |
| $\Delta$ | discriminant of quadratic polynomial |
| $\epsilon_{ijk}$, $\epsilon^{ijk}$ | covariant and contravariant relative permutation tensor in 3D space |
| $\underline{\epsilon}_{ijk}$, $\underline{\epsilon}^{ijk}$ | covariant and contravariant absolute permutation tensor in 3D space |
| $\epsilon_{\alpha\beta}$, $\epsilon^{\alpha\beta}$ | covariant and contravariant relative permutation tensor in 2D space |
| $\underline{\epsilon}_{\alpha\beta}$, $\underline{\epsilon}^{\alpha\beta}$ | covariant and contravariant absolute permutation tensor in 2D space |
| $\theta$ | angle or parameter |
| $\theta_s$ | sum of interior angles of polygon |
| $\kappa$ | curvature of curve |
| $\kappa_1$, $\kappa_2$ | principal curvatures of surface at a given point |
| $\kappa_{\mathbf{B}}$, $\kappa_{\mathbf{T}}$ | curvature of binormal and tangent spherical indicatrices |
| $\kappa_g$, $\kappa_n$ | geodesic and normal curvatures |
| $\kappa_{gu}$, $\kappa_{gv}$ | geodesic curvatures of $u$ and $v$ coordinate curves |
| $\kappa_{nu}$, $\kappa_{nv}$ | normal curvatures of $u$ and $v$ coordinate curves |
| $\mathbf{K}$ | curvature vector of curve |
| $\mathbf{K}_g$, $\mathbf{K}_n$ | geodesic and normal components of curvature vector of curve |
| $\lambda$ | direction parameter of surface |
| $\xi$ | real parameter |
| $\rho$ | pseudo-radius of pseudo-sphere |
| $\rho, \phi$ | polar coordinates of plane |
| $\rho, \phi, z$ | cylindrical coordinates of 3D space |
| $\sigma$ | area of surface patch |
| $\tau$ | torsion of curve |
| $\tau_{\mathbf{B}}$, $\tau_{\mathbf{T}}$ | torsion of binormal and tangent spherical indicatrices |
| $\tau_g$ | geodesic torsion |
| $\phi$ | angle or parameter |
| $\chi$ | Euler characteristic |
| $\omega$ | real parameter |

# Chapter 1
# Preliminaries

1. Give a brief definition of differential geometry indicating the other disciplines of mathematics to which differential geometry is intimately linked.
   **Answer**: Differential geometry is a branch of mathematics that largely employs methods and techniques of other branches of mathematics such as differential and integral calculus, topology and tensor calculus to investigate geometric issues related to abstract objects, such as space curves and surfaces, and their properties where these investigations are mostly focused on these properties at small scales. The investigations of differential geometry also include characterizing categories of these objects.
   There is an intimate relation between differential geometry and the disciplines of differential topology and differential equations.

2. A surface embedded in a 3D space can be regarded as a 2D and as a 3D object at the same time. Discuss this briefly. From the same perspective, discuss also the state of a curve embedded in a surface which in its turn is embedded in a 3D space.
   **Answer**: The surface can be seen as a 2D object when viewed internally from inside its space and can be seen as a 3D object when viewed externally from the embedding 3D space.
   Following the example of surface, a curve can be seen internally as a 1D object. It can also be seen externally as a 2D object considering the embedding 2D surface and hence it is characterized as a surface curve. The curve may also be seen externally as a 3D object considering the embedding 3D space and hence it is characterized as a space curve.

3. What are the following symbols: $[\alpha\beta, \gamma]$, $[ij, k]$, $\Gamma^{\gamma}_{\alpha\beta}$ and $\Gamma^{k}_{ij}$? What is the difference between those with Greek indices and those with Latin indices?
   **Answer**:
   $[\alpha\beta, \gamma]$ is the Christoffel symbol of the first kind for surface.
   $[ij, k]$ is the Christoffel symbol of the first kind for space.
   $\Gamma^{\gamma}_{\alpha\beta}$ is the Christoffel symbol of the second kind for surface.
   $\Gamma^{k}_{ij}$ is the Christoffel symbol of the second kind for space.
   Those with Greek indices represent surface while those with Latin indices represent space.

4. What is the relation between the coefficients of the surface covariant metric tensor and the surface covariant curvature tensor on one hand and the coefficients of the first and second fundamental forms on the other? What are the symbols representing all these coefficients?
   **Answer**:

The coefficients of the first fundamental form are equal to the coefficients of the surface covariant metric tensor.
The coefficients of the second fundamental form are equal to the coefficients of the surface covariant curvature tensor.
The symbols representing these coefficients are:
$E, F, G$ symbolize the coefficients of the first fundamental form.
$a_{11}, a_{12}, a_{22}$ symbolize the coefficients of the surface covariant metric tensor.
$e, f, g$ symbolize the coefficients of the second fundamental form.[1]
$b_{11}, b_{12}, b_{22}$ symbolize the coefficients of the surface covariant curvature tensor.[2]
Accordingly, we have:

$$(E, F, G, e, f, g) = (a_{11}, a_{12}, a_{22}, b_{11}, b_{12}, b_{22})$$

5. What is the difference between the local and global properties of a manifold? Give an example for each. What are the colloquial terms used to label these two categories?
**Answer**: The local properties correspond to the characteristics of the manifold in the immediate neighborhood of a point in the manifold such as the curvature of a curve or surface at that point, while the global properties correspond to the characteristics of the manifold on a large scale and over extended regions of the manifold such as the number of stationary points of a curve or a surface or being a one-side surface like Mobius strip which is locally a double-side surface.
These two categories are referred to colloquially as *in the small* and *in the large.*

6. What is the meaning of "intrinsic" and "extrinsic" properties of a manifold? Give an example for each.
**Answer**: Intrinsic properties are those properties which are independent in their existence and definition from the ambient space that embraces the object such as the distance along a given curve or the Gaussian curvature of a surface at a given point, while extrinsic properties are those properties which depend in their existence and definition on the external embedding space such as having a normal vector at a point on the curve or the surface.

7. Explain the concept of "2D inhabitant" and how it is used to classify the properties of a space surface.
**Answer**: A 2D inhabitant is a creature whose perception is restricted to the 2D manifold in which he lives and hence he has no conception of any $n$D ($n > 2$) embedding space. Accordingly, all the properties that are conceived by this creature are intrinsic to the 2D manifold (or surface) while all the other properties that are beyond the perception of this creature are extrinsic to the 2D manifold.

8. Find the equation of the plane passing through the points: $(1, 2, 0)$, $(0, -3, 1.5)$ and $(1, 0, -1)$. What is the normal unit vector to this plane?

[1] We should note that the coefficients of the second fundamental form may also be symbolized by $L, M, N$ (see Exercise 84 of § 3).
[2] Due to symmetry, we have: $a_{21} = a_{12}$ and $b_{21} = b_{12}$.

**Answer**: In the following points we outline the method for solving this problem:
(a) We obtain two vectors embedded in the plane from these three points, that is:

$$\begin{aligned}(1,2,0)-(1,0,-1) &= (0,2,1)\\(0,-3,1.5)-(1,0,-1) &= (-1,-3,2.5)\end{aligned}$$

(b) We obtain a normal vector to the plane by taking the cross product of these two vectors, that is:

$$\begin{vmatrix}\mathbf{i} & \mathbf{j} & \mathbf{k}\\0 & 2 & 1\\-1 & -3 & 2.5\end{vmatrix} = \mathbf{i}\,(5+3)-\mathbf{j}\,[0+1]+\mathbf{k}\,(0+2)=8\mathbf{i}-\mathbf{j}+2\mathbf{k}$$

where $\mathbf{i},\mathbf{j},\mathbf{k}$ are the Cartesian basis vectors.
(c) The equation of a plane surface is:

$$ax+by+cz=d$$

where $(a,b,c)$ is the normal vector to the plane and $d$ is a constant. Hence, the equation of the plane is:

$$8x-y+2z=d$$

(d) To determine $d$ we substitute the coordinates of one of the three points (say the first point) in the last equation, that is:

$$\begin{aligned}(8\times 1)-2+(2\times 0) &= d\\d &= 6\end{aligned}$$

Hence, the equation of the plane is:

$$8x-y+2z=6$$

(e) We verify the above equation by substituting the three points in the equation to obtain an identity:
First point $(1,2,0)$:

$$\begin{aligned}8\,(1)-(2)+2\,(0) &= 6\\6 &= 6\end{aligned}$$

Second point $(0,-3,1.5)$:

$$\begin{aligned}8\,(0)-(-3)+2\,(1.5) &= 6\\6 &= 6\end{aligned}$$

Third point $(1,0,-1)$:

$$8\,(1)-(0)+2\,(-1) = 6$$

$$6 = 6$$

Hence, the equation $8x - y + 2z = 6$ (which is an equation of a plane) is satisfied by the three points $(1, 2, 0)$, $(0, -3, 1.5)$ and $(1, 0, -1)$. Now, since only one plane can pass through three points then the equation $8x - y + 2z = 6$ is the equation of the plane that passes through these three points, as required.
The normal unit vector to this plane is:

$$\mathbf{n} = \frac{8\mathbf{i} - \mathbf{j} + 2\mathbf{k}}{|8\mathbf{i} - \mathbf{j} + 2\mathbf{k}|} = \frac{8\mathbf{i} - \mathbf{j} + 2\mathbf{k}}{\sqrt{8^2 + (-1)^2 + 2^2}} = \frac{8\mathbf{i} - \mathbf{j} + 2\mathbf{k}}{\sqrt{69}}$$

9. Is the normal unit vector of a plane surface unique?
**Answer**: No. It can be in one of two opposite directions.

10. Define briefly each one of the following terms: surface of revolution, meridians and parallels.
**Answer**:
Surface of revolution is an axially symmetric surface generated by a plane curve $C$ (called the profile of the surface) revolving around a straight line $L$ (called the axis of revolution or the axis of symmetry of the surface) contained in the plane of the curve but not intersecting the curve.
Meridians are plane curves on the surface of revolution formed by the intersection of a plane containing the axis of revolution with the surface, and hence the meridians are identical versions of the profile curve $C$.
Parallels are circles generated by intersecting the surface of revolution by planes perpendicular to the axis of revolution, and hence they represent the paths of specific points on the profile curve $C$.

11. Prove that the meridians and parallels of a surface of revolution are mutually perpendicular at their points of intersection.
**Answer**: Let assume that the surface of revolution is generated by a profile curve $C$ that is in the $xz$ plane and hence it can be $t$-parameterized by $x = g$ and $z = h$ where $g$ and $h$ are scalar functions of $t$, i.e. $g = g(t)$ and $h = h(t)$. Hence, any point on the profile curve $C$ is given by:

$$\mathbf{i}g + \mathbf{j}0 + \mathbf{k}h = \mathbf{i}g + \mathbf{k}h$$

where $\mathbf{i}, \mathbf{j}, \mathbf{k}$ are the Cartesian basis vectors. Now, if we assume that the surface of revolution is generated by rotating $C$ around the $z$-axis (i.e. the $z$-axis is the axis of revolution) then the position vector of any point on the surface should be given in the 3D space by the following form:

$$\mathbf{r}(t, \phi) = \mathbf{i}g\cos\phi + \mathbf{j}g\sin\phi + \mathbf{k}h$$

where $\phi$ is the angle through which the profile curve that passes through that point is rotated around the $z$-axis (in the positive sense) starting from its initial position in the

$xz$ plane. Now, the meridians are the $\phi$-independent curves on the surface and hence the tangent to any meridian that passes through a given point on the surface should be given by:

$$\mathbf{T}_m = \partial_t \mathbf{r} = \mathbf{i}\,[\partial_t g]\cos\phi + \mathbf{j}\,[\partial_t g]\sin\phi + \mathbf{k}\partial_t h$$

Regarding the parallels, they are $z$-centered circles on the surface and hence they are the $t$-independent curves on the surface. Therefore, the tangent to any parallel that passes through a given point on the surface should be given by:

$$\mathbf{T}_p = \partial_\phi \mathbf{r} = -\mathbf{i}g\sin\phi + \mathbf{j}g\cos\phi + \mathbf{k}0$$

By taking the dot product of $\mathbf{T}_m$ and $\mathbf{T}_p$ we get (noting that the Cartesian basis vectors are orthonormal set):

$$\begin{aligned}\mathbf{T}_m \cdot \mathbf{T}_p &= (\mathbf{i}\,[\partial_t g]\cos\phi + \mathbf{j}\,[\partial_t g]\sin\phi + \mathbf{k}\partial_t h)\cdot(-\mathbf{i}g\sin\phi + \mathbf{j}g\cos\phi + \mathbf{k}0)\\ &= -[\partial_t g]\,g\cos\phi\sin\phi + [\partial_t g]\,g\sin\phi\cos\phi + 0\\ &= 0\end{aligned}$$

i.e. $\mathbf{T}_m$ and $\mathbf{T}_p$ (which are the tangents to the meridian and parallel that pass through any given point on the surface) are mutually perpendicular and hence the meridians and parallels of a surface of revolution are mutually perpendicular at their points of intersection, as required.

12. State the parametric equations of the following geometric shapes: torus, hyperboloid of one sheet, and hyperbolic paraboloid.
**Answer**:
Torus:

$$\begin{aligned}x &= (R + r\cos\phi)\cos\theta\\ y &= (R + r\cos\phi)\sin\theta\\ z &= r\sin\phi\end{aligned}$$

where $R$ is the torus radius, $r$ is the radius of the generating circle ($r < R$), $\phi \in [0, 2\pi)$ is the angle of variation of $r$, and $\theta \in [0, 2\pi)$ is the angle of variation of $R$.
Hyperboloid of one sheet:

$$\begin{aligned}x &= a\,\cosh\xi\,\cos\theta\\ y &= b\,\cosh\xi\,\sin\theta\\ z &= c\,\sinh\xi\end{aligned}$$

where $a, b, c$ are non-zero real constants and $\xi$, $\theta$ are real parameters with $-\infty < \xi < +\infty$ and $0 \le \theta < 2\pi$.
Hyperbolic paraboloid:

$$x = a\xi$$

$$\begin{aligned} y &= b\omega \\ z &= c\xi\omega \end{aligned}$$

where $a, b, c$ are non-zero real constants and $\xi$, $\omega$ are real parameters with $-\infty < \xi < +\infty$ and $-\infty < \omega < +\infty$.

13. Write down the parametric equations of a circle in the $uv$ plane centered at point $(a, b)$ with radius $r = c$ where $a, b, c$ are real constants and $c > 0$.
**Answer**:

$$\begin{aligned} u &= a + c\cos\theta \\ v &= b + c\sin\theta \end{aligned}$$

where $0 \le \theta < 2\pi$.

14. Find the parametric equations of an ellipse in the $xy$ plane centered at the origin of coordinates with $A = 5$ and $B = 3$ where $A, B$ are the semi-major and semi-minor axes.
**Answer**: If we assume that the major and minor axes are along the $x$ and $y$ axes respectively then we have:

$$\begin{aligned} x &= 5\cos t \\ y &= 3\sin t \end{aligned}$$

where $0 \le t < 2\pi$.

15. A surface of revolution may be represented locally in a 3D space by the following form: $\mathbf{r}(u, v) = (u\cos v,\, u\sin v,\, f(u, v))$ where $f$ is a continuous function. Determine the equations representing the parallels and meridians of this surface.
**Answer**: If we replace the generic $u, v$ parameters with more specific parameters $\phi, t$ (where $v \equiv \phi$ and $f \equiv t$), then a surface of revolution can be represented spatially as: $\mathbf{r}(\phi, t) = (\rho\cos\phi,\, \rho\sin\phi,\, t)$ where $\rho$ and $\phi$ are cylindrical coordinates and $\rho$ is a function of $t$, i.e. $\rho = \rho(t)$. Now, according to this parameterization the parallels are characterized by having a constant $t$ and hence they are given by:

$$\mathbf{r}(\phi, t_o) = (\rho_o\cos\phi,\, \rho_o\sin\phi,\, t_o)$$

where $t_o$ is a constant and $\rho_o = \rho(t_o)$. Similarly, the meridians are characterized by having constant $\phi$ and hence they are given by:

$$\mathbf{r}(\phi_o, t) = (\rho\cos\phi_o,\, \rho\sin\phi_o,\, t)$$

where $\phi_o$ is a constant.

16. Find the parametric equations of a curve formed by the intersection of the surfaces represented by: $\mathbf{r}_1(u,v) = (u, u^2, v)$ and $\mathbf{r}_2(u,v) = (u, v, u^2)$ where $-\infty < u, v < +\infty$.
**Answer**: The curve should satisfy both representations and hence we should have:

$$\begin{aligned} u &= u \\ u^2 &= v \\ v &= u^2 \end{aligned}$$

The first equation is a trivial identity while the second and third equations are the same and hence the intersection curve is given by:

$$\mathbf{r}_C(u) = \left(u, u^2, u^2\right)$$

that is:

$$\begin{aligned} x &= u \\ y &= u^2 \\ z &= u^2 \end{aligned}$$

17. Write down the general form of the parametric equations of each of the following surfaces: hyperboloid of two sheets, parabolic cylinder, catenoid, monkey saddle and pseudo-sphere.
**Answer**:
Hyperboloid of two sheets:

$$\begin{aligned} x &= a \sinh\xi \cos\theta \\ y &= b \sinh\xi \sin\theta \\ z &= c \cosh\xi \end{aligned}$$

where $a, b, c$ are non-zero real constants and $\xi$, $\theta$ are real parameters with $0 \le \xi < \infty$ and $0 \le \theta < 2\pi$. We note that the two parts of the surface correspond to $\pm c$.
Parabolic cylinder:

$$\begin{aligned} x &= \xi \\ y &= a\,\xi^2 \\ z &= b\,\omega \end{aligned}$$

where $a, b$ are non-zero real constants and $\xi$, $\omega$ are real parameters with $-\infty < \xi < +\infty$ and $-\infty < \omega < +\infty$.
Catenoid:

$$x = a \cosh\left(\frac{\xi}{a}\right) \cos\theta$$

$$\begin{aligned} y &= a \cosh\left(\frac{\xi}{a}\right) \sin\theta \\ z &= \xi \end{aligned}$$

where $a$ is a non-zero real constant and $\xi$, $\theta$ are real parameters with $-\infty < \xi < +\infty$ and $0 \le \theta < 2\pi$.
Monkey saddle:

$$\begin{aligned} x &= \xi \\ y &= \omega \\ z &= \xi^3 - 3\xi\omega^2 \end{aligned}$$

where $\xi$, $\omega$ are real parameters with $-\infty < \xi < +\infty$ and $-\infty < \omega < +\infty$.
Pseudo-sphere:

$$x = a \sin\theta \cos\phi \tag{1}$$

$$y = a \sin\theta \sin\phi \tag{2}$$

$$z = a\left[\cos\theta + \ln\left(\tan\frac{\theta}{2}\right)\right] \tag{3}$$

where $a$ is a real constant and $\theta$, $\phi$ are real parameters with $0 < \theta < \pi$ and $0 \le \phi < 2\pi$.

18. Sketch the following (using a 3D computer graphic package if available): (a) a straight line passing through the point $(11, -5, 6.3)$ and parallel to the vector $(-3, -1.8, 6.5)$ (b) a plane passing through the point $(6, -8.2, -7)$ with a normal vector $(3, -1.6, -2.5)$.
**Answer**:
(a) The sketch should look something like Figure 1.
(b) The sketch should look something like Figure 2.

19. A surface is parameterized by: $\mathbf{r}(u, v) = (a \sinh u \cos v,\ b \sinh u \sin v,\ c \cosh u)$. What is the name of this surface? What is the condition for this surface to be a surface of revolution around the third spatial axis?
**Answer**: This is a hyperboloid of two sheets with $(\xi, \theta) \equiv (u, v)$.[3] The condition is $a = b$.

20. Find the equation of the straight line passing through the point $(-6, 3.1, 8.4)$ and the point $(1, 0, -3)$.
**Answer**: We outline the solution in the following points:
(a) We obtain the direction vector $\mathbf{d}$ from the two points, that is:

$$\mathbf{d} = (-6, 3.1, 8.4) - (1, 0, -3) = (-7, 3.1, 11.4)$$

(b) Let $P(x, y, z)$ be an arbitrary point on the line. The vector that connects $P$ to one of the given points [say point $(1, 0, -3)$] should be parallel to the direction vector, that is:

$$(x, y, z) - (1, 0, -3) = t\mathbf{d}$$

[3] As noted earlier, the two parts of the surface correspond to $\pm c$.

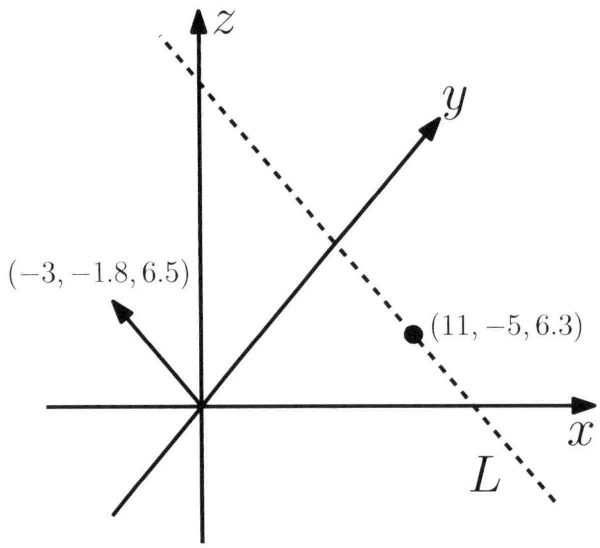


Figure 1: A sketch of a straight line $L$ passing through the point $(11, -5, 6.3)$ and parallel to the vector $(-3, -1.8, 6.5)$.

where $t$ is a real parameter.
(c) Hence the equation of the line is:

$$(x-1, y, z+3) = t\,(-7, 3.1, 11.4)$$

which is usually given in the form:

$$\frac{x-1}{-7} = \frac{y}{3.1} = \frac{z+3}{11.4} = t$$

(d) We note that the point $(-6, 3.1, 8.4)$ corresponds to $t = 1$ while the point $(1, 0, -3)$ corresponds to $t = 0$ and hence they are on the line. This may serve as verification.

21. Classify the following as curves or surfaces: ellipsoid, elliptic paraboloid, catenary, helicoid, enneper and tractrix.
**Answer**: Catenary and tractrix are curves while ellipsoid, elliptic paraboloid, helicoid and enneper are surfaces.

22. Make a simple sketch for each one of the geometric shapes in the previous question. Use a computer graphic package if convenient.
**Answer**: The sketches of all these geometric shapes (except catenary) are given in the book and hence the required answer is no more than a simple replica of these sketches. Regarding the catenary, the sketch should look something like Figure 3.

23. Prove that $\mathbf{v} \times \frac{d\mathbf{v}}{dt} = \mathbf{0}$ *iff* the direction of the vector $\mathbf{v}(t)$ is constant.
**Answer**: There are two parts to this proof:

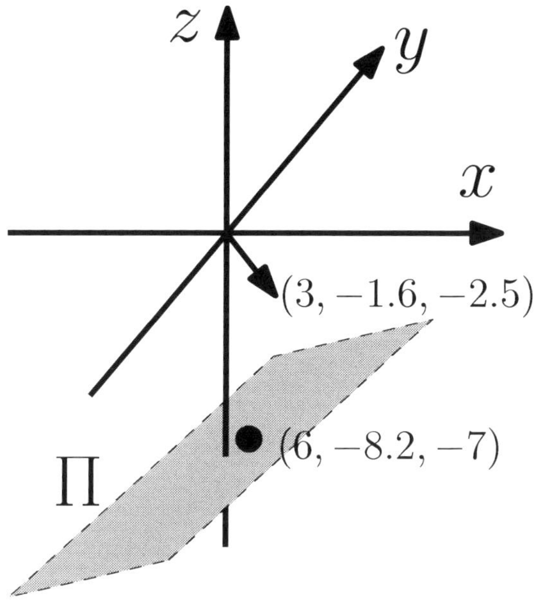


Figure 2: A sketch of a plane Π passing through the point $(6, -8.2, -7)$ with a normal vector $(3, -1.6, -2.5)$.

(a) If the direction of the vector $\mathbf{v}(t)$ is constant then $\mathbf{v} \times \frac{d\mathbf{v}}{dt} = \mathbf{0}$: this is because in this case $\mathbf{v}$ can be expressed as:

$$\mathbf{v} = |\mathbf{v}| \frac{\mathbf{v}}{|\mathbf{v}|} = |\mathbf{v}| \, \mathbf{d}$$

where $\mathbf{d}$ is a constant unit vector since the direction of the vector $\mathbf{v}(t)$ is constant. Hence, by the product rule of differentiation we have:

$$\frac{d\mathbf{v}}{dt} = \frac{d}{dt} (|\mathbf{v}| \, \mathbf{d}) = \frac{d \, |\mathbf{v}|}{dt} \mathbf{d} + |\mathbf{v}| \frac{d\mathbf{d}}{dt} = \frac{d \, |\mathbf{v}|}{dt} \mathbf{d} + |\mathbf{v}| \, \mathbf{0} = \frac{d \, |\mathbf{v}|}{dt} \mathbf{d}$$

where $\frac{d\mathbf{d}}{dt} = \mathbf{0}$ is justified by the fact that $\mathbf{d}$ is a constant vector. Therefore, we have:

$$\mathbf{v} \times \frac{d\mathbf{v}}{dt} = |\mathbf{v}| \, \mathbf{d} \times \frac{d \, |\mathbf{v}|}{dt} \mathbf{d} = |\mathbf{v}| \frac{d \, |\mathbf{v}|}{dt} (\mathbf{d} \times \mathbf{d}) = \mathbf{0}$$

where the last step is justified by the fact that the cross product of a vector by itself is zero.[4]

(b) If $\mathbf{v} \times \frac{d\mathbf{v}}{dt} = \mathbf{0}$ then the direction of the vector $\mathbf{v}(t)$ is constant: because if $\mathbf{v} \times \frac{d\mathbf{v}}{dt} = \mathbf{0}$ then either $\frac{d\mathbf{v}}{dt}$ is co-directional to $\mathbf{v}$ (and hence $\mathbf{v}$ is constant in direction) or $\frac{d\mathbf{v}}{dt}$ is zero

[4] In fact, the proof of this part can be reduced to the statement that since the direction of the vector $\mathbf{v}$ is constant then $\frac{d\mathbf{v}}{dt}$ must be co-directional to $\mathbf{v}$ (since a change can occur only to the magnitude of the vector) including being zero (if $\mathbf{v}$ is constant in magnitude as well) and hence $\mathbf{v} \times \frac{d\mathbf{v}}{dt} = \mathbf{0}$ because the cross product of two co-directional vectors is identically zero. However, we preferred the above proof for more clarity.

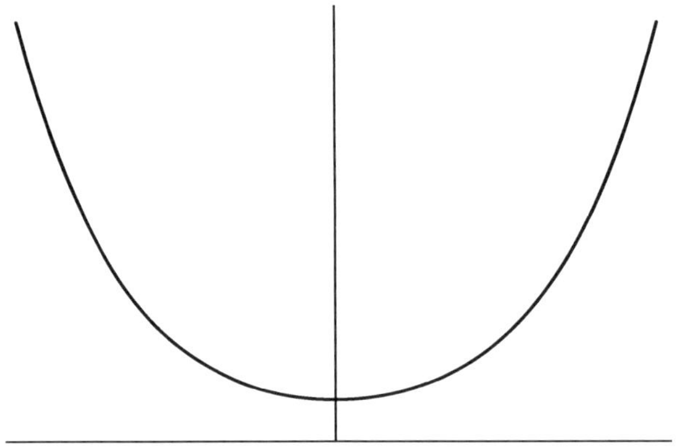

Figure 3: A simple sketch of a catenary where the horizontal and vertical lines may represent the $x$ and $y$ coordinate axes.

(and hence $\mathbf{v}$ is constant in magnitude and direction). So, in both cases the vector $\mathbf{v}$ must have constant direction. In fact, the second case is a special case of the first case. Note: it is noteworthy that if a vector is constant in magnitude only then its derivative is perpendicular to the vector, and if it is constant in direction only then its derivative has the same constant direction, and if it is constant in magnitude and direction then its derivative is zero.

24. Define "Euler characteristic" stating the equation that links it to the number of vertices, faces and edges of a polyhedron.
**Answer**: The Euler characteristic is a parameter of closed surfaces that topologically characterizes these surfaces by establishing a connection between the number of vertices, edges and faces that identify the surface shape. For polyhedral surfaces, the Euler characteristic is given by:
$$\chi = \mathcal{V} + \mathcal{F} - \mathcal{E}$$
where $\chi$ is the Euler characteristic of the surface, and $\mathcal{V}, \mathcal{F}, \mathcal{E}$ are the numbers of vertices, faces and edges of the polyhedron. The Euler characteristic can also be defined for more general types of surface as we will see in the next question.

25. Explain how the Euler characteristic is defined for non-polyhedral compact orientable surfaces such as ellipsoids.
**Answer**: The Euler characteristic of a compact orientable non-polyhedral surface, like ellipsoid and torus, can be obtained by polygonal decomposition based on dividing the entire surface into a finite number of non-overlapping curvilinear polygons which share at most edges or vertices, as seen in Figure 4 where an ellipsoid is polygonally decomposed into four non-overlapping curvilinear polygons.

26. What is the topological meaning of "genus of a surface"?
**Answer**: The topological genus of a surface is the number of handles or topological holes on the surface.

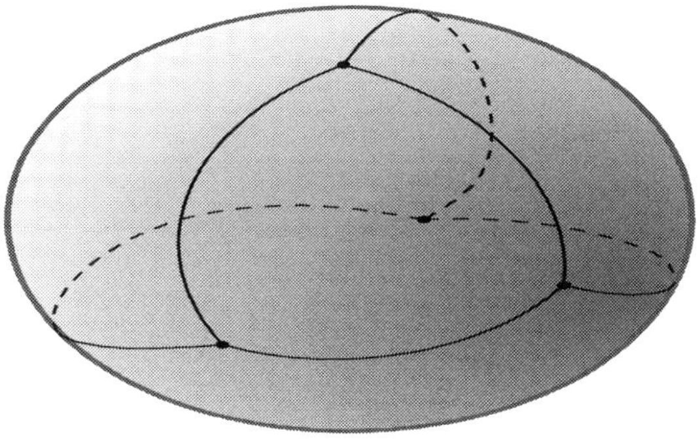

Figure 4: Polygonal decomposition of an ellipsoid into four non-overlapping curvilinear polygons.

27. Give examples for surfaces of genus 0, 1, 2 and 3 from common geometric shapes other than those given in the text.
**Answer**: The surface of ball is of genus 0. The surface of coffee cup (i.e. a cup with a single closed handle) is of genus 1. The surface of traditional trophy (i.e. what looks like a cup with 2 closed handles) is of genus 2. The surface of some antique jars (i.e. those with 3 closed handles) is of genus 3.

28. By using the Euler formula, calculate the Euler characteristic $\chi$ of the following surfaces: (a) parallelepiped (b) dodecahedron (c) icosahedron. Show your work in detail.
**Answer**:
(a) Parallelepiped: we have 8 vertices, 6 faces and 12 edges. Hence:
$$\chi = \mathcal{V} + \mathcal{F} - \mathcal{E} = 8 + 6 - 12 = 2$$
(b) Dodecahedron: we have 20 vertices, 12 faces and 30 edges. Hence:
$$\chi = \mathcal{V} + \mathcal{F} - \mathcal{E} = 20 + 12 - 30 = 2$$
(c) Icosahedron: we have 12 vertices, 20 faces and 30 edges. Hence:
$$\chi = \mathcal{V} + \mathcal{F} - \mathcal{E} = 12 + 20 - 30 = 2$$

29. By using polygonal decomposition, calculate the Euler characteristic $\chi$ of the following surfaces: (a) sphere (b) ellipsoid (c) torus. Show your work in detail with simple sketches to demonstrate the polygonal decomposition in each case.
**Answer**:
(a) Sphere: in Figure 5, a sphere is decomposed into four curvilinear polygons. As we see, we have 4 vertices, 4 faces and 6 edges. Hence:
$$\chi = \mathcal{V} + \mathcal{F} - \mathcal{E} = 4 + 4 - 6 = 2$$

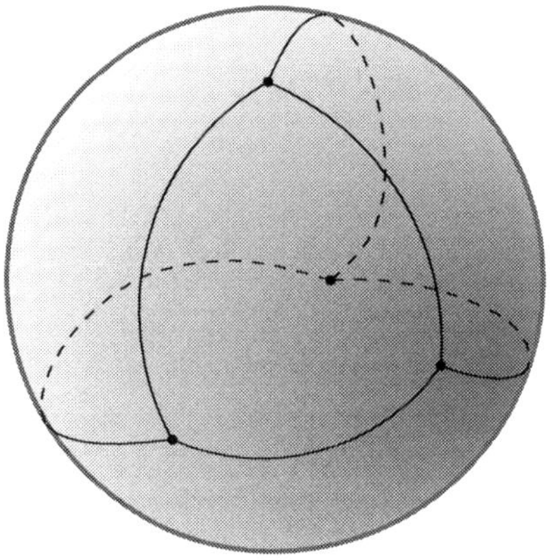

Figure 5: Polygonal decomposition of a sphere into four non-overlapping curvilinear polygons.

(b) Ellipsoid: in Figure 4, an ellipsoid is decomposed into four curvilinear polygons. As we see, we have 4 vertices, 4 faces and 6 edges. Hence:

$$\chi = \mathcal{V} + \mathcal{F} - \mathcal{E} = 4 + 4 - 6 = 2$$

(c) Torus: in Figure 6, a torus is decomposed into four curvilinear polygons. As we see, we have 4 vertices, 4 faces and 8 edges. Hence:

$$\chi = \mathcal{V} + \mathcal{F} - \mathcal{E} = 4 + 4 - 8 = 0$$

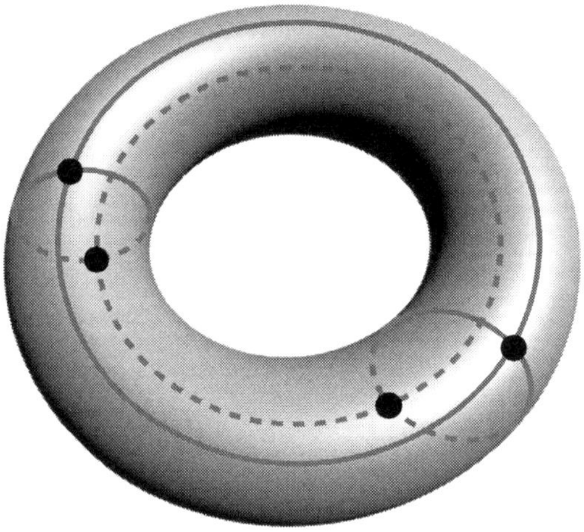

Figure 6: Polygonal decomposition of a torus into four non-overlapping curvilinear polygons.

30. What is the genus of the surfaces in the previous question?
**Answer**: We have:

$$\chi = 2\,(1 - \mathfrak{g})$$

where $\mathfrak{g}$ is the genus of the surface. Hence:

$$\mathfrak{g} = 1 - \frac{\chi}{2}$$

Accordingly:
(a) Sphere:

$$\mathfrak{g} = 1 - \frac{\chi}{2} = 1 - \frac{2}{2} = 0$$

(b) Ellipsoid:

$$\mathfrak{g} = 1 - \frac{\chi}{2} = 1 - \frac{2}{2} = 0$$

(c) Torus:

$$\mathfrak{g} = 1 - \frac{\chi}{2} = 1 - \frac{0}{2} = 1$$

31. What is the Cartesian form of the equation of a sphere centered at point $(a, b, c)$ with radius $r = d$ where $a, b, c, d$ are real constants and $d > 0$?
**Answer**:

$$(x - a)^2 + (y - b)^2 + (z - c)^2 = d^2$$

32. Explain briefly the meaning of the following terms: bicontinuous function, surface of class $C^n$, and sufficiently smooth curve.
**Answer**:
Bicontinuous function is a continuous function with a continuous inverse.
Surface of class $C^n$ is a surface that is mathematically represented by a function of class $C^n$ (i.e. the function and all its first $n$ partial derivatives do exist and they are continuous).
Sufficiently smooth curve is a curve whose functional relation is sufficiently differentiable for the intended purpose, being of class $C^n$ at least where $n$ is the minimum requirement for the differentiability index to satisfy the required conditions.

33. Explain in detail, using equations and simple sketches, the concept of "deleted neighborhood" in 1D and 2D flat and curved spaces as seen from the ambient space.[5]
**Answer**:
1D: a deleted neighborhood of a point $P$ on a 1D interval on a straight line is defined as the set of all points $x \in \mathbb{R}$ in the interval such that:

$$0 < |x - x_P| < \epsilon$$

where $x_P$ is the coordinate of $P$ on the real line and $\epsilon$ is a positive real number. For a space curve represented by $\mathbf{r} = \mathbf{r}(t)$, where $\mathbf{r}$ is the spatial representation of the curve and $t$ is a general parameter in the curve representation, the definition applies to the neighborhood of $t_P$ where $t_P$ is the value of $t$ corresponding to the point $P$ on the curve. This is outlined in Figure 7.

[5] In this context, "flat" and "curved" are used in a generic rather than technical sense.

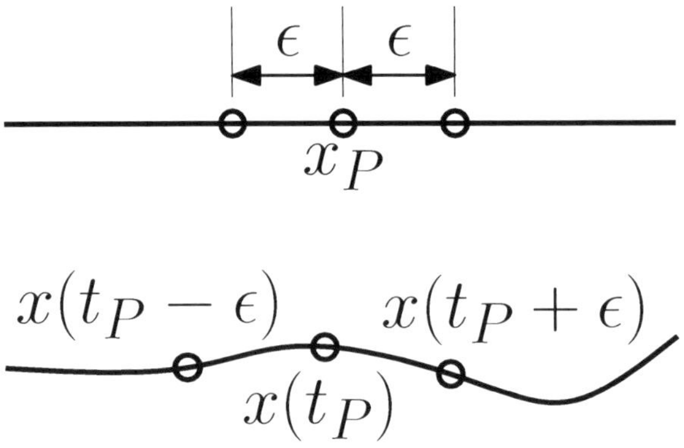


Figure 7: Deleted neighborhood on a straight line (top) and on a space curve (bottom).

2D: a deleted neighborhood of a point $P$ on a 2D flat surface is defined as the set of all points $(x, y) \in \mathbb{R}^2$ on the surface such that:

$$0 < \sqrt{(x - x_P)^2 + (y - y_P)^2} < \epsilon$$

where $(x_P, y_P)$ are the coordinates of $P$ on the plane and $\epsilon$ is a positive real number. For a space surface (which is not flat in general) represented by $\mathbf{r} = \mathbf{r}(u, v)$, where $\mathbf{r}$ is the spatial representation of the surface and $u, v$ are the surface coordinates on the $uv$ plane that map on the surface, the definition applies to the neighborhood of $(u_P, v_P)$ where $(u_P, v_P)$ are the coordinates on the 2D $uv$ plane corresponding to the point $P$ on the surface. This is outlined in Figure 8.

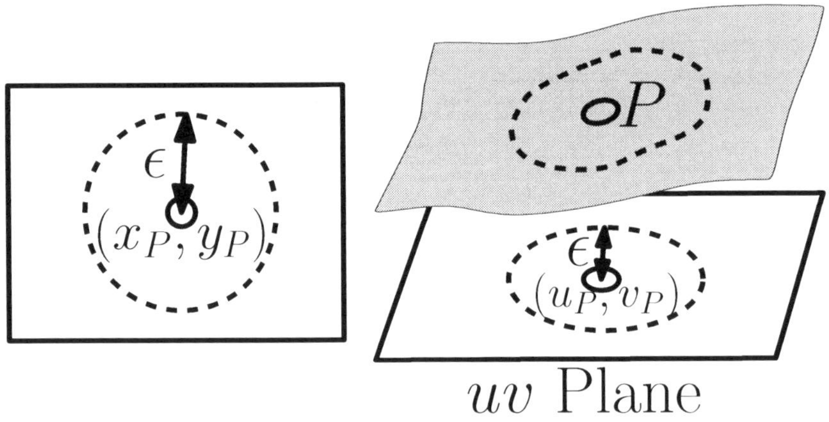


Figure 8: Deleted neighborhood on a plane (left) and on a space surface (right).

34. How can we extend the concept of "deleted neighborhood" to spaces of dimensionality higher than 2?
**Answer**: This is done by extending the neighborhood of the point to higher dimensionality where this neighborhood is defined by a sphere centered at the point with a

given radius $\epsilon$. It should be obvious that for $n$D spaces with $n > 3$ the concepts of "sphere" and "radius" are generalizations of these concepts from their 3D meanings.

35. Find the equation of a cone generated by rotating the line $z = -2x$ around the $z$-axis.
**Answer**: The equation of quadric cone is given by:

$$\frac{x^2}{a^2} + \frac{y^2}{b^2} - \frac{z^2}{c^2} = 0$$

It is obvious that the point $(1, 0, -2)$ is on the cone. On substituting this points into the above equation we obtain:

$$\begin{aligned}\frac{1}{a^2} - \frac{4}{c^2} &= 0 \\ c^2 &= 4a^2\end{aligned}$$

Hence, the above equation becomes:

$$\frac{x^2}{a^2} + \frac{y^2}{b^2} - \frac{z^2}{4a^2} = 0$$

Also, from the axial symmetry around the $z$-axis we should have $b^2 = a^2$. Hence, the last equation becomes:

$$\frac{x^2}{a^2} + \frac{y^2}{a^2} - \frac{z^2}{4a^2} = 0$$

On multiplying the two sides with $4a^2$ we obtain:

$$4x^2 + 4y^2 - z^2 = 0$$

which is the required equation.

36. Derive the parametric equations of a helix rotating around the $z$-axis, passing through the point $(3, 0, 0)$ and climbing (or descending) 5.3 units in the $z$-direction as it makes a $4\pi$ turn around the $z$-axis.
**Answer**: The parametric equations of helix are:

$$\begin{aligned}x &= a\cos\theta \\ y &= a\sin\theta \\ z &= b\theta\end{aligned}$$

From the point $(3, 0, 0)$ which correspond to $\theta = 0$ we conclude that $a = 3$. Also, from the fact that it climbs 5.3 units in the $z$-direction as it makes a $4\pi$ turn around the $z$-axis we conclude that $b = \frac{5.3}{4\pi}$ (i.e. $z = b\theta \quad \to \quad 5.3 = b4\pi$). Hence, the parametric equations of this helix are:

$$x = 3\cos\theta$$

$$
\begin{aligned}
y &= 3\sin\theta \\
z &= \frac{5.3}{4\pi}\theta
\end{aligned}
$$

As indicated earlier, the point $(3, 0, 0)$ corresponds to $\theta = 0$ while the point $(3, 0, 5.3)$ corresponds to $\theta = 4\pi$.

37. Define "positive definite" in words and by stating the mathematical conditions for a quadratic expression to be positive definite.
**Answer**: A variable is described as positive definite if it possesses positive values ($> 0$) over the whole of its domain. The mathematical conditions for a quadratic expression $a_1x^2 + 2a_2xy + a_3y^2$ to be positive definite is:

$$a_1 > 0 \qquad \text{and} \qquad (a_1a_3 - a_2a_2) > 0$$

38. Describe orthogonal coordinate transformations and how they are characterized by their Jacobian.
**Answer**: Orthogonal coordinate transformations are characterized by being made of combinations of translation, rotation and reflection of axes. The Jacobian of orthogonal transformations is unity, that is $J = \pm 1$.

39. State the difference between positive and negative orthogonal transformations.
**Answer**: Positive orthogonal transformations have positive Jacobian (i.e. $J = +1$) while negative orthogonal transformations have negative Jacobian (i.e. $J = -1$). Also, positive orthogonal transformations consist solely of translation and rotation while negative orthogonal transformations include reflection, by applying an odd number of axes reversal, as well.

40. Find a set of parametric equations representing a cylinder generated by rotating a straight line parallel to the $z$-axis and passing through the point $(2.5, 0, 0)$ around the $z$-axis.
**Answer**:

$$
\begin{aligned}
x &= 2.5\cos\theta \\
y &= 2.5\sin\theta \\
z &= t
\end{aligned}
$$

where $0 \le \theta < 2\pi$ and $-\infty < t < +\infty$. The point $(2.5, 0, 0)$ corresponds to $\theta = 0$ and $t = 0$.

41. What is the unit normal vector to the surface of a sphere, centered on the origin of coordinates with radius $r$, at a point on its surface with coordinates $(x_P, y_P, z_P)$? Consider the possibility of having more than one normal vector at that point.
**Answer**: The unit normal vector is given by:

$$\mathbf{n} = \pm\frac{x_P\mathbf{i} + y_P\mathbf{j} + z_P\mathbf{k}}{\sqrt{(x_P)^2 + (y_P)^2 + (z_P)^2}} = \pm\frac{x_P\mathbf{i} + y_P\mathbf{j} + z_P\mathbf{k}}{r}$$

where $\mathbf{i}, \mathbf{j}, \mathbf{k}$ are the Cartesian basis vectors. The $\pm$ sign refers to the two possibilities about the direction of the vector.

42. What "coordinate curves" means? What are the other names given to these curves?
**Answer**: "Coordinate curves" on a surface are the map or projection of the $uv$ curves of the $uv$ grid of the parameters plane onto the space surface. Hence, they are curves along which only one coordinate variable ($u$ or $v$) varies while the other coordinate variable ($v$ or $u$) remains constant. So, along the $u$ coordinate curves $v$ is held constant while along the $v$ coordinate curves $u$ is held constant. Coordinate curves are also known as parametric curves or parametric lines.

43. Define regular representation of a class $C^n$ surface patch in a 3D Euclidean space stating its mathematical conditions.
**Answer**: A regular representation of class $C^n$ ($n > 0$) of a surface patch $S$ in a 3D Euclidean space is defined as a functional mapping of an open set $\Omega$ in the $uv$ plane onto $S$ that satisfies the following two conditions:
• The functional mapping relation is of class $C^n$ over the entire $\Omega$.
• The Jacobian matrix of the transformation between the representation of the surface in the 3D space and its 2D domain is of rank 2 for all the points in $\Omega$.

44. Using a parametric representation of the elliptic paraboloid, show that it is a regular surface.
**Answer**: The elliptic paraboloid is defined parametrically by:

$$\begin{aligned} x &= a\sqrt{\xi}\cos\theta \\ y &= b\sqrt{\xi}\sin\theta \\ z &= c\xi \end{aligned}$$

It is obvious that this representation is continuous and differentiable of all orders over the entire $(u, v) \equiv (\xi, \theta)$ ($-\infty < \xi < +\infty$ and $-\infty < \theta < +\infty$). Moreover, the Jacobian matrix is:

$$\begin{aligned} \mathbf{J} &= \begin{bmatrix} \partial_u S_1 & \partial_v S_1 \\ \partial_u S_2 & \partial_v S_2 \\ \partial_u S_3 & \partial_v S_3 \end{bmatrix} \\ &= \begin{bmatrix} \partial_\xi x & \partial_\theta x \\ \partial_\xi y & \partial_\theta y \\ \partial_\xi z & \partial_\theta z \end{bmatrix} \\ &= \begin{bmatrix} a\left(2\sqrt{\xi}\right)^{-1}\cos\theta & -a\sqrt{\xi}\sin\theta \\ b\left(2\sqrt{\xi}\right)^{-1}\sin\theta & +b\sqrt{\xi}\cos\theta \\ c & 0 \end{bmatrix} \end{aligned}$$

which is of rank 2 over the entire $(u, v)$ plane.[6] Hence, the elliptic paraboloid is a

[6] We note that we should exclude the line $\xi = 0$ but this is a trivial problem since it is accidental caused by this particular representation.

regular surface.
The problem can also be solved by using the condition $\mathbf{E}_1 \times \mathbf{E}_2 \neq \mathbf{0}$, as indicated in the book. So, if we parameterize the elliptic paraboloid as (see § 6 in the book):[7]

$$\begin{aligned} x &= u \\ y &= v \\ z &= \frac{u^2}{a^2} + \frac{v^2}{b^2} \end{aligned}$$

then we have:

$$\mathbf{E}_1 \times \mathbf{E}_2 = \begin{vmatrix} \mathbf{i} & \mathbf{j} & \mathbf{k} \\ 1 & 0 & \frac{2u}{a^2} \\ 0 & 1 & \frac{2v}{b^2} \end{vmatrix} = -\frac{2u}{a^2}\mathbf{i} - \frac{2v}{b^2}\mathbf{j} + \mathbf{k} \neq \mathbf{0}$$

As we see, $\mathbf{E}_1 \times \mathbf{E}_2$ does not vanish at any point on the $(u, v)$ plane.

45. What are the reasons for having a singular point on a space surface?
**Answer**: Singularity occurs either because of an intrinsic geometric reason or because of the particular parametric representation of the surface. The first type of singularity is inherent and hence it cannot be removed, while the second type is accidental and hence it can be removed by changing the representation.

46. A sphere centered at the origin of coordinates can be represented parametrically by: $\mathbf{r}(\theta, \phi) = a(\sin\theta\cos\phi, \sin\theta\sin\phi, \cos\theta)$ where $a > 0$ is a constant (which is its radius), $0 \le \theta \le \pi$ and $0 \le \phi < 2\pi$. At what points, if any, this representation is not regular?
**Answer**: This representation has continuous partial derivatives of all orders. Using the parameterization $(u, v) \equiv (\theta, \phi)$, the cross product $\mathbf{E}_1 \times \mathbf{E}_2$ is given by:

$$\begin{aligned} \mathbf{E}_1 \times \mathbf{E}_2 &= a^2 \begin{vmatrix} \mathbf{i} & \mathbf{j} & \mathbf{k} \\ \cos\theta\cos\phi & \cos\theta\sin\phi & -\sin\theta \\ -\sin\theta\sin\phi & \sin\theta\cos\phi & 0 \end{vmatrix} \\ &= a^2\left[\mathbf{i}\sin^2\theta\cos\phi + \mathbf{j}\sin^2\theta\sin\phi + \mathbf{k}\left(\cos\theta\sin\theta\cos^2\phi + \cos\theta\sin\theta\sin^2\phi\right)\right] \\ &= a^2\left[\mathbf{i}\sin^2\theta\cos\phi + \mathbf{j}\sin^2\theta\sin\phi + \mathbf{k}\cos\theta\sin\theta\left(\cos^2\phi + \sin^2\phi\right)\right] \\ &= a^2\left[\mathbf{i}\sin^2\theta\cos\phi + \mathbf{j}\sin^2\theta\sin\phi + \mathbf{k}\cos\theta\sin\theta\right] \end{aligned}$$

Hence:

$$\begin{aligned} |\mathbf{E}_1 \times \mathbf{E}_2| &= a^2\sqrt{\sin^4\theta\cos^2\phi + \sin^4\theta\sin^2\phi + \cos^2\theta\sin^2\theta} \\ &= a^2\sqrt{\sin^4\theta\left(\cos^2\phi + \sin^2\phi\right) + \cos^2\theta\sin^2\theta} \\ &= a^2\sqrt{\sin^4\theta + \cos^2\theta\sin^2\theta} \end{aligned}$$

[7] This representation has partial derivatives of all orders.

$$\begin{aligned} &= a^2\sqrt{\sin^2\theta\left(\sin^2\theta+\cos^2\theta\right)} \\ &= a^2\sqrt{\sin^2\theta} \\ &= a^2\left|\sin\theta\right| \end{aligned}$$

Hence, $|\mathbf{E}_1 \times \mathbf{E}_2| = 0$ when $\theta = 0$ and $\theta = \pi$. Therefore, this representation is regular except at the two poles.

47. How a mathematical correspondence can be established between points on two different curves and two different surfaces?
**Answer**: Establishing a mathematical correspondence between the points on two curves with common parameterization is straightforward since the common parameterization can be used to make this correspondence. However, when the two curves have different parameterizations then a one-to-one correspondence between the two parameters should be established and the corresponding points then refer to two points with corresponding values of the two parameters. Corresponding points on two surfaces can be similarly defined considering that surfaces require two independent parameters in their characterization.

48. State the mathematical conditions that are satisfied by the intrinsic distance between two points on a smooth connected surface.
**Answer**: The intrinsic distance $d$ between two points, $P_1$ and $P_2$, on a smooth connected surface satisfies the following conditions:
• Symmetry: $d(P_1, P_2) = d(P_2, P_1)$.
• Triangle inequality: $d(P_1, P_2) \le d(P_1, P_3) + d(P_2, P_3)$ where $P_3$ is a point on the surface.
• Positive definiteness: $d(P_1, P_2) > 0$ with $d(P_1, P_2) = 0$ *iff* $P_1$ and $P_2$ are the same point.

49. Is it guaranteed that an arc of minimum length between two specific points on a surface does exist and it is unique?
**Answer**: No, it is not guaranteed in general (i.e. it may not exit and if it does exist it may not be unique) although this may be the case for certain types of surface as explained in the book.

50. Prove the three properties of intrinsic distance, i.e. symmetry, triangle inequality, and positive definiteness.
**Answer**: Referring to the mathematical conditions that have been stated in the answer of Exercise 48, we have:
• Symmetry: the length of any regular curve connecting two given points $P_1$ and $P_2$ is independent of the orientation of the curve. Hence, the set of all numbers representing the lengths of all regular curves connecting $P_1$ and $P_2$ is independent of orientation. Now, since the intrinsic distance $d$ is one of these numbers (since it is the infimum of these numbers), then $d$ is independent of orientation, i.e. $d(P_1, P_2) = d(P_2, P_1)$, as required.

• Triangle inequality: it is obvious that $d(P_1, P_2)$ is less than or equal to the length $L$ of any regular curve $C$ that connects $P_1$ and $P_2$ and passes through $P_3$. Now, if $L_1$ is the length of the segment of $C$ that connects $P_1$ and $P_3$, and $L_2$ is the length of the segment of $C$ that connects $P_2$ and $P_3$, then we have:

$$\begin{aligned} d(P_1, P_2) &\leq L \\ d(P_1, P_2) &\leq L_1 + L_2 \end{aligned}$$

Now, since $d(P_1, P_3) \leq L_1$ and $d(P_2, P_3) \leq L_2$ then we can express $L_1$ and $L_2$ as:

$$\begin{aligned} L_1 &= d(P_1, P_3) + \epsilon_1 \\ L_2 &= d(P_2, P_3) + \epsilon_2 \end{aligned}$$

where $\epsilon_1$ and $\epsilon_2$ are arbitrary positive numbers. Hence, we should have:

$$\begin{aligned} d(P_1, P_2) &\leq L_1 + L_2 \\ d(P_1, P_2) &\leq d(P_1, P_3) + \epsilon_1 + d(P_2, P_3) + \epsilon_2 \\ d(P_1, P_2) &\leq d(P_1, P_3) + d(P_2, P_3) + \epsilon \end{aligned}$$

where $\epsilon = \epsilon_1 + \epsilon_2$. Now, since $\epsilon_1$ and $\epsilon_2$ are arbitrary then $\epsilon$ is also arbitrary and hence we should have:

$$d(P_1, P_2) \leq d(P_1, P_3) + d(P_2, P_3)$$

as required.
• Positive definiteness: the length between any two distinct points is a positive number. Hence, the set of all numbers representing the lengths of all regular curves connecting $P_1$ and $P_2$ consists of positive numbers. Now, since the intrinsic distance $d$ is one of these numbers (since it is the infimum of these numbers) then $d$ is positive and it is zero only when $P_1$ and $P_2$ are identical, i.e. $d(P_1, P_2) > 0$ with $d(P_1, P_2) = 0$ *iff* $P_1$ and $P_2$ are the same point, as required.

51. Show that the intrinsic distance $d$ between two points is invariant under a local isometric mapping $f$, i.e. $d(f(P_1), f(P_2)) = d(P_1, P_2)$.
**Answer**: The defining property of isometric mapping is that it preserves distances. Now, if surface $S_1$ is isometrically mapped onto surface $S_2$ then by definition the length $L_1$ of any regular curve $C_1$ connecting two points $P_1$ and $P_2$ on $S_1$ should be equal to the length $L_2$ of the mapped curve $C_2$ that connects the corresponding points $f(P_1)$ and $f(P_2)$ on $S_2$. This means that the set of all numbers representing the lengths of all regular curves that connect $P_1$ and $P_2$ on $S_1$ will be mapped correspondingly onto the set of all numbers representing the lengths of all regular curves that connect $f(P_1)$ and $f(P_2)$ on $S_2$ where the corresponding numbers in the two sets are equal. In other words, the two sets are identical and hence the infimum of the second set should be equal to the infimum of the first set, that is:

$$d(f(P_1), f(P_2)) = d(P_1, P_2)$$

as required.

52. Find the intrinsic distance on the unit sphere centered at the origin of coordinates between the point $(1, 0, 0)$ and the point $(\frac{1}{\sqrt{3}}, \frac{1}{\sqrt{3}}, \frac{1}{\sqrt{3}})$.
**Answer**: The intrinsic distance is the length of the smaller arc of the great circle that passes through these points. Now, the angle $\theta$ $(0 \le \theta \le \pi)$ between the two vectors $(1, 0, 0)$ and $(\frac{1}{\sqrt{3}}, \frac{1}{\sqrt{3}}, \frac{1}{\sqrt{3}})$ is given by:

$$\begin{aligned}\cos\theta &= \frac{(1,0,0)\cdot\left(1/\sqrt{3},1/\sqrt{3},1/\sqrt{3}\right)}{|(1,0,0)|\left|\left(1/\sqrt{3},1/\sqrt{3},1/\sqrt{3}\right)\right|} \\ &= \frac{1/\sqrt{3}}{1} \\ &= \frac{1}{\sqrt{3}}\end{aligned}$$

and hence $\theta \simeq 0.9553$ rad. Accordingly:

$$d = r\theta \simeq 1 \times 0.9553 = 0.9553$$

53. What is the intrinsic distance between the two points of the last question in a 3D Euclidean space that encloses the sphere?
**Answer**: If we extend the concept of intrinsic distance to spaces of higher dimensionality than 2D, then the intrinsic distance between the point $(1, 0, 0)$ and the point $(\frac{1}{\sqrt{3}}, \frac{1}{\sqrt{3}}, \frac{1}{\sqrt{3}})$ in a 3D Euclidean space is the straight line segment that connects these two points, that is:

$$\begin{aligned}d &= \sqrt{\left(1-\frac{1}{\sqrt{3}}\right)^2+\left(0-\frac{1}{\sqrt{3}}\right)^2+\left(0-\frac{1}{\sqrt{3}}\right)^2} \\ &= \sqrt{\frac{3-2\sqrt{3}+1}{3}+\frac{1}{3}+\frac{1}{3}} \\ &= \sqrt{\frac{6-2\sqrt{3}}{3}} \\ &\simeq 0.9194\end{aligned}$$

54. Define the basis vectors and state their roles.
**Answer**: The basis vectors of a given space with a given coordinate system are vectors that describe how the coordinates vary with motion along trajectories in the space. For example, they can be tangents to the coordinate curves or gradients of coordinate surfaces and hence they describe and quantify the variation of coordinates at any point in the space. The basis vectors have crucial roles in characterizing the space and describing its geometry. For example, they are used in the definition and construction of essential abstract objects and concepts such as the metric tensor of the space. The basis vectors may also be used in differential geometry as a moving coordinate frame for the enveloping space of their underlying constructions, as described in the book.

55. Describe in detail the two main sets of space basis vectors in differential geometry related to curves and surfaces. Are there any other sets of basis vectors?
**Answer**: The differential geometry of curves and surfaces employs two main sets of basis vectors:
• One set is constructed on space curves and consists of three unit vectors: the tangent $\mathbf{T}$, the normal $\mathbf{N}$ and the binormal $\mathbf{B}$ to the curve.
• Another set is constructed on surfaces and consists of two linearly independent vectors which are the tangents to the coordinate curves of the surface, $\mathbf{E}_1$ and $\mathbf{E}_2$, plus the unit normal vector to the surface $\mathbf{n}$.
Each one of these basis sets is defined on each regular point of the curve or the surface and hence the vectors in each one of these basis sets vary in general from one point to another, i.e. they are functions of position.
Yes, there are other sets of basis vectors that are defined and used in differential geometry such as the Darboux basis set $(\mathbf{d}_1, \mathbf{d}_2, \mathbf{n})$ as described in detail in the book (also see Exercises 11, 12 and 47 of § 4).

56. Are the basis vectors necessarily of unit length and/or mutually orthogonal? If not, give examples of basis vectors which are not of unit length and/or mutually orthogonal.
**Answer**: No, the basis vectors are not necessarily of unit length and/or mutually orthogonal. For example, the basis vectors $\mathbf{E}_1$ and $\mathbf{E}_2$ of the elliptic paraboloid of Exercise 44 are:

$$\mathbf{E}_1 = \mathbf{i} + \frac{2u}{a^2}\mathbf{k} \qquad \text{and} \qquad \mathbf{E}_2 = \mathbf{j} + \frac{2v}{b^2}\mathbf{k}$$

As we see:

$$|\mathbf{E}_1| = \sqrt{1 + \frac{4u^2}{a^4}} \qquad \text{and} \qquad |\mathbf{E}_2| = \sqrt{1 + \frac{4v^2}{b^4}}$$

and hence in general they are not necessarily of unit length. Moreover:

$$\mathbf{E}_1 \cdot \mathbf{E}_2 = \frac{4uv}{a^2b^2}$$

which is not zero in general and hence they are not necessarily orthogonal.
Another example is the basis vectors $\mathbf{E}_1$ and $\mathbf{E}_2$ of the sphere of Exercise 46 which are:

$$\begin{aligned}
\mathbf{E}_1 &= +\mathbf{i}a\cos\theta\cos\phi + \mathbf{j}a\cos\theta\sin\phi - \mathbf{k}a\sin\theta \\
\mathbf{E}_2 &= -\mathbf{i}a\sin\theta\sin\phi + \mathbf{j}a\sin\theta\cos\phi
\end{aligned}$$

Therefore:

$$\begin{aligned}
|\mathbf{E}_1| &= a\sqrt{\cos^2\theta\cos^2\phi + \cos^2\theta\sin^2\phi + \sin^2\theta} \\
&= a\sqrt{\cos^2\theta\left(\cos^2\phi + \sin^2\phi\right) + \sin^2\theta} \\
&= a\sqrt{\cos^2\theta + \sin^2\theta}
\end{aligned}$$

$$\begin{aligned}
&= a \\
|\mathbf{E}_2| &= a\sqrt{\sin^2\theta\sin^2\phi + \sin^2\theta\cos^2\phi} \\
&= a\sqrt{\sin^2\theta\left(\sin^2\phi + \cos^2\phi\right)} \\
&= a\sqrt{\sin^2\theta} \\
&= a\,|\sin\theta|
\end{aligned}$$

Hence, $\mathbf{E}_1$ and $\mathbf{E}_2$ are not necessarily of unit length. However:

$$\begin{aligned}
\mathbf{E}_1\cdot\mathbf{E}_2 &= (\mathbf{i}a\cos\theta\cos\phi + \mathbf{j}a\cos\theta\sin\phi - \mathbf{k}a\sin\theta)\cdot(-\mathbf{i}a\sin\theta\sin\phi + \mathbf{j}a\sin\theta\cos\phi) \\
&= -a^2\cos\theta\cos\phi\sin\theta\sin\phi + a^2\cos\theta\cos\phi\sin\theta\sin\phi \\
&= 0
\end{aligned}$$

and hence they are orthogonal.
On the other hand, the basis vectors of the orthonormal Cartesian system (i.e. $\mathbf{i},\mathbf{j},\mathbf{k}$) are orthogonal and of unit length, that is:

$$\mathbf{i}\cdot\mathbf{j} = \mathbf{i}\cdot\mathbf{k} = \mathbf{j}\cdot\mathbf{k} = 0 \qquad \text{and} \qquad \mathbf{i}\cdot\mathbf{i} = \mathbf{j}\cdot\mathbf{j} = \mathbf{k}\cdot\mathbf{k} = 1$$

57. Define, in mathematical terms, flat and curved spaces giving examples for each.
**Answer**: A space is flat if it is possible to find a coordinate system for the space with a diagonal metric tensor whose all diagonal elements are $\pm 1$; otherwise the space is curved. In mathematical terms, an $n$D space is flat *iff* it is possible to find a coordinate system for which the line element $ds$ is given by:

$$(ds)^2 = \zeta_1(dx^1)^2 + \zeta_2(dx^2)^2 + \ldots + \zeta_n(dx^n)^2 = \sum_{i=1}^{n}\zeta_i(dx^i)^2$$

where the indexed $\zeta$ are $\pm 1$, the indexed $x$ are the coordinates of the space and the above condition applies over the entire space. Examples of flat space are plane and 3D Euclidean space. Examples of curved space are sphere and ellipsoid.

58. State a sufficient and necessary condition for an $n$D space to be flat.
**Answer**: It is the condition that we stated in the answer of the previous question, i.e. the space is flat *iff* it can be coordinated by a system whose metric tensor is diagonal with all the diagonal elements being $+1$ or $-1$ and hence the line element $ds$ is given by $(ds)^2 = \sum_{i=1}^{n}\zeta_i(dx^i)^2$ throughout the space.

59. Is it necessary that an $n$D curved space possesses universally constant curvature? If not, give an example of a space with variable curvature in sign and magnitude.
**Answer**: No. A curved space can have constant curvature such as sphere whose curvature is $1/R^2$ (where $R$ is the radius) throughout the space and can have variable curvature such as torus whose curvature varies in sign and magnitude from point to point over the surface and hence it has points with positive curvature, points with negative curvature and points with zero curvature (see Exercise 85 of § 4).

60. What is the locus of the points (if any) which are shared between the $xy$ plane and the following surfaces: (a) a sphere centered at $(0, 0, 5)$ with radius $r = 6$ (b) a sphere centered at $(1, 1, 1)$ with radius $r = 1.5$ (c) a plane passing through the point $(5, -9.6, 0)$ with a unit normal vector $(0, 0, -1)$?
**Answer**:
(a) The equation of a sphere centered at $(0, 0, 5)$ with radius $r = 6$ is $x^2+y^2+(z-5)^2 = 36$ while the equation of the $xy$ plane is $z = 0$. On substituting from the second equation into the first equation, we obtain:

$$\begin{aligned} x^2 + y^2 + (0-5)^2 &= 36 \\ x^2 + y^2 + 25 &= 36 \\ x^2 + y^2 &= 11 \end{aligned}$$

and hence the locus is a circle in the $xy$ plane that is centered on the origin with radius $r = \sqrt{11}$.
(b) The equation of a sphere centered at $(1, 1, 1)$ with radius $r = 1.5$ is $(x-1)^2 + (y-1)^2 + (z-1)^2 = 2.25$ while the equation of the $xy$ plane is $z = 0$. On substituting from the second equation into the first equation, we obtain:

$$\begin{aligned} (x-1)^2 + (y-1)^2 + (0-1)^2 &= 2.25 \\ (x-1)^2 + (y-1)^2 + 1 &= 2.25 \\ (x-1)^2 + (y-1)^2 &= 1.25 \end{aligned}$$

and hence the locus is a circle in the $xy$ plane that is centered on the point $(1, 1, 0)$ with radius $r = \sqrt{1.25}$.
(c) The equation of a plane passing through the point $(5, -9.6, 0)$ with a unit normal vector $(0, 0, -1)$ is:

$$\begin{aligned} 0\,(x-5) + 0\,(y+9.6) - 1\,(z-0) &= 0 \\ z &= 0 \end{aligned}$$

and the equation of the $xy$ plane is also $z = 0$. Hence, the two planes are identical.

61. Describe the commonly used approach for investigating the Riemannian geometry of curved manifolds.
**Answer**: The common approach is to embed the manifold in a Euclidean space of higher dimensionality and inspect the properties of the manifold from this perspective. For example, the geometry of curved surfaces (2D spaces) is investigated by immersing the surfaces in a 3D Euclidean space and examining their properties from this external enveloping 3D space.

62. Give a brief definition of homogeneous coordinate systems giving a common example of such systems.
**Answer**: Homogeneous coordinate system is a system associated with the unity tensor

(i.e. diagonal with all the diagonal elements being +1) as the metric of its underlying space. The common example of homogeneous coordinate systems is orthonormal Cartesian systems in 2D or 3D spaces.

63. What is the relation between the Christoffel symbols of the first kind and the Christoffel symbols of the second kind?
**Answer**: The Christoffel symbols of the second kind are obtained from the Christoffel symbols of the first kind by raising the third index of the first kind, that is:

$$\Gamma^k_{ij} = g^{kl}\,[ij,l]$$

where $[ij,l]$ and $\Gamma^k_{ij}$ are the Christoffel symbols of the first and second kind and $g^{kl}$ is the contravariant metric tensor. Hence, the Christoffel symbols of the first kind are obtained from the Christoffel symbols of the second kind by lowering that index, that is:

$$[ij,l] = g_{kl}\Gamma^k_{ij}$$

where $g_{kl}$ is the covariant metric tensor.

64. Find the mathematical expressions for the symbols $[12,1]$ and $\Gamma^1_{22}$ of a surface in terms of the coefficients of the surface metric tensor.
**Answer**: The Christoffel symbols of the first and second kind are given by:

$$\begin{aligned}
[\alpha\beta,\gamma] &= \frac{1}{2}\left(\partial_\beta a_{\alpha\gamma} + \partial_\alpha a_{\beta\gamma} - \partial_\gamma a_{\alpha\beta}\right) \\
\Gamma^\gamma_{\alpha\beta} &= \frac{a^{\gamma\delta}}{2}\left(\partial_\beta a_{\alpha\delta} + \partial_\alpha a_{\beta\delta} - \partial_\delta a_{\alpha\beta}\right)
\end{aligned}$$

where the indexed $a$ are the coefficients of the surface metric tensor. Hence:

$$[12,1] = \frac{1}{2}\left(\partial_2 a_{11} + \partial_1 a_{21} - \partial_1 a_{12}\right) = \frac{1}{2}\partial_2 a_{11} = \frac{1}{2}\partial_v a_{11}$$

where the symmetry of the metric tensor is used in the second equality. Similarly:

$$\begin{aligned}
\Gamma^1_{22} &= \frac{a^{1\delta}}{2}\left(\partial_2 a_{2\delta} + \partial_2 a_{2\delta} - \partial_\delta a_{22}\right) \\
&= \frac{a^{1\delta}}{2}\left(2\partial_2 a_{2\delta} - \partial_\delta a_{22}\right) \\
&= \frac{a^{11}}{2}\left(2\partial_2 a_{21} - \partial_1 a_{22}\right) + \frac{a^{12}}{2}\left(2\partial_2 a_{22} - \partial_2 a_{22}\right) \\
&= \frac{2a^{11}\partial_2 a_{21}}{2} - \frac{a^{11}\partial_1 a_{22}}{2} + \frac{2a^{12}\partial_2 a_{22}}{2} - \frac{a^{12}\partial_2 a_{22}}{2} \\
&= \frac{2a^{11}\partial_2 a_{21}}{2} - \frac{a^{11}\partial_1 a_{22}}{2} + \frac{a^{12}\partial_2 a_{22}}{2}
\end{aligned}$$

Now, since the contravariant metric tensor is the inverse of the covariant metric tensor then we have:

$$\begin{bmatrix} a^{11} & a^{12} \\ a^{21} & a^{22} \end{bmatrix} = \frac{1}{a}\begin{bmatrix} a_{22} & -a_{12} \\ -a_{21} & a_{11} \end{bmatrix}$$

where $a$ is the determinant of the surface covariant metric tensor. On substituting from the last equation into the previous equation we obtain:

$$\begin{aligned}\Gamma^1_{22} &= \frac{2a^{11}\partial_2 a_{21}}{2} - \frac{a^{11}\partial_1 a_{22}}{2} + \frac{a^{12}\partial_2 a_{22}}{2} \\ &= \frac{2a_{22}\partial_2 a_{21}}{2a} - \frac{a_{22}\partial_1 a_{22}}{2a} - \frac{a_{12}\partial_2 a_{22}}{2a} \\ &= \frac{2a_{22}\partial_v a_{21} - a_{22}\partial_u a_{22} - a_{12}\partial_v a_{22}}{2a}\end{aligned}$$

65. Using the definition of the Christoffel symbols of the first kind and the rules of tensors, derive the following equations:

$$\begin{aligned}[11,2] &= \partial_u a_{12} - \frac{\partial_v a_{11}}{2} = F_u - \frac{E_v}{2} \\ [22,1] &= \partial_v a_{12} - \frac{\partial_u a_{22}}{2} = F_v - \frac{G_u}{2}\end{aligned}$$

**Answer**: The Christoffel symbols of the first kind are given by:

$$[\alpha\beta,\gamma] = \frac{1}{2}\left(\partial_\beta a_{\alpha\gamma} + \partial_\alpha a_{\beta\gamma} - \partial_\gamma a_{\alpha\beta}\right)$$

where the indexed $a$ are the coefficients of the surface covariant metric tensor. Hence:

$$\begin{aligned}[11,2] &= \frac{1}{2}\left(\partial_1 a_{12} + \partial_1 a_{12} - \partial_2 a_{11}\right) \\ &= \frac{1}{2}\left(2\partial_1 a_{12} - \partial_2 a_{11}\right) \\ &= \partial_1 a_{12} - \frac{\partial_2 a_{11}}{2} \\ &= \partial_u a_{12} - \frac{\partial_v a_{11}}{2} \\ &= F_u - \frac{E_v}{2}\end{aligned}$$

Similarly:

$$\begin{aligned}[22,1] &= \frac{1}{2}\left(\partial_2 a_{21} + \partial_2 a_{21} - \partial_1 a_{22}\right) \\ &= \frac{1}{2}\left(2\partial_2 a_{21} - \partial_1 a_{22}\right) \\ &= \partial_2 a_{21} - \frac{\partial_1 a_{22}}{2} \\ &= \partial_v a_{12} - \frac{\partial_u a_{22}}{2} \\ &= F_v - \frac{G_u}{2}\end{aligned}$$

where the symmetry of the metric tensor is used in line 4.

66. Using the definition of the Christoffel symbols of the second kind and the rules of tensors, derive the following equations:

$$\Gamma^2_{12} = \frac{a_{11}\partial_u a_{22} - a_{12}\partial_v a_{11}}{2a} = \frac{EG_u - FE_v}{2a} = \Gamma^2_{21}$$

$$\Gamma^2_{22} = \frac{a_{11}\partial_v a_{22} - 2a_{12}\partial_v a_{12} + a_{12}\partial_u a_{22}}{2a} = \frac{EG_v - 2FF_v + FG_u}{2a}$$

**Answer**: The Christoffel symbols of the second kind are given by:

$$\Gamma^\gamma_{\alpha\beta} = \frac{a^{\gamma\delta}}{2}\left(\partial_\beta a_{\alpha\delta} + \partial_\alpha a_{\beta\delta} - \partial_\delta a_{\alpha\beta}\right)$$

where the indexed $a$ are the coefficients of the surface metric tensor. Hence:

$$\begin{aligned}
\Gamma^2_{12} &= \frac{a^{2\delta}}{2}\left(\partial_2 a_{1\delta} + \partial_1 a_{2\delta} - \partial_\delta a_{12}\right)\\
&= \frac{a^{21}}{2}\left(\partial_2 a_{11} + \partial_1 a_{21} - \partial_1 a_{12}\right) + \frac{a^{22}}{2}\left(\partial_2 a_{12} + \partial_1 a_{22} - \partial_2 a_{12}\right)\\
&= \frac{a^{21}}{2}\partial_2 a_{11} + \frac{a^{21}}{2}\partial_1 a_{21} - \frac{a^{21}}{2}\partial_1 a_{12} + \frac{a^{22}}{2}\partial_2 a_{12} + \frac{a^{22}}{2}\partial_1 a_{22} - \frac{a^{22}}{2}\partial_2 a_{12}\\
&= \frac{a^{21}}{2}\partial_2 a_{11} + \frac{a^{22}}{2}\partial_1 a_{22}\\
&= \frac{a^{21}\partial_2 a_{11} + a^{22}\partial_1 a_{22}}{2}\\
&= \frac{-a_{21}\partial_2 a_{11} + a_{11}\partial_1 a_{22}}{2a}\\
&= \frac{a_{11}\partial_u a_{22} - a_{12}\partial_v a_{11}}{2a}\\
&= \frac{EG_u - FE_v}{2a}\\
&= \Gamma^2_{21}
\end{aligned}$$

where the symmetry of the metric tensor is used in lines 4 and 7 while the fact that the contravariant and covariant forms of the metric tensor are inverses of each other is used in line 6 (see Exercise 64). The equality $\Gamma^2_{12} = \Gamma^2_{21}$ is based on the symmetry of the Christoffel symbols in their paired indices.
Similarly:

$$\begin{aligned}
\Gamma^2_{22} &= \frac{a^{2\delta}}{2}\left(\partial_2 a_{2\delta} + \partial_2 a_{2\delta} - \partial_\delta a_{22}\right)\\
&= \frac{a^{21}}{2}\left(\partial_2 a_{21} + \partial_2 a_{21} - \partial_1 a_{22}\right) + \frac{a^{22}}{2}\left(\partial_2 a_{22} + \partial_2 a_{22} - \partial_2 a_{22}\right)\\
&= \frac{2a^{21}}{2}\partial_2 a_{21} - \frac{a^{21}}{2}\partial_1 a_{22} + \frac{a^{22}}{2}\partial_2 a_{22}
\end{aligned}$$

$$\begin{aligned}
&= \frac{2a^{21}\partial_2 a_{21} - a^{21}\partial_1 a_{22} + a^{22}\partial_2 a_{22}}{2} \\
&= \frac{-2a_{21}\partial_2 a_{21} + a_{21}\partial_1 a_{22} + a_{11}\partial_2 a_{22}}{2a} \\
&= \frac{a_{11}\partial_2 a_{22} - 2a_{21}\partial_2 a_{21} + a_{21}\partial_1 a_{22}}{2a} \\
&= \frac{a_{11}\partial_v a_{22} - 2a_{12}\partial_v a_{12} + a_{12}\partial_u a_{22}}{2a} \\
&= \frac{EG_v - 2FF_v + FG_u}{2a}
\end{aligned}$$

where line 5 is justified by the fact that the contravariant and covariant forms of the metric tensor are inverses of each other (see Exercise 64) while line 7 is justified by the symmetry of the metric tensor.

67. State the mathematical relations that correlate the Christoffel symbols of the first and second kind to the surface basis vectors and their derivatives.
**Answer**: They are:

$$\begin{aligned}
[\alpha\beta,\gamma] &= \frac{\partial \mathbf{E}_\alpha}{\partial u^\beta} \cdot \mathbf{E}_\gamma \\
\Gamma^\gamma_{\alpha\beta} &= \frac{\partial \mathbf{E}_\alpha}{\partial u^\beta} \cdot \mathbf{E}^\gamma
\end{aligned}$$

where $\alpha, \beta, \gamma = 1, 2$.

68. What is the relation between the Riemann-Christoffel curvature tensor and the Gaussian curvature of a surface?
**Answer**: It is:

$$R_{\alpha\beta\gamma\delta} = K \underline{\epsilon}_{\alpha\beta} \underline{\epsilon}_{\gamma\delta}$$

where $R_{\alpha\beta\gamma\delta}$ is the Riemann-Christoffel curvature tensor, $K$ is the Gaussian curvature, $\underline{\epsilon}_{\alpha\beta}$ and $\underline{\epsilon}_{\gamma\delta}$ are the absolute permutation tensor for surface, and $\alpha, \beta, \gamma, \delta = 1, 2$.

69. How many independent non-vanishing components the 2D Riemann-Christoffel curvature tensor possesses?
**Answer**: It possesses only one independent non-vanishing component.

70. What is the significance of having an identically vanishing Riemann-Christoffel curvature tensor on a 2D surface?
**Answer**: The significance is that the surface is intrinsically flat and hence it is a plane or isometric to plane.

71. Show that the Riemann-Christoffel curvature tensor of the first kind is anti-symmetric in its first two indices and in its last two indices and block symmetric with respect to these two sets of indices.
**Answer**: Riemann-Christoffel curvature tensor of the first kind is given by:

$$R_{ijkl} = \partial_k [jl, i] - \partial_l [jk, i] + [il, r]\,\Gamma^r_{jk} - [ik, r]\,\Gamma^r_{jl}$$

$$= \frac{1}{2}\left(\partial_k\partial_j g_{li} + \partial_l\partial_i g_{jk} - \partial_k\partial_i g_{jl} - \partial_l\partial_j g_{ki}\right) + g^{rs}\left([il,r]\,[jk,s] - [ik,r]\,[jl,s]\right) \quad (4)$$

where the second equality is justified in the text books of tensor calculus.[8] Hence, if we take Eq. 4 as definition for the Riemann-Christoffel curvature tensor of the first kind then we have:

• Anti-symmetry in the first two indices:

$$\begin{aligned}
R_{jikl} &= \frac{1}{2}\left(\partial_k\partial_i g_{lj} + \partial_l\partial_j g_{ik} - \partial_k\partial_j g_{il} - \partial_l\partial_i g_{kj}\right) + g^{rs}\left([jl,r]\,[ik,s] - [jk,r]\,[il,s]\right) \\
&= -\left[\frac{1}{2}\left(\partial_k\partial_j g_{il} + \partial_l\partial_i g_{kj} - \partial_k\partial_i g_{lj} - \partial_l\partial_j g_{ik}\right) + g^{rs}\left([jk,r]\,[il,s] - [jl,r]\,[ik,s]\right)\right] \\
&= -\left[\frac{1}{2}\left(\partial_k\partial_j g_{li} + \partial_l\partial_i g_{jk} - \partial_k\partial_i g_{jl} - \partial_l\partial_j g_{ki}\right) + g^{sr}\left([il,s]\,[jk,r] - [ik,s]\,[jl,r]\right)\right] \\
&= -\left[\frac{1}{2}\left(\partial_k\partial_j g_{li} + \partial_l\partial_i g_{jk} - \partial_k\partial_i g_{jl} - \partial_l\partial_j g_{ki}\right) + g^{rs}\left([il,r]\,[jk,s] - [ik,r]\,[jl,s]\right)\right] \\
&= -R_{ijkl}
\end{aligned}$$

where line 1 is obtained from the given definition (i.e Eq. 4) with exchanging the indices $ij$, line 2 is obtained by taking a common factor of $-1$, line 3 is based on the symmetry of metric tensor plus reordering, line 4 is relabeling of dummy indices, and line 5 is the definition of $R_{ijkl}$ as given by Eq. 4.

• Anti-symmetry in the last two indices:

$$\begin{aligned}
R_{ijlk} &= \frac{1}{2}\left(\partial_l\partial_j g_{ki} + \partial_k\partial_i g_{jl} - \partial_l\partial_i g_{jk} - \partial_k\partial_j g_{li}\right) + g^{rs}\left([ik,r]\,[jl,s] - [il,r]\,[jk,s]\right) \\
&= -\left[\frac{1}{2}\left(\partial_l\partial_i g_{jk} + \partial_k\partial_j g_{li} - \partial_l\partial_j g_{ki} - \partial_k\partial_i g_{jl}\right) + g^{rs}\left([il,r]\,[jk,s] - [ik,r]\,[jl,s]\right)\right] \\
&= -\left[\frac{1}{2}\left(\partial_k\partial_j g_{li} + \partial_l\partial_i g_{jk} - \partial_k\partial_i g_{jl} - \partial_l\partial_j g_{ki}\right) + g^{rs}\left([il,r]\,[jk,s] - [ik,r]\,[jl,s]\right)\right] \\
&= -R_{ijkl}
\end{aligned}$$

where line 1 is obtained from the given definition (i.e Eq. 4) with exchanging the indices $kl$, line 2 is obtained by taking a common factor of $-1$, line 3 is reordering of terms, and line 4 is the definition of $R_{ijkl}$ as given by Eq. 4.

• Block symmetry:

$$\begin{aligned}
R_{klij} &= \frac{1}{2}\left(\partial_i\partial_l g_{jk} + \partial_j\partial_k g_{li} - \partial_i\partial_k g_{lj} - \partial_j\partial_l g_{ik}\right) + g^{rs}\left([kj,r]\,[li,s] - [ki,r]\,[lj,s]\right) \\
&= \frac{1}{2}\left(\partial_l\partial_i g_{jk} + \partial_k\partial_j g_{li} - \partial_k\partial_i g_{lj} - \partial_l\partial_j g_{ik}\right) + g^{rs}\left([kj,r]\,[li,s] - [ki,r]\,[lj,s]\right) \\
&= \frac{1}{2}\left(\partial_k\partial_j g_{li} + \partial_l\partial_i g_{jk} - \partial_k\partial_i g_{lj} - \partial_l\partial_j g_{ik}\right) + g^{rs}\left([li,s]\,[kj,r] - [ki,r]\,[lj,s]\right)
\end{aligned}$$

[8] See for example: Principles of Tensor Calculus by the author.

$$
\begin{aligned}
&= \frac{1}{2}\left(\partial_k \partial_j g_{li} + \partial_l \partial_i g_{jk} - \partial_k \partial_i g_{jl} - \partial_l \partial_j g_{ki}\right) + g^{rs}\left([il,s]\,[jk,r] - [ik,r]\,[jl,s]\right) \\
&= \frac{1}{2}\left(\partial_k \partial_j g_{li} + \partial_l \partial_i g_{jk} - \partial_k \partial_i g_{jl} - \partial_l \partial_j g_{ki}\right) + g^{rs}\left([il,r]\,[jk,s] - [ik,r]\,[jl,s]\right) \\
&= R_{ijkl}
\end{aligned}
$$

where line 1 is obtained from the given definition (i.e Eq. 4) with relabeling the indices ($ijkl \to klij$), line 2 is the commutativity of partial differential operators, line 3 is reordering of terms and factors, line 4 is based on the symmetry of the metric tensor and the symmetry of the Christoffel symbols in their paired indices, line 5 is relabeling of dummy indices plus the symmetry of the metric tensor, and line 6 is the definition of $R_{ijkl}$ as given by Eq. 4.

72. Prove the following equations:

$$
\begin{aligned}
[\alpha\beta,\gamma] &= \frac{\partial \mathbf{E}_\alpha}{\partial u^\beta}\cdot \mathbf{E}_\gamma \\
\Gamma^\gamma_{\alpha\beta} &= \frac{\partial \mathbf{E}_\alpha}{\partial u^\beta}\cdot \mathbf{E}^\gamma
\end{aligned}
$$

**Answer**:

• First equation:

$$
\begin{aligned}
[\alpha\beta,\gamma] &= \frac{1}{2}\left(\partial_\beta a_{\alpha\gamma} + \partial_\alpha a_{\beta\gamma} - \partial_\gamma a_{\alpha\beta}\right) \\
2\,[\alpha\beta,\gamma] &= \partial_\beta a_{\alpha\gamma} + \partial_\alpha a_{\beta\gamma} - \partial_\gamma a_{\alpha\beta} \\
2\,[\alpha\beta,\gamma] &= \partial_\beta\left(\mathbf{E}_\alpha\cdot\mathbf{E}_\gamma\right) + \partial_\alpha\left(\mathbf{E}_\beta\cdot\mathbf{E}_\gamma\right) - \partial_\gamma\left(\mathbf{E}_\alpha\cdot\mathbf{E}_\beta\right) \\
2\,[\alpha\beta,\gamma] &= \left(\partial_\beta\mathbf{E}_\alpha\right)\cdot\mathbf{E}_\gamma + \mathbf{E}_\alpha\cdot\left(\partial_\beta\mathbf{E}_\gamma\right) + \left(\partial_\alpha\mathbf{E}_\beta\right)\cdot\mathbf{E}_\gamma + \\
&\quad \mathbf{E}_\beta\cdot\left(\partial_\alpha\mathbf{E}_\gamma\right) - \left(\partial_\gamma\mathbf{E}_\alpha\right)\cdot\mathbf{E}_\beta - \mathbf{E}_\alpha\cdot\left(\partial_\gamma\mathbf{E}_\beta\right) \\
2\,[\alpha\beta,\gamma] &= \left(\partial_\beta\mathbf{E}_\alpha\right)\cdot\mathbf{E}_\gamma + \mathbf{E}_\alpha\cdot\left(\partial_\beta\mathbf{E}_\gamma\right) + \left(\partial_\beta\mathbf{E}_\alpha\right)\cdot\mathbf{E}_\gamma + \\
&\quad \mathbf{E}_\beta\cdot\left(\partial_\alpha\mathbf{E}_\gamma\right) - \left(\partial_\alpha\mathbf{E}_\gamma\right)\cdot\mathbf{E}_\beta - \mathbf{E}_\alpha\cdot\left(\partial_\beta\mathbf{E}_\gamma\right) \\
2\,[\alpha\beta,\gamma] &= 2\left(\partial_\beta\mathbf{E}_\alpha\right)\cdot\mathbf{E}_\gamma \\
[\alpha\beta,\gamma] &= \left(\partial_\beta\mathbf{E}_\alpha\right)\cdot\mathbf{E}_\gamma \\
[\alpha\beta,\gamma] &= \frac{\partial \mathbf{E}_\alpha}{\partial u^\beta}\cdot\mathbf{E}_\gamma
\end{aligned}
$$

where equality 1 is the definition of the Riemann-Christoffel curvature tensor of the first kind, equality 3 is based on the relation between the coefficients of the metric tensor and the basis vectors, equality 4 is the product rule of differentiation, and equality 5 is based on the identity $\partial_\alpha\mathbf{E}_\beta = \partial_\beta\mathbf{E}_\alpha$ which can be justified as follows:

$$
\partial_\alpha\mathbf{E}_\beta = \partial_\alpha\partial_\beta\mathbf{r} = \partial_\beta\partial_\alpha\mathbf{r} = \partial_\beta\mathbf{E}_\alpha
$$

while the remaining equalities are simple algebraic manipulation and notation.

• Second equation:

$$
\Gamma^\gamma_{\alpha\beta} = a^{\gamma\delta}\,[\alpha\beta,\delta]
$$

$$= a^{\gamma\delta}\left(\frac{\partial \mathbf{E}_\alpha}{\partial u^\beta}\cdot \mathbf{E}_\delta\right)$$

$$= \frac{\partial \mathbf{E}_\alpha}{\partial u^\beta}\cdot \mathbf{E}^\gamma$$

where equality 1 is the definition of the Riemann-Christoffel curvature tensor of the second kind, equality 2 is based on the first equation of this question which was proved in the first part, and equality 3 is an index raising operation.

73. What is the rank of the Ricci curvature tensor?
**Answer**: It is 2.

74. State the mathematical relation that links the Ricci curvature tensor of the first kind to the Christoffel symbols of the second kind and their partial derivatives.
**Answer**: It is:

$$R_{ij} = \partial_j \Gamma^a_{ia} - \partial_a \Gamma^a_{ij} + \Gamma^a_{bj}\Gamma^b_{ia} - \Gamma^a_{ba}\Gamma^b_{ij}$$

where $R_{ij}$ is the Ricci curvature tensor of the first kind and the indexed $\Gamma$ are the Christoffel symbols of the second kind.

75. What is the relation between the elements of the Ricci curvature tensor of the first kind and the Gaussian curvature?
**Answer**: It is:

$$\frac{R_{11}}{a_{11}} = \frac{R_{12}}{a_{12}} = \frac{R_{21}}{a_{21}} = \frac{R_{22}}{a_{22}} = -K$$

where the indexed $R$ are the elements of the Ricci curvature tensor of the first kind, the indexed $a$ are the elements of the covariant metric tensor and $K$ is the Gaussian curvature.

76. How do you obtain the Ricci scalar from the Riemann-Christoffel curvature tensor of the first kind? Explain your answer step by step.
**Answer**: We do the following:
(a) We obtain the Riemann-Christoffel curvature tensor of the second kind by raising the first index of the Riemann-Christoffel curvature tensor of the first kind, that is:

$$R^{\omega}{}_{\alpha\beta\gamma} = a^{\omega\phi} R_{\phi\alpha\beta\gamma}$$

(b) We obtain the Ricci curvature tensor of the first kind by contracting the contravariant index with the last covariant index of the Riemann-Christoffel curvature tensor of the second kind, that is:

$$R_{\alpha\beta} = \delta^{\gamma}_{\omega} R^{\omega}{}_{\alpha\beta\gamma} = R^{\omega}{}_{\alpha\beta\omega}$$

(c) We obtain the Ricci curvature tensor of the second kind by raising the first index of the Ricci curvature tensor of the first kind, that is:

$$R^{\gamma}{}_{\beta} = a^{\gamma\alpha} R_{\alpha\beta}$$

(d) We obtain the Ricci curvature scalar $\mathcal{R}$ by contracting the indices of the Ricci curvature tensor of the second kind, that is:

$$\mathcal{R} = \delta^{\beta}_{\gamma} R^{\gamma}{}_{\beta} = R^{\beta}{}_{\beta}$$

# Chapter 2
# Curves in Space

1. State the technical definition of space curve outlining the difference between a curve and its trace.
   **Answer**: Space curve is technically defined as a differentiable parameterized mapping between an interval of the real line and a connected subset of the embedding space, that is:
   $$C(t) : I \rightarrow \mathbb{R}^n$$
   where $C$ represents a space curve defined on the interval $I \subseteq \mathbb{R}$ and parameterized by the variable $t \in I$.
   The difference between a curve and its trace is that the curve is a mathematical relation (or mapping) while the trace is the image of this mapping in the embedding space.

2. What is the most common way of defining space curves mathematically? Give an example from simple curves like circle and helix.
   **Answer**: The most common way of defining space curves mathematically is by parameterization. For example, a circle in the $xy$ plane centered on the origin is defined parametrically as:
   $$\begin{aligned} x(\theta) &= R\cos\theta \\ y(\theta) &= R\sin\theta \\ z(\theta) &= 0 \end{aligned}$$
   where $R$ is a constant (i.e. radius) while $\theta$ $(0 \le \theta < 2\pi)$ is a real parameter (which may be seen as an angle of rotation).
   Similarly, a regular helix spinning around the $z$-axis may be defined parametrically as:
   $$\begin{aligned} x(t) &= a\cos t \\ y(t) &= a\sin t \\ z(t) &= bt \end{aligned}$$
   where $a$ and $b$ are non-zero real constants while $t$ $(-\infty < t < +\infty)$ is a real parameter.

3. Make a clear distinction between general and natural parameterization of space curves.
   **Answer**: Natural parameterization of a curve is parameterization by arc length while general parameterization is parameterization by any legitimate parameter whether arc length or not. General parameterization may also be restricted to parameterization by parameters other than arc length.

4. State a mathematical condition for a space curve to be parameterized naturally.
**Answer**: The condition for a space curve $C(t) : I \rightarrow \mathbb{R}^3$ (where $t \in I$ is the curve parameter and $I \subseteq \mathbb{R}$ is an interval over which the curve is defined) to be parameterized naturally is that for all values of $t$ we have:

$$\left|\frac{d\mathbf{r}}{dt}\right| = 1$$

where $\mathbf{r}(t)$ is the position vector representing the curve in the ambient space.

5. What is the relation between two natural parameters of a given space curve?
**Answer**: The relation between two natural parameters, $s$ and $\check{s}$, is:

$$\check{s} = \pm s + c$$

where $c$ is a real constant.

6. The following equation: $\mathbf{r}(t) = \left(\frac{t}{2}, -\frac{t}{2}, \frac{t}{\sqrt{2}}\right)$ is a parametric representation of a space curve. Is this a natural parameterization? Justify your answer.
**Answer**: The condition for natural parameterization is:

$$\left|\frac{d\mathbf{r}}{dt}\right| = 1$$

Hence:

$$\begin{aligned}
\left|\frac{d\mathbf{r}(t)}{dt}\right| &= \left|\frac{d}{dt}\left(\frac{t}{2}, -\frac{t}{2}, \frac{t}{\sqrt{2}}\right)\right| \\
&= \left|\left(\frac{1}{2}, -\frac{1}{2}, \frac{1}{\sqrt{2}}\right)\right| \\
&= \sqrt{\left(\frac{1}{2}\right)^2 + \left(-\frac{1}{2}\right)^2 + \left(\frac{1}{\sqrt{2}}\right)^2} \\
&= \sqrt{\frac{1}{4} + \frac{1}{4} + \frac{1}{2}} \\
&= 1
\end{aligned}$$

Therefore, this a natural parameterization.

7. Prove that two natural parameters, $s$ and $\check{s}$, of a curve are related by the equation $\check{s} = \pm s + c$ where $c$ is a real constant.
**Answer**: If we correlate $\check{s}$ to $s$ $\left[\text{i.e. } \check{s} = \check{s}(s)\right]$ then by the chain rule of differentiation we have:

$$\frac{d\mathbf{r}}{ds} = \frac{d\mathbf{r}}{d\check{s}}\frac{d\check{s}}{ds}$$

Hence:

$$\left|\frac{d\mathbf{r}}{ds}\right| = \left|\frac{d\mathbf{r}}{d\check{s}}\right|\left|\frac{d\check{s}}{ds}\right|$$

But since $s$ and $\check{s}$ are natural parameters then we have:

$$\left|\frac{d\mathbf{r}}{ds}\right| = 1 \qquad \text{and} \qquad \left|\frac{d\mathbf{r}}{d\check{s}}\right| = 1$$

Therefore:

$$\begin{aligned}
\left|\frac{d\mathbf{r}}{ds}\right| &= \left|\frac{d\mathbf{r}}{d\check{s}}\right|\left|\frac{d\check{s}}{ds}\right| \\
1 &= 1 \times \left|\frac{d\check{s}}{ds}\right| \\
\left|\frac{d\check{s}}{ds}\right| &= 1 \\
\frac{d\check{s}}{ds} &= \pm 1 \\
\check{s} &= \pm s + c
\end{aligned}$$

where $c$ is a real constant, as required.

8. Using tensor notation, write down the equation of the tangent vector to a surface curve represented parametrically by $C(u^1(t), u^2(t))$ where $t$ is a general parameter.
**Answer**: It is:

$$\frac{dx^i}{dt} = x^i_\alpha \frac{du^\alpha}{dt}$$

where $x^i$ represent spatial coordinates, $x^i_\alpha \equiv \frac{\partial x^i}{\partial u^\alpha}$ is the surface basis vector, and $i = 1, 2, 3$ and $\alpha = 1, 2$.

9. What is the meaning of having "regular curve at a specific point"? What "regular curve" means?
**Answer**: A space curve $C(t) : I \to \mathbb{R}^3$ (where $I \subseteq R$ and $t \in I$ is a general parameter) is described as "regular curve at a specific point" $P$ *iff* $\dot{\mathbf{r}}(t_0)$ exists and $\dot{\mathbf{r}}(t_0) \neq \mathbf{0}$ where $\mathbf{r}(t)$ is the spatial representation of $C$, the overdot stands for differentiation with respect to the general parameter $t$, and $t_0$ is the value of the parameter corresponding to the specific point $P$. The curve is described as "regular curve" *iff* it is regular at each interior point in $I$.

10. Prove that the parametric representation: $\mathbf{r}(t) = (t^2, e^t, t+1)$ of a space curve is regular for all $t$.
**Answer**: We have:

$$\dot{\mathbf{r}} \equiv \frac{d\mathbf{r}}{dt} = (2t, e^t, 1)$$

and hence $\dot{\mathbf{r}}$ exists for all $t$ $(-\infty < t < +\infty)$. Moreover:

$$\dot{\mathbf{r}} = (2t, e^t, 1) \neq (0, 0, 0) \equiv \mathbf{0}$$

for any $t$. Hence, the parametric representation of this space curve is regular for all $t$, as required.

11. State the condition for a vector to be tangent to a regular surface at a given point on the surface.
**Answer**: A vector $\mathbf{v}$ is described as tangent to a regular surface $S$ at a given point $P$ on $S$ if there is a regular curve $C$ on $S$ passing through $P$ such that $\mathbf{v} = \alpha \frac{d\mathbf{r}(t)}{dt}$ where $\alpha$ is a non-zero scalar, $\mathbf{r}(t)$ is the spatial representation of $C$ and $\frac{d\mathbf{r}(t)}{dt}$ is evaluated at $P$.

12. Find the unit tangent vector, $\mathbf{T}(t)$, for a space curve represented by: $\mathbf{r}(t) = (t^2, t, \sin t)$.
**Answer**: The tangent of this space curve is:

$$\frac{d\mathbf{r}}{dt} = (2t, 1, \cos t)$$

Hence:

$$\mathbf{T} = \frac{(2t, 1, \cos t)}{|(2t, 1, \cos t)|} = \frac{(2t, 1, \cos t)}{\sqrt{4t^2 + 1 + \cos^2 t}}$$

13. Find the arc length of a space curve given by: $\mathbf{r}(t) = (5t, 7\cosh t, 2\sinh t)$ for $1 \leq t \leq 4$.
**Answer**: If $L$ is the required arc length and $I$ is the interval $[1, 4]$ then we have:

$$\begin{aligned} L &= \int_I |d\mathbf{r}| \\ &= \int_1^4 \left| \frac{d\mathbf{r}}{dt} \right| dt \\ &= \int_1^4 |(5, 7\sinh t, 2\cosh t)|\, dt \\ &= \int_1^4 \sqrt{25 + 49\sinh^2 t + 4\cosh^2 t}\, dt \\ &\simeq 188.995 \end{aligned}$$

where the last step is obtained by numerical integration.[9]

14. Find the curvature, as a function of $t$, of a space curve represented by:

$$\mathbf{r}(t) = \left(\cos t - 1, \sin t + t, t^2\right)$$

**Answer**: We have:

$$\dot{\mathbf{r}} = \frac{d\mathbf{r}}{dt} = (-\sin t, \cos t + 1, 2t)$$

[9] It may also be obtained analytically using elliptic integrals of the second kind.

$$\begin{aligned}
\ddot{\mathbf{r}} &= \frac{d\dot{\mathbf{r}}}{dt} = (-\cos t, -\sin t, 2) \\
\dot{\mathbf{r}} \cdot \dot{\mathbf{r}} &= \sin^2 t + \cos^2 t + 2\cos t + 1 + 4t^2 = 2 + 2\cos t + 4t^2 \\
\ddot{\mathbf{r}} \cdot \ddot{\mathbf{r}} &= \cos^2 t + \sin^2 t + 4 = 5 \\
\dot{\mathbf{r}} \cdot \ddot{\mathbf{r}} &= \sin t \cos t - \sin t \cos t - \sin t + 4t = -\sin t + 4t
\end{aligned}$$

Hence, the curvature $\kappa$ is:

$$\begin{aligned}
\kappa &= \frac{\sqrt{(\dot{\mathbf{r}} \cdot \dot{\mathbf{r}})(\ddot{\mathbf{r}} \cdot \ddot{\mathbf{r}}) - (\dot{\mathbf{r}} \cdot \ddot{\mathbf{r}})^2}}{(\dot{\mathbf{r}} \cdot \dot{\mathbf{r}})^{3/2}} \\
&= \frac{\sqrt{(2 + 2\cos t + 4t^2)(5) - (-\sin t + 4t)^2}}{(2 + 2\cos t + 4t^2)^{3/2}} \\
&= \frac{\sqrt{(10 + 10\cos t + 20t^2) - \left(\sin^2 t - 8t\sin t + 16t^2\right)}}{(2 + 2\cos t + 4t^2)^{3/2}} \\
&= \frac{\sqrt{10 + 10\cos t + 20t^2 - \sin^2 t + 8t\sin t - 16t^2}}{(2 + 2\cos t + 4t^2)^{3/2}} \\
&= \frac{\sqrt{10 + 10\cos t + 4t^2 - \sin^2 t + 8t\sin t}}{(2 + 2\cos t + 4t^2)^{3/2}}
\end{aligned}$$

15. Define "periodic curve" giving two common examples other than those given in the text.
**Answer**: A curve $C$ is described as periodic if it can be represented parametrically by a continuous function of the form $\mathbf{r}(t + T) = \mathbf{r}(t)$ where $\mathbf{r}$ is the spatial representation of $C$, $t$ is a general parameter and $T$ is a constant called the function period.
Polar curves provide many common examples of periodic curves. One example[10] is rose curves which are defined parametrically by:

$$\rho = A\cos(n\phi) \qquad \text{or} \qquad \rho = A\sin(n\phi)$$

where $\rho$ and $\phi$ are plane polar coordinates, $A$ is a constant (represents the magnitude of petals) and $n$ is an integer that determines the number of petals[11] (see Figure 9). Common types of limacon are also good examples of periodic curves.

16. Should a periodic curve be a plane curve?
**Answer**: No. For example, a curve defined parametrically by $\mathbf{r}(\theta) = \left(\cos\theta, 2\sin\theta, 3\sin^2\theta\right)$ is a periodic curve (with period $T = 2\pi$) but it is not a plane curve (see Figure 10).

17. Should a periodic continuous curve be a closed curve? Should a periodic smooth curve be a closed curve?

[10] In fact, this is a group of examples.

[11] The number of petals is $n$ for odd $n$ and $2n$ for even $n$.

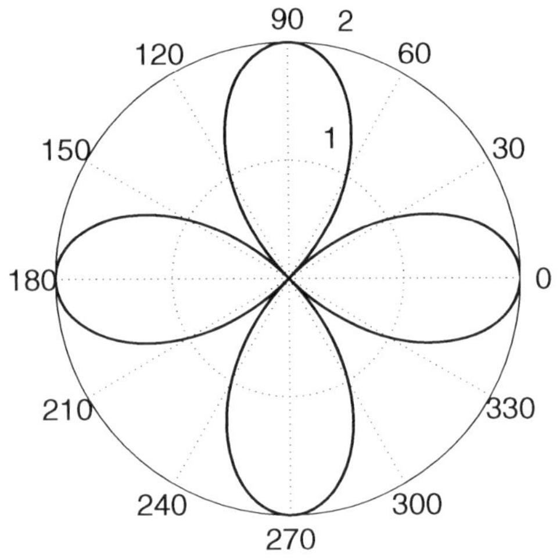


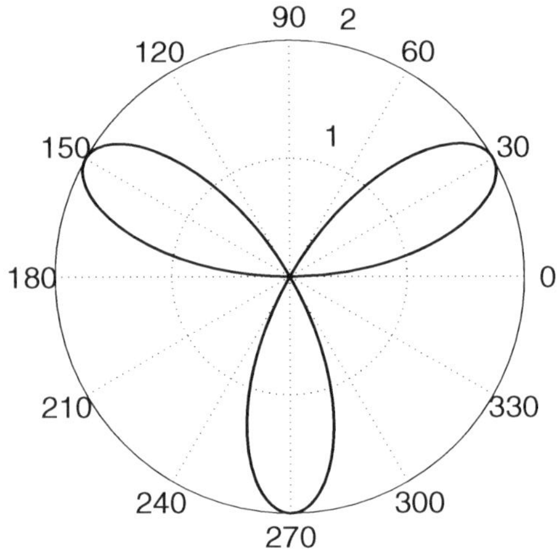


Figure 9: Examples of periodic rose curves: (a) the polar curve $\rho = 2\cos(2\phi)$ which is of period $T = 2\pi$ and (b) the polar curve $\rho = 2\sin(3\phi)$ which is of period $T = \pi$. The numbers on the perimeter are the polar angle $\phi$ in degrees.

**Answer**: Yes, a continuous periodic curve should be a closed curve (assuming it is defined over a minimum of one period).
Yes, a periodic smooth curve should be a closed curve (assuming it is defined over a minimum of one period).

18. A plane curve called *cissoid of Diocles* is given in polar coordinates by: $\rho = 2\sin\phi\tan\phi$.[12]
Find the equation of the curve in a rectangular Cartesian coordinate system.
**Answer**: We have:

$$\begin{aligned}
\rho &= 2\sin\phi\tan\phi \\
\rho &= 2\sin\phi\frac{\sin\phi}{\cos\phi} \\
\rho &= 2\frac{\sin^2\phi}{\cos\phi} \\
\rho\cos\phi &= 2\sin^2\phi \\
x &= 2\sin^2\phi \\
x &= 2\frac{y^2}{x^2+y^2} \\
x\left(x^2+y^2\right) &= 2y^2
\end{aligned}$$

where in line 5 we used the polar to Cartesian transformation $x = \rho\cos\phi$ while in line

[12] In fact, this curve in its more general form is given by $\rho = 2A\sin\phi\tan\phi$ where $A$ is a constant and hence the form given in this question is a special case corresponding to $A = 1$.

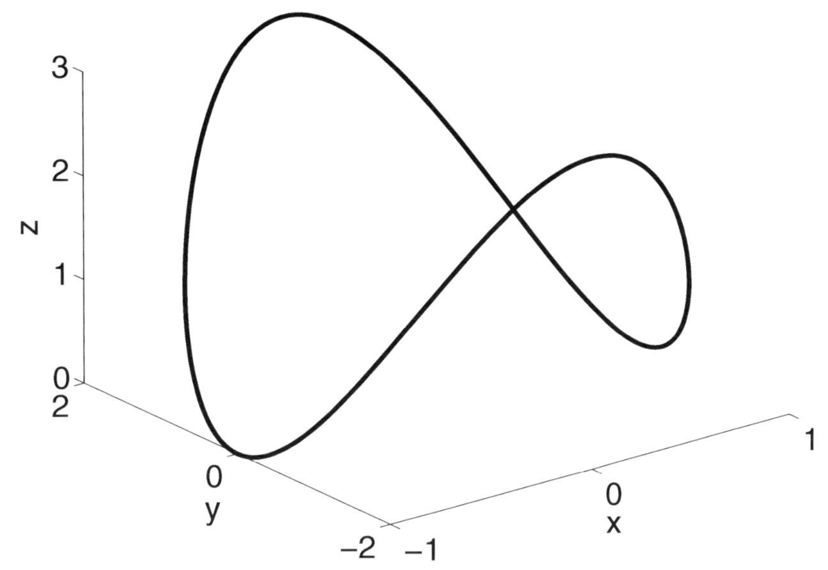


Figure 10: The periodic twisted curve $\mathbf{r}(\theta) = \left(\cos\theta, 2\sin\theta, 3\sin^2\theta\right)$.

6 we used the polar to Cartesian transformation $y = \rho \sin\phi$ (i.e. $\sin\phi = y/\rho$) plus the Cartesian to polar transformation $\rho = \sqrt{x^2 + y^2}$.

19. Sketch the curve of the previous question for $0 \le |\phi| \le \frac{\pi}{4}$ (notice the two branches).
**Answer**: The polar plot of this curve is shown in Figure 11.

20. Show that for a curve represented spatially by $\mathbf{r}$ and parameterized naturally by $s$ and generally by $t$, the relation between $s$ and $t$ is given by: $\left|\frac{ds}{dt}\right| = \left|\frac{d\mathbf{r}}{dt}\right|$.
**Answer**: If we correlate $s$ to $t$ then we have $s = s(t)$ and hence:

$$\begin{aligned} \frac{d\mathbf{r}}{dt} &= \frac{d\mathbf{r}}{ds}\frac{ds}{dt} \\ \left|\frac{d\mathbf{r}}{dt}\right| &= \left|\frac{d\mathbf{r}}{ds}\right|\left|\frac{ds}{dt}\right| \\ \left|\frac{d\mathbf{r}}{dt}\right| &= \left|\frac{ds}{dt}\right| \end{aligned}$$

where line 1 is justified by the chain rule of differentiation while line 3 is justified by the identity $\left|\frac{d\mathbf{r}}{ds}\right| = 1$ since $s$ is a natural parameter.

21. Define Frenet frame with a simple sketch showing the basis vectors at a given point on an arbitrary space curve.
**Answer**: Frenet frame is a set of orthonormal basis vectors for the embedding space that consists of the triad $\mathbf{T}, \mathbf{N}, \mathbf{B}$ which are the unit tangent, the unit normal and the unit binormal vectors of the space curve to which the frame belongs. The required sketch should be similar to Figure 12.

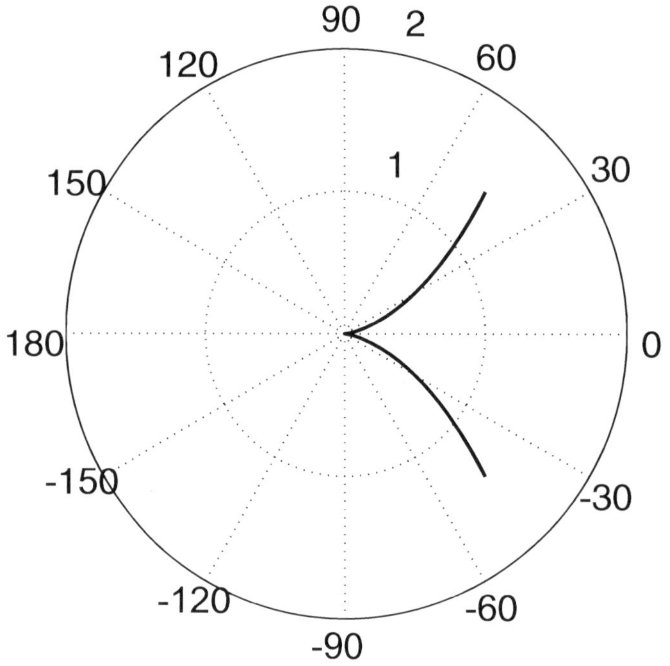


Figure 11: The polar plot of the *cissoid of Diocles* curve $\rho = 2 \sin \phi \tan \phi$ for polar angle $0 \le |\phi| \le \frac{\pi}{4}$. The numbers on the perimeter are the polar angle $\phi$ in degrees.

22. Write down the mathematical equations of the unit vectors $\mathbf{T}, \mathbf{N}, \mathbf{B}$ for a curve parameterized by a general parameter $t$ and a natural parameter $s$.
**Answer**: For a curve parameterized by a general parameter $t$ we have:

$$\begin{aligned} \mathbf{T} &= \frac{\dot{\mathbf{r}}}{|\dot{\mathbf{r}}|} \\ \mathbf{N} &= \frac{\dot{\mathbf{r}} \times (\ddot{\mathbf{r}} \times \dot{\mathbf{r}})}{|\dot{\mathbf{r}}|\,|\ddot{\mathbf{r}} \times \dot{\mathbf{r}}|} \\ \mathbf{B} &= \frac{\dot{\mathbf{r}} \times \ddot{\mathbf{r}}}{|\dot{\mathbf{r}} \times \ddot{\mathbf{r}}|} \end{aligned}$$

where $\mathbf{r} = \mathbf{r}(t)$ is the spatial representation of the curve and the overdot represents derivative with respect to $t$.
For a curve parameterized by a natural parameter $s$ we have:

$$\mathbf{T} = \mathbf{r}' \tag{5}$$

$$\kappa\mathbf{N} = \mathbf{T}' \tag{6}$$

$$\tau\mathbf{B} = \kappa\mathbf{T} + \mathbf{N}' \tag{7}$$

where $\mathbf{r} = \mathbf{r}(s)$ is the spatial representation of the curve, $\kappa$ and $\tau$ are the curvature and torsion of the curve and the prime represents derivative with respect to $s$.

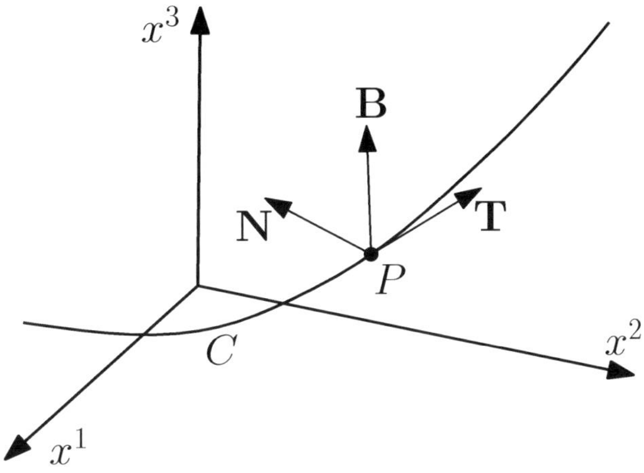


Figure 12: Frenet frame that consists of the basis vectors $\mathbf{T}, \mathbf{N}, \mathbf{B}$ at point $P$ on a space curve $C$.

23. State the mathematical definition of the curvature $\kappa$ and the torsion $\tau$ of an $s$-parameterized space curve.
**Answer**:

$$\begin{aligned} \kappa &= \mathbf{N} \cdot \mathbf{T}' \\ \tau &= \mathbf{N}' \cdot \mathbf{B} \end{aligned}$$

where $\mathbf{T}, \mathbf{N}, \mathbf{B}$ are the unit tangent, the unit normal and the unit binormal vectors to the curve and the prime represents derivative with respect to $s$.

24. Show that the sufficient and necessary condition for a space curve to be a plane curve is that its torsion vanishes identically.
**Answer**: Let Assume first that the space curve is naturally parameterized by $s$. We have two parts to this proof:
(a) If the torsion vanishes identically then the space curve is a plane curve: since $\mathbf{T}, \mathbf{N}, \mathbf{B}$ are orthonormal right handed vector set then we have:

$$\mathbf{B} = \mathbf{T} \times \mathbf{N}$$

On taking the derivative with respect to $s$ we get:

$$\begin{aligned} \mathbf{B}' &= (\mathbf{T}' \times \mathbf{N}) + (\mathbf{T} \times \mathbf{N}') \qquad (8) \\ &= (\kappa\mathbf{N} \times \mathbf{N}) + (\mathbf{T} \times \mathbf{N}') \\ &= \mathbf{0} + (\mathbf{T} \times \mathbf{N}') \\ &= \mathbf{T} \times \mathbf{N}' \\ &= \mathbf{T} \times (\tau\mathbf{B} - \kappa\mathbf{T}) \\ &= (\mathbf{T} \times \tau\mathbf{B}) - \kappa (\mathbf{T} \times \mathbf{T}) \\ &= (\mathbf{T} \times \tau\mathbf{B}) - \mathbf{0} \end{aligned}$$

$$\begin{aligned} &= \tau \mathbf{T} \times \mathbf{B} \\ &= -\tau \mathbf{N} \end{aligned}$$

where line 1 is the product rule of differentiation, line 2 is from Eq. 6, line 3 is the fact that the cross product of parallel vectors is zero, line 5 is from Eq. 7, line 6 is the distributivity of cross product over sum, and line 9 is because $\mathbf{T}, \mathbf{N}, \mathbf{B}$ are orthonormal right handed vector set. Accordingly, if $\tau = 0$ identically then $\mathbf{B}' = \mathbf{0}$ identically and hence $\mathbf{B}$ is a constant vector, say $\mathbf{B}_c$ where the subscript indicates its constancy. Now, since the curve is represented spatially by $\mathbf{r}(s)$ then we have:

$$\begin{aligned} \frac{d}{ds}(\mathbf{r} \cdot \mathbf{B}_c) &= \mathbf{r}' \cdot \mathbf{B}_c + \mathbf{r} \cdot \mathbf{B}_c' \\ &= \mathbf{r}' \cdot \mathbf{B}_c + \mathbf{0} \\ &= \mathbf{r}' \cdot \mathbf{B}_c \\ &= \mathbf{T} \cdot \mathbf{B}_c \\ &= 0 \end{aligned}$$

where line 4 is from Eq. 5 while line 5 is because $\mathbf{T}$ and $\mathbf{B}_c$ are orthogonal. Hence:

$$\mathbf{r} \cdot \mathbf{B}_c = d$$

where $d$ is a constant. As we see, this equation looks like an equation of a plane represented spatially by $\mathbf{r}$ with a normal vector $\mathbf{B}_c$ (refer to Exercise 8) and hence the curve should be confined to a plane (which is the osculating plane of the curve), i.e. the curve is a plane curve, as required.
(b) If the space curve is a plane curve then the torsion vanishes identically: since the space curve is a plane curve then $\mathbf{T}$ and $\mathbf{N}$ are confined to a plane and hence $\mathbf{B}$ $(= \mathbf{T} \times \mathbf{N})$ is constant because it is a unit vector normal to the plane.[13] Accordingly:

$$\begin{aligned} \mathbf{B}' &= \mathbf{0} \\ -\tau \mathbf{N} &= \mathbf{0} \\ \tau &= 0 \end{aligned}$$

where line 2 is from Eq. 8 which is obtained in part (a) while line 3 is because $\mathbf{N} \neq \mathbf{0}$ in general for a twisted curve.

25. What are the curvature $\kappa$ and torsion $\tau$ of (a) a straight line (b) a circle with radius $R = 3.2$ (c) a curve parameterized by: $\mathbf{r}(t) = (3\cos t, 3\sin t, 1.9t)$?
**Answer**:
(a) Straight line: $\kappa = 0$ and $\tau = 0$.

[13] Being unit vector means it is constant in magnitude while being normal to a plane means it is constant in direction and hence it is constant vector since it is constant in magnitude and direction.

(b) Circle with radius $R = 3.2$: $\kappa = 1/R = 0.3125$ and $\tau = 0$.
(c) Curve parameterized by $\mathbf{r}(t) = (3\cos t, 3\sin t, 1.9t)$: we have:

$$
\begin{aligned}
\dot{\mathbf{r}} &= \frac{d\mathbf{r}}{dt} = (-3\sin t, 3\cos t, 1.9) \\
\ddot{\mathbf{r}} &= \frac{d\dot{\mathbf{r}}}{dt} = (-3\cos t, -3\sin t, 0) \\
\dddot{\mathbf{r}} &= \frac{d\ddot{\mathbf{r}}}{dt} = (3\sin t, -3\cos t, 0) \\
\dot{\mathbf{r}} \times \ddot{\mathbf{r}} &= \begin{bmatrix} \mathbf{i} & \mathbf{j} & \mathbf{k} \\ -3\sin t & 3\cos t & 1.9 \\ -3\cos t & -3\sin t & 0 \end{bmatrix} = (5.7\sin t, -5.7\cos t, 9) \\
\ddot{\mathbf{r}} \times \dddot{\mathbf{r}} &= \begin{bmatrix} \mathbf{i} & \mathbf{j} & \mathbf{k} \\ -3\cos t & -3\sin t & 0 \\ 3\sin t & -3\cos t & 0 \end{bmatrix} = (0, 0, 9)
\end{aligned}
$$

Hence:

$$
\begin{aligned}
\kappa &= \frac{|\dot{\mathbf{r}} \times \ddot{\mathbf{r}}|}{|\dot{\mathbf{r}}|^3} \\
&= \frac{\left(32.49\sin^2 t + 32.49\cos^2 t + 81\right)^{1/2}}{\left(9\sin^2 t + 9\cos^2 t + 3.61\right)^{3/2}} \\
&= \frac{(113.49)^{1/2}}{(12.61)^{3/2}} \\
&\simeq 0.2379
\end{aligned}
$$

$$
\begin{aligned}
\tau &= \frac{\dot{\mathbf{r}} \cdot (\ddot{\mathbf{r}} \times \dddot{\mathbf{r}})}{|\dot{\mathbf{r}} \times \ddot{\mathbf{r}}|^2} \\
&= \frac{(-3\sin t, 3\cos t, 1.9) \cdot (0, 0, 9)}{32.49\sin^2 t + 32.49\cos^2 t + 81} \\
&= \frac{17.1}{113.49} \\
&\simeq 0.1507
\end{aligned}
$$

We should remark that from the given parametric equation of the curve we can easily see that this curve is a helix with $a = 3$ and $b = 1.9$ and hence we can easily find the curvature and torsion from the formulae:

$$
\begin{aligned}
\kappa &= \frac{a}{a^2 + b^2} = \frac{3}{3^2 + 1.9^2} \simeq 0.2379 \\
\tau &= \frac{b}{a^2 + b^2} = \frac{1.9}{3^2 + 1.9^2} \simeq 0.1507
\end{aligned}
$$

However, we preferred to use the lengthy method for demonstration and verification.

26. Find the curvature and torsion, as functions of $t$, of (a) a curve represented by: $\mathbf{r}(t) = (\sin t + t, \cos t - 3, t + 2)$ (b) a curve represented by: $\mathbf{r}(t) = (t, 2t^2, t^3)$ (c) a curve represented by: $\mathbf{r}(t) = (at, bt^3, ct^2)$ where $a, b, c$ are non-vanishing real constants.
**Answer**:
(a) We have:

$$\begin{aligned}
\dot{\mathbf{r}} &= \frac{d\mathbf{r}}{dt} = (\cos t + 1, -\sin t, 1) \\
\ddot{\mathbf{r}} &= \frac{d\dot{\mathbf{r}}}{dt} = (-\sin t, -\cos t, 0) \\
\dddot{\mathbf{r}} &= \frac{d\ddot{\mathbf{r}}}{dt} = (-\cos t, \sin t, 0) \\
\dot{\mathbf{r}} \times \ddot{\mathbf{r}} &= \begin{bmatrix} \mathbf{i} & \mathbf{j} & \mathbf{k} \\ \cos t + 1 & -\sin t & 1 \\ -\sin t & -\cos t & 0 \end{bmatrix} = (\cos t, -\sin t, -1 - \cos t) \\
\ddot{\mathbf{r}} \times \dddot{\mathbf{r}} &= \begin{bmatrix} \mathbf{i} & \mathbf{j} & \mathbf{k} \\ -\sin t & -\cos t & 0 \\ -\cos t & \sin t & 0 \end{bmatrix} = (0, 0, -1)
\end{aligned}$$

Hence:

$$\begin{aligned}
\kappa &= \frac{|\dot{\mathbf{r}} \times \ddot{\mathbf{r}}|}{|\dot{\mathbf{r}}|^3} \\
&= \frac{\left(\cos^2 t + \sin^2 t + 1 + 2\cos t + \cos^2 t\right)^{1/2}}{\left(\cos^2 t + 2\cos t + 1 + \sin^2 t + 1\right)^{3/2}} \\
&= \frac{\left(2 + 2\cos t + \cos^2 t\right)^{1/2}}{\left(2\cos t + 3\right)^{3/2}} \\
\tau &= \frac{\dot{\mathbf{r}} \cdot (\ddot{\mathbf{r}} \times \dddot{\mathbf{r}})}{|\dot{\mathbf{r}} \times \ddot{\mathbf{r}}|^2} \\
&= \frac{(\cos t + 1, -\sin t, 1) \cdot (0, 0, -1)}{\cos^2 t + \sin^2 t + 1 + 2\cos t + \cos^2 t} \\
&= \frac{-1}{2 + 2\cos t + \cos^2 t}
\end{aligned}$$

(b) We have:

$$\begin{aligned}
\dot{\mathbf{r}} &= \frac{d\mathbf{r}}{dt} = \left(1, 4t, 3t^2\right) \\
\ddot{\mathbf{r}} &= \frac{d\dot{\mathbf{r}}}{dt} = (0, 4, 6t) \\
\dddot{\mathbf{r}} &= \frac{d\ddot{\mathbf{r}}}{dt} = (0, 0, 6)
\end{aligned}$$

$$
\begin{aligned}
\dot{\mathbf{r}} \times \ddot{\mathbf{r}} &= \begin{bmatrix} \mathbf{i} & \mathbf{j} & \mathbf{k} \\ 1 & 4t & 3t^2 \\ 0 & 4 & 6t \end{bmatrix} = \big(12t^2, -6t, 4\big) \\
\ddot{\mathbf{r}} \times \dddot{\mathbf{r}} &= \begin{bmatrix} \mathbf{i} & \mathbf{j} & \mathbf{k} \\ 0 & 4 & 6t \\ 0 & 0 & 6 \end{bmatrix} = (24, 0, 0)
\end{aligned}
$$

Hence:

$$
\begin{aligned}
\kappa &= \frac{|\dot{\mathbf{r}} \times \ddot{\mathbf{r}}|}{|\dot{\mathbf{r}}|^3} \\
&= \frac{(144t^4 + 36t^2 + 16)^{1/2}}{(1 + 16t^2 + 9t^4)^{3/2}} \\
\tau &= \frac{\dot{\mathbf{r}} \cdot (\ddot{\mathbf{r}} \times \dddot{\mathbf{r}})}{|\dot{\mathbf{r}} \times \ddot{\mathbf{r}}|^2} \\
&= \frac{(1, 4t, 3t^2) \cdot (24, 0, 0)}{144t^4 + 36t^2 + 16} \\
&= \frac{24}{144t^4 + 36t^2 + 16} \\
&= \frac{6}{36t^4 + 9t^2 + 4}
\end{aligned}
$$

(c) We have:

$$
\begin{aligned}
\dot{\mathbf{r}} &= \frac{d\mathbf{r}}{dt} = \big(a, 3bt^2, 2ct\big) \\
\ddot{\mathbf{r}} &= \frac{d\dot{\mathbf{r}}}{dt} = (0, 6bt, 2c) \\
\dddot{\mathbf{r}} &= \frac{d\ddot{\mathbf{r}}}{dt} = (0, 6b, 0) \\
\dot{\mathbf{r}} \times \ddot{\mathbf{r}} &= \begin{bmatrix} \mathbf{i} & \mathbf{j} & \mathbf{k} \\ a & 3bt^2 & 2ct \\ 0 & 6bt & 2c \end{bmatrix} = \big(-6bct^2, -2ac, 6abt\big) \\
\ddot{\mathbf{r}} \times \dddot{\mathbf{r}} &= \begin{bmatrix} \mathbf{i} & \mathbf{j} & \mathbf{k} \\ 0 & 6bt & 2c \\ 0 & 6b & 0 \end{bmatrix} = (-12bc, 0, 0)
\end{aligned}
$$

Hence:

$$
\begin{aligned}
\kappa &= \frac{|\dot{\mathbf{r}} \times \ddot{\mathbf{r}}|}{|\dot{\mathbf{r}}|^3} \\
&= \frac{(36b^2c^2t^4 + 4a^2c^2 + 36a^2b^2t^2)^{1/2}}{(a^2 + 9b^2t^4 + 4c^2t^2)^{3/2}}
\end{aligned}
$$

$$\begin{aligned}
\tau &= \frac{\dot{\mathbf{r}} \cdot (\ddot{\mathbf{r}} \times \dddot{\mathbf{r}})}{|\dot{\mathbf{r}} \times \ddot{\mathbf{r}}|^2} \\
&= \frac{(a, 3bt^2, 2ct) \cdot (-12bc, 0, 0)}{36b^2c^2t^4 + 4a^2c^2 + 36a^2b^2t^2} \\
&= \frac{-12abc}{36b^2c^2t^4 + 4a^2c^2 + 36a^2b^2t^2} \\
&= \frac{-3abc}{9b^2c^2t^4 + a^2c^2 + 9a^2b^2t^2}
\end{aligned}$$

27. What are the mathematical conditions that represent the fact that the vectors $\mathbf{T}, \mathbf{N}, \mathbf{B}$ are mutually orthogonal and of unit length?
**Answer**:
Orthogonality:

$$\begin{aligned}
\mathbf{T} \cdot \mathbf{N} &= 0 \\
\mathbf{T} \cdot \mathbf{B} &= 0 \\
\mathbf{N} \cdot \mathbf{B} &= 0
\end{aligned}$$

Unit length:

$$\begin{aligned}
\mathbf{T} \cdot \mathbf{T} &= 1 \\
\mathbf{N} \cdot \mathbf{N} &= 1 \\
\mathbf{B} \cdot \mathbf{B} &= 1
\end{aligned}$$

28. Prove the theorem of Lancret which states that a space curve of class $C^3$ with non-vanishing curvature is a helix *iff* the ratio of its torsion to its curvature is constant along the curve.
**Answer**: There are two parts to this proof:
(a) If the curve is a helix then the ratio of torsion to curvature is constant: a circular helix is generally parameterized by:

$$\mathbf{r}(t) = (a \cos t, a \sin t, bt)$$

where $a$ and $b$ are real constants and $t$ is a general real parameter. Accordingly:

$$\begin{aligned}
\dot{\mathbf{r}} &= \frac{d\mathbf{r}}{dt} = (-a \sin t, a \cos t, b) \\
\ddot{\mathbf{r}} &= \frac{d\dot{\mathbf{r}}}{dt} = (-a \cos t, -a \sin t, 0) \\
\dddot{\mathbf{r}} &= \frac{d\ddot{\mathbf{r}}}{dt} = (a \sin t, -a \cos t, 0) \\
\dot{\mathbf{r}} \times \ddot{\mathbf{r}} &= \begin{bmatrix} \mathbf{i} & \mathbf{j} & \mathbf{k} \\ -a \sin t & a \cos t & b \\ -a \cos t & -a \sin t & 0 \end{bmatrix} = \left(ab \sin t, -ab \cos t, a^2\right)
\end{aligned}$$

$$\ddot{\mathbf{r}} \times \dddot{\mathbf{r}} = \begin{bmatrix} \mathbf{i} & \mathbf{j} & \mathbf{k} \\ -a\cos t & -a\sin t & 0 \\ a\sin t & -a\cos t & 0 \end{bmatrix} = \left(0, 0, a^2\right)$$

Hence:

$$\begin{aligned} \kappa &= \frac{|\dot{\mathbf{r}} \times \ddot{\mathbf{r}}|}{|\dot{\mathbf{r}}|^3} \qquad (9) \\ &= \frac{\left(a^2b^2\sin^2 t + a^2b^2\cos^2 t + a^4\right)^{1/2}}{\left(a^2\sin^2 t + a^2\cos^2 t + b^2\right)^{3/2}} \\ &= \frac{\left(a^2b^2 + a^4\right)^{1/2}}{\left(a^2 + b^2\right)^{3/2}} \\ &= \frac{a\left(b^2 + a^2\right)^{1/2}}{\left(a^2 + b^2\right)^{3/2}} \\ &= \frac{a}{a^2 + b^2} \end{aligned}$$

$$\begin{aligned} \tau &= \frac{\dot{\mathbf{r}} \cdot \left(\ddot{\mathbf{r}} \times \dddot{\mathbf{r}}\right)}{|\dot{\mathbf{r}} \times \ddot{\mathbf{r}}|^2} \qquad (10) \\ &= \frac{\left(-a\sin t, a\cos t, b\right) \cdot \left(0, 0, a^2\right)}{a^2b^2\sin^2 t + a^2b^2\cos^2 t + a^4} \\ &= \frac{ba^2}{a^2b^2 + a^4} \\ &= \frac{b}{a^2 + b^2} \end{aligned}$$

Therefore:

$$\frac{\tau}{\kappa} = \frac{b}{a}$$

which is constant (since $a$ and $b$ are constants), as required.
(b) If the ratio of torsion to curvature is constant then the curve is a helix: in this part of the proof we use the defining property of circular helix that is its tangent makes a constant angle with its axis of rotation (see Exercise 16 of § 6), i.e. a space curve is a helix *iff* its tangent makes a constant angle with a given straight line. For simplicity, let assume (with no loss of generality) that the curve is naturally parameterized and hence it is represented by $\mathbf{r}(s)$. Now, for a naturally parameterized curve we have:

$$\begin{aligned} \tau &= \mathbf{N}' \cdot \mathbf{B} = |\mathbf{N}'| \, |\mathbf{B}| \cos\phi = |\mathbf{N}'| \cos\phi \\ \kappa &= -\mathbf{N}' \cdot \mathbf{T} = -|\mathbf{N}'| \, |\mathbf{T}| \cos\theta = -|\mathbf{N}'| \cos\theta \end{aligned}$$

where $\phi$ is the angle between $\mathbf{N}'$ and $\mathbf{B}$ and $\theta$ is the angle between $\mathbf{N}'$ and $\mathbf{T}$ and where the last equalities are because $\mathbf{B}$ and $\mathbf{T}$ are unit vectors. Hence:

$$\frac{\tau}{\kappa} = \frac{|\mathbf{N}'| \cos\phi}{-|\mathbf{N}'| \cos\theta} = -\frac{\cos\phi}{\cos\theta}$$

Now, since $\mathbf{N}'$ is a linear combination of $\mathbf{B}$ and $\mathbf{T}$ (i.e. $\mathbf{N}' = \tau\mathbf{B} - \kappa\mathbf{T}$, see Eq. 7), and $\mathbf{B}$ and $\mathbf{T}$ are orthonormal then $\cos\theta = \sin\phi$ and hence:

$$\frac{\tau}{\kappa} = -\frac{\cos\phi}{\sin\phi}$$

Now, because the ratio of torsion to curvature is constant then $\phi$ should be constant, that is:

$$\frac{\tau}{\kappa} = -\frac{\cos\phi_c}{\sin\phi_c}$$

where the subscript $c$ indicates its constancy. Accordingly:

$$\begin{aligned}
\tau\sin\phi_c + \kappa\cos\phi_c &= 0 \\
\tau\mathbf{N}\sin\phi_c + \kappa\mathbf{N}\cos\phi_c &= \mathbf{0} \\
-\mathbf{B}'\sin\phi_c + \mathbf{T}'\cos\phi_c &= \mathbf{0} \\
-\mathbf{B}\sin\phi_c + \mathbf{T}\cos\phi_c &= \mathbf{v}
\end{aligned}$$

where in line 2 we multiplied both sides by $\mathbf{N}$, in line 3 we used Eqs. 8 and 6, and in line 4 we integrated both sides (with $\mathbf{v}$ being the constant of integration). Now, since $\mathbf{B}$ and $\mathbf{T}$ are orthonormal then $\mathbf{v}$ is of unity magnitude, i.e. $\mathbf{v}$ is a constant unit vector. On inner multiplying both sides of the last equation with $\mathbf{T}$ we obtain:

$$\begin{aligned}
-(\mathbf{T}\cdot\mathbf{B})\sin\phi_c + (\mathbf{T}\cdot\mathbf{T})\cos\phi_c &= \mathbf{T}\cdot\mathbf{v} \\
\cos\phi_c &= \mathbf{T}\cdot\mathbf{v} \\
\mathbf{T}\cdot\mathbf{v} &= \text{constant}
\end{aligned}$$

where line 2 is because $\mathbf{T}$ and $\mathbf{B}$ are orthogonal and $\mathbf{T}$ is a unit vector, while line 3 is because $\phi_c$ is constant. The last equality (i.e. $\mathbf{T}\cdot\mathbf{v} = \text{constant}$) is the defining property of circular helix where $\mathbf{T}$ is its tangent, $\mathbf{v}$ is the unit vector that defines the orientation of its axis of rotation and $\phi_c$ is the constant angle between its tangent and its axis.

29. Show that if two space curves, which have an injective association, possess parallel tangent vectors at their corresponding points, then their normal and binormal vectors at these points are parallel as well.
**Answer**: Let the curves be represented as $\mathbf{r}_1(s_1)$ and $\mathbf{r}_2(s_2)$ where $s_1$ is a natural parameter for $\mathbf{r}_1$ and $s_2$ is a natural parameter for $\mathbf{r}_2$. Now, since $\mathbf{T}_1$ and $\mathbf{T}_2$ are parallel at their corresponding points and they are unit vectors then we have $\mathbf{T}_1 = \mathbf{T}_2$ and hence:

$$\begin{aligned}
\frac{d\mathbf{T}_1}{ds_1} &= \frac{d\mathbf{T}_2}{ds_1} \\
\frac{d\mathbf{T}_1}{ds_1} &= \frac{d\mathbf{T}_2}{ds_2}\frac{ds_2}{ds_1} \\
\kappa_1\mathbf{N}_1 &= \kappa_2\mathbf{N}_2\frac{ds_2}{ds_1}
\end{aligned}$$

(where we used the chain rule in line 2 and Eq. 6 in line 3) and hence their principal normal vectors are also parallel. Now, since $\mathbf{B} = \mathbf{T} \times \mathbf{N}$ and $\mathbf{T}_1$ is parallel to $\mathbf{T}_2$ and $\mathbf{N}_1$ is parallel to $\mathbf{N}_2$ then $\mathbf{B}_1$ must also be parallel to $\mathbf{B}_2$.

30. Prove the following equation: $|\kappa\tau| = |\mathbf{T}' \cdot \mathbf{B}'|$.
**Answer**: From Eqs. 6 and 8 we have:

$$\begin{aligned} \mathbf{T}' &= \kappa\mathbf{N} \\ \mathbf{B}' &= -\tau\mathbf{N} \end{aligned}$$

Hence:

$$\begin{aligned} \mathbf{T}' \cdot \mathbf{B}' &= -\kappa\mathbf{N} \cdot \tau\mathbf{N} \\ \mathbf{T}' \cdot \mathbf{B}' &= -\kappa\tau\mathbf{N} \cdot \mathbf{N} \\ \mathbf{T}' \cdot \mathbf{B}' &= -\kappa\tau \\ |\mathbf{T}' \cdot \mathbf{B}'| &= |\kappa\tau| \end{aligned}$$

31. Give a parametric representation of a circular helix using a natural parameter $s$.
**Answer**: A circular helix is generally parameterized by:

$$\mathbf{r}(t) = (a\cos t, a\sin t, bt)$$

where $a$ and $b$ are real constants and $t$ is a general parameter. Now, it was shown in Exercise 20 that for a curve represented spatially by $\mathbf{r}$ and parameterized naturally by $s$ and generally by $t$ the relation between $s$ and $t$ is given by $\left|\frac{ds}{dt}\right| = \left|\frac{d\mathbf{r}}{dt}\right|$. Hence:

$$\begin{aligned} \left|\frac{ds}{dt}\right| &= \left|\frac{d\mathbf{r}}{dt}\right| \\ \left|\frac{ds}{dt}\right| &= \left|\frac{d}{dt}(a\cos t, a\sin t, bt)\right| \\ \left|\frac{ds}{dt}\right| &= |(-a\sin t, a\cos t, b)| \\ \left|\frac{ds}{dt}\right| &= \sqrt{a^2\sin^2 t + a^2\cos^2 t + b^2} \\ \left|\frac{ds}{dt}\right| &= \sqrt{a^2 + b^2} \\ \frac{ds}{dt} &= \pm\sqrt{a^2 + b^2} \\ s &= \pm t\sqrt{a^2 + b^2} + c \end{aligned}$$

where $c$ is a constant. Now, if we take only the positive root and assume $c = 0$ then we have:

$$t = \frac{s}{\sqrt{a^2 + b^2}}$$

On substituting from the last equation into the above parametric equation of helix we get:

$$\mathbf{r}(s) = \left( a\cos\frac{s}{\sqrt{a^2+b^2}},\ a\sin\frac{s}{\sqrt{a^2+b^2}},\ \frac{bs}{\sqrt{a^2+b^2}} \right) \tag{11}$$

As we see:

$$\begin{aligned} \left|\frac{d\mathbf{r}}{ds}\right| &= \left|\left( -\frac{a}{\sqrt{a^2+b^2}}\sin\frac{s}{\sqrt{a^2+b^2}},\ \frac{a}{\sqrt{a^2+b^2}}\cos\frac{s}{\sqrt{a^2+b^2}},\ \frac{b}{\sqrt{a^2+b^2}} \right)\right| \\ &= \left[ \frac{a^2}{a^2+b^2}\sin^2\frac{s}{\sqrt{a^2+b^2}} + \frac{a^2}{a^2+b^2}\cos^2\frac{s}{\sqrt{a^2+b^2}} + \frac{b^2}{a^2+b^2} \right]^{1/2} \\ &= \left[ \frac{a^2}{a^2+b^2} + \frac{b^2}{a^2+b^2} \right]^{1/2} \\ &= \left[ \frac{a^2+b^2}{a^2+b^2} \right]^{1/2} \\ &= 1 \end{aligned}$$

as it should be.

32. For a plane curve in a 3D space given by the equation: $y = 2x^2 - x + 3$ $(0 \le x \le 10)$, find the equations of the osculating, normal and rectifying planes at point $(1, 4)$.
**Answer**: This curve can be represented by $\mathbf{r}(t) = (t, 2t^2 - t + 3, 0)$. Hence:

$$\begin{aligned} \dot{\mathbf{r}} &= \frac{d\mathbf{r}}{dt} = (1, 4t-1, 0) \\ \mathbf{T} &= \frac{\dot{\mathbf{r}}}{|\dot{\mathbf{r}}|} = \frac{(1, 4t-1, 0)}{\sqrt{16t^2 - 8t + 2}} \\ \mathbf{B} &= \mathbf{k} = (0, 0, 1) \\ \mathbf{N} &= \mathbf{B} \times \mathbf{T} = \frac{(1-4t, 1, 0)}{\sqrt{16t^2 - 8t + 2}} \end{aligned}$$

The osculating plane is the plane passing through the point $(1, 4, 0)$ with normal vector $\mathbf{B}$, that is:

$$\begin{aligned} (0, 0, 1) \cdot (x-1, y-4, , z) &= 0 \\ z &= 0 \end{aligned}$$

i.e. it is the $xy$ plane which should be obvious.
The normal plane is the plane passing through the point $(1, 4, 0)$ with normal vector $\mathbf{T}$ corresponding to $t = 1$, that is:

$$\begin{aligned} \frac{(1, 3, 0)}{\sqrt{10}} \cdot (x-1, y-4, , z) &= 0 \\ x - 1 + 3y - 12 &= 0 \end{aligned}$$

$$x + 3y = 13$$

The rectifying plane is the plane passing through the point $(1, 4, 0)$ with normal vector $\mathbf{N}$ corresponding to $t = 1$, that is:

$$\begin{aligned}
\frac{(-3, 1, 0)}{\sqrt{10}} \cdot (x - 1, y - 4, , z) &= 0 \\
-3x + 3 + y - 4 &= 0 \\
-3x + y &= 1
\end{aligned}$$

33. For a space curve parameterized as: $(x, y, z) = (t, t^3, 3t^2)$, find the equations of the tangent, principal normal and binormal lines passing through the point $(1, 1, 3)$ on the curve.
**Answer**: the point $(1, 1, 3)$ corresponds to $t = 1$ and hence we have:

$$\begin{aligned}
\dot{\mathbf{r}} &= \frac{d\mathbf{r}}{dt} = \left(1, 3t^2, 6t\right) = (1, 3, 6) \\
\ddot{\mathbf{r}} &= \frac{d\dot{\mathbf{r}}}{dt} = (0, 6t, 6) = (0, 6, 6) \\
\dot{\mathbf{r}} \times \ddot{\mathbf{r}} &= (-18, -6, 6) \\
\mathbf{T} &= \frac{\dot{\mathbf{r}}}{|\dot{\mathbf{r}}|} = \frac{1}{\sqrt{46}}(1, 3, 6) \\
\mathbf{B} &= \frac{\dot{\mathbf{r}} \times \ddot{\mathbf{r}}}{|\dot{\mathbf{r}} \times \ddot{\mathbf{r}}|} = \frac{1}{\sqrt{396}}(-18, -6, 6) \\
\mathbf{N} &= \mathbf{B} \times \mathbf{T} = \frac{1}{\sqrt{18216}}(-54, 114, -48)
\end{aligned}$$

The tangent line is the line passing through the point $(1, 1, 3)$ and oriented along $\mathbf{T}$, that is:

$$(x - 1, y - 1, z - 3) = \lambda_t (1, 3, 6)$$

where $\lambda_t$ is a real parameter.
The principal normal line is the line passing through the point $(1, 1, 3)$ and oriented along $\mathbf{N}$, that is:

$$(x - 1, y - 1, z - 3) = \lambda_n (-54, 114, -48)$$

where $\lambda_n$ is a real parameter.
The binormal line is the line passing through the point $(1, 1, 3)$ and oriented along $\mathbf{B}$, that is:

$$(x - 1, y - 1, z - 3) = \lambda_b (-18, -6, 6)$$

where $\lambda_b$ is a real parameter.

34. Give an example of a non-planar space curve whose principal normal vector at all points of the curve is parallel to a particular plane.

**Answer**: Circular helix. This can be demonstrated by a circular helix whose axis of rotation is the $z$-axis and hence it is $t$-parameterized by:

$$\mathbf{r}(t) = (a\cos t, a\sin t, bt)$$

where $a$ and $b$ are real constants and $t$ is a general real parameter. Accordingly:

$$\begin{aligned}
\dot{\mathbf{r}} &= \frac{d\mathbf{r}}{dt} = (-a\sin t, a\cos t, b) \\
\ddot{\mathbf{r}} &= \frac{d\dot{\mathbf{r}}}{dt} = (-a\cos t, -a\sin t, 0) \\
\ddot{\mathbf{r}} \times \dot{\mathbf{r}} &= \begin{bmatrix} \mathbf{i} & \mathbf{j} & \mathbf{k} \\ -a\cos t & -a\sin t & 0 \\ -a\sin t & a\cos t & b \end{bmatrix} = \left(-ab\sin t, ab\cos t, -a^2\right) \\
\dot{\mathbf{r}} \times (\ddot{\mathbf{r}} \times \dot{\mathbf{r}}) &= \begin{bmatrix} \mathbf{i} & \mathbf{j} & \mathbf{k} \\ -a\sin t & a\cos t & b \\ -ab\sin t & ab\cos t & -a^2 \end{bmatrix} \\
&= \left(-a^3\cos t - ab^2\cos t, -ab^2\sin t - a^3\sin t, 0\right) \\
\mathbf{N} &= c\dot{\mathbf{r}} \times (\ddot{\mathbf{r}} \times \dot{\mathbf{r}}) \\
&= c\left(-a^3\cos t - ab^2\cos t, -ab^2\sin t - a^3\sin t, 0\right)
\end{aligned}$$

where $c = |\dot{\mathbf{r}} \times (\ddot{\mathbf{r}} \times \dot{\mathbf{r}})|^{-1}$. Now, since the $z$ component of $\mathbf{N}$ is zero then it is parallel to the $xy$ plane (and indeed to any plane parallel to the $xy$ plane). So, we can conclude that the principal normal vector of circular helix (which is not a plane curve) at any point of the curve is parallel to any plane perpendicular to its axis of rotation.

35. For a space curve parameterized as: $(x, y, z) = (\cos t, \sin t, 5t)$, find the equations of the osculating, rectifying and normal planes at the point on the curve with $t = 1.3$.
**Answer**: We have:

$$\begin{aligned}
\dot{\mathbf{r}} &= \frac{d\mathbf{r}}{dt} = (-\sin t, \cos t, 5) \\
\ddot{\mathbf{r}} &= \frac{d\dot{\mathbf{r}}}{dt} = (-\cos t, -\sin t, 0) \\
\dot{\mathbf{r}} \times \ddot{\mathbf{r}} &= \begin{bmatrix} \mathbf{i} & \mathbf{j} & \mathbf{k} \\ -\sin t & \cos t & 5 \\ -\cos t & -\sin t & 0 \end{bmatrix} = (5\sin t, -5\cos t, 1) \\
\mathbf{T} &= \frac{\dot{\mathbf{r}}}{|\dot{\mathbf{r}}|} = \frac{(-\sin t, \cos t, 5)}{\sqrt{26}} \\
\mathbf{B} &= \frac{\dot{\mathbf{r}} \times \ddot{\mathbf{r}}}{|\dot{\mathbf{r}} \times \ddot{\mathbf{r}}|} = \frac{(5\sin t, -5\cos t, 1)}{\sqrt{26}} \\
\mathbf{N} &= \mathbf{B} \times \mathbf{T} = (-\cos t, -\sin t, 0)
\end{aligned}$$

The osculating plane is the plane passing through the point $(\cos 1.3, \sin 1.3, 6.5)$ $\big[$i.e. the point corresponding to $t = 1.3\big]$ with a normal vector $\mathbf{B}$ corresponding to $t = 1.3$, that is:

$$\begin{aligned}
\frac{(5\sin 1.3, -5\cos 1.3, 1)}{\sqrt{26}} \cdot (x - \cos 1.3, y - \sin 1.3, z - 6.5) &= 0 \\
5\sin 1.3\,(x - \cos 1.3) - 5\cos 1.3\,(y - \sin 1.3) + (z - 6.5) &= 0 \\
5x\sin 1.3 - 5\sin 1.3\cos 1.3 - 5y\cos 1.3 + 5\cos 1.3\sin 1.3 + z - 6.5 &= 0 \\
5x\sin 1.3 - 5y\cos 1.3 + z &= 6.5 \\
4.8178x - 1.3375y + z &\simeq 6.5
\end{aligned}$$

The rectifying plane is the plane passing through the point $(\cos 1.3, \sin 1.3, 6.5)$ with a normal vector $\mathbf{N}$ corresponding to $t = 1.3$, that is:

$$\begin{aligned}
(-\cos 1.3, -\sin 1.3, 0) \cdot (x - \cos 1.3, y - \sin 1.3, z - 6.5) &= 0 \\
-\cos 1.3\,(x - \cos 1.3) - \sin 1.3\,(y - \sin 1.3) + 0 &= 0 \\
-x\cos 1.3 + \cos^2 1.3 - y\sin 1.3 + \sin^2 1.3 &= 0 \\
x\cos 1.3 + y\sin 1.3 &= 1 \\
0.2675x + 0.9636y &\simeq 1
\end{aligned}$$

The normal plane is the plane passing through the point$(\cos 1.3, \sin 1.3, 6.5)$ with a normal vector $\mathbf{T}$ corresponding to $t = 1.3$, that is:

$$\begin{aligned}
\frac{(-\sin 1.3, \cos 1.3, 5)}{\sqrt{26}} \cdot (x - \cos 1.3, y - \sin 1.3, z - 6.5) &= 0 \\
-\sin 1.3\,(x - \cos 1.3) + \cos 1.3\,(y - \sin 1.3) + 5\,(z - 6.5) &= 0 \\
-x\sin 1.3 + \sin 1.3\cos 1.3 + y\cos 1.3 - \cos 1.3\sin 1.3 + 5z - 32.5 &= 0 \\
-x\sin 1.3 + y\cos 1.3 + 5z &= 32.5 \\
-0.9636x + 0.2675y + 5z &\simeq 32.5
\end{aligned}$$

36. Which of the three vectors $\mathbf{T}, \mathbf{N}, \mathbf{B}$ is not affected by reversing the sense of traversing the space curve?
**Answer**: It is $\mathbf{N}$.

37. What are the tangent vector $\mathbf{T}$, the principal normal vector $\mathbf{N}$ and the binormal vector $\mathbf{B}$ of a helix represented by: $\mathbf{r}(t) = (3\cos t, 3\sin t, 3t)$?
**Answer**: We have:

$$\begin{aligned}
\dot{\mathbf{r}} &= \frac{d\mathbf{r}}{dt} = (-3\sin t, 3\cos t, 3) \\
\ddot{\mathbf{r}} &= \frac{d\dot{\mathbf{r}}}{dt} = (-3\cos t, -3\sin t, 0)
\end{aligned}$$

$$\begin{aligned}
\dot{\mathbf{r}} \times \ddot{\mathbf{r}} &= \begin{bmatrix} \mathbf{i} & \mathbf{j} & \mathbf{k} \\ -3\sin t & 3\cos t & 3 \\ -3\cos t & -3\sin t & 0 \end{bmatrix} = (9\sin t, -9\cos t, 9) \\
\mathbf{T} &= \frac{\dot{\mathbf{r}}}{|\dot{\mathbf{r}}|} = \frac{(-\sin t, \cos t, 1)}{\sqrt{2}} \\
\mathbf{B} &= \frac{\dot{\mathbf{r}} \times \ddot{\mathbf{r}}}{|\dot{\mathbf{r}} \times \ddot{\mathbf{r}}|} = \frac{(\sin t, -\cos t, 1)}{\sqrt{2}} \\
\mathbf{N} &= \mathbf{B} \times \mathbf{T} = (-\cos t, -\sin t, 0)
\end{aligned}$$

38. Find the three vectors $\mathbf{T}, \mathbf{N}, \mathbf{B}$ along a curve represented by: $\mathbf{r}(t) = (2t-3, t^3+t, 5-t^2)$.
**Answer**: We have:

$$\begin{aligned}
\dot{\mathbf{r}} &= \frac{d\mathbf{r}}{dt} = \left(2, 3t^2 + 1, -2t\right) \\
\ddot{\mathbf{r}} &= \frac{d\dot{\mathbf{r}}}{dt} = (0, 6t, -2) \\
\dot{\mathbf{r}} \times \ddot{\mathbf{r}} &= \begin{bmatrix} \mathbf{i} & \mathbf{j} & \mathbf{k} \\ 2 & 3t^2+1 & -2t \\ 0 & 6t & -2 \end{bmatrix} = \left(6t^2 - 2, 4, 12t\right) \\
\mathbf{T} &= \frac{\dot{\mathbf{r}}}{|\dot{\mathbf{r}}|} = \frac{(2, 3t^2 + 1, -2t)}{\sqrt{9t^4 + 10t^2 + 5}} \\
\mathbf{B} &= \frac{\dot{\mathbf{r}} \times \ddot{\mathbf{r}}}{|\dot{\mathbf{r}} \times \ddot{\mathbf{r}}|} = \frac{(3t^2 - 1, 2, 6t)}{\sqrt{9t^4 + 30t^2 + 5}} \\
\mathbf{N} &= \mathbf{B} \times \mathbf{T} = \frac{(-18t^3 - 10t,\ 6t^3 + 10t,\ 9t^4 - 5)}{\sqrt{(9t^4 + 30t^2 + 5)\,(9t^4 + 10t^2 + 5)}}
\end{aligned}$$

39. Make a simple sketch for the osculating, rectifying and normal planes at a point on an arbitrary space curve. Use a computer graphic package if convenient.
**Answer**: The sketch should look like Figure 13.

40. Write down the equation of Lancret that is related to the third curvature of space curves and discuss its significance.
**Answer**: The equation of Lancret states that:

$$(ds_{\mathbf{N}})^2 = (ds_{\mathbf{T}})^2 + (ds_{\mathbf{B}})^2$$

where $ds_{\mathbf{N}}$, $ds_{\mathbf{T}}$ and $ds_{\mathbf{B}}$ are the lengths of the line element components in the principal normal direction, the tangent direction and the binormal direction. Its significance is that it determines the third curvature $\sqrt{(ds_{\mathbf{T}})^2 + (ds_{\mathbf{B}})^2}$ of the curve and links it to one of the line element components (i.e. $ds_{\mathbf{N}}$) of the curve.

41. Discuss, in detail, the fundamental theorem of space curves in differential geometry outlining its application and significance.

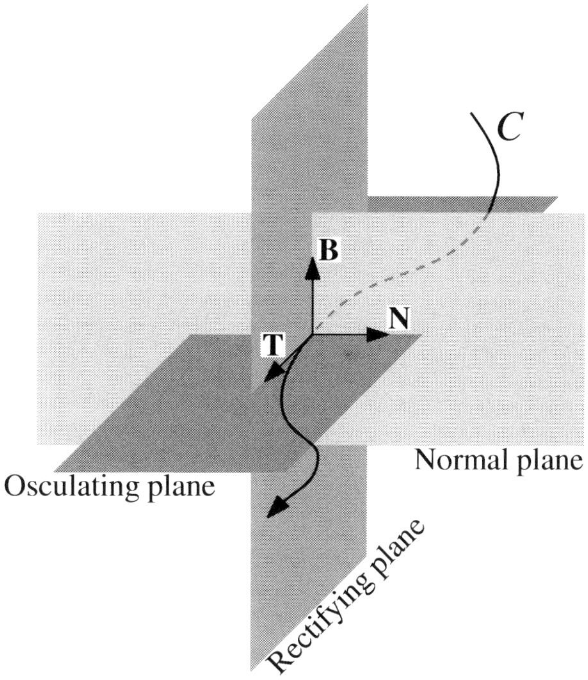


Figure 13: Osculating, rectifying and normal planes at a point on a space curve $C$.

**Answer**: The essence of the fundamental theorem of space curves in differential geometry is that any space curve is completely determined by its curvature and torsion. In technical terms, given a real interval $I \subseteq \mathbb{R}$ and two differentiable real functions: $\kappa(s) > 0$ and $\tau(s)$ where $s \in I$, there is a uniquely defined parameterized regular space curve $C(s)$: $I \to \mathbb{R}^3$ of class $C^2$ with $\kappa(s)$ and $\tau(s)$ being the curvature and torsion of $C$ respectively and $s$ is its arc length. Hence, any other curve meeting these conditions will be different from $C$ only by a rigid motion transformation (i.e. translation and rotation) which determines its position and orientation in space. On the other hand, any curve with the above-described properties possesses uniquely defined $\kappa(s)$ and $\tau(s)$. The significance of the fundamental theorem of space curves is that it provides the existence and uniqueness conditions for curves. Apart from its theoretical significance and value it can have practical applications such as determining and identifying space curves from their characteristics.

42. Given that the curvature of a plane curve is given by: $\kappa = \frac{1}{3s+5}$ where $s > 0$ is a natural parameter, find the equation of this curve.
**Answer**: Because it is a plane curve then its torsion is zero, i.e. $\tau = 0$. So, let assume (with no loss of generality) that the curve is confined to the $xy$ plane and its tangent and normal vectors are given by:

$$\mathbf{T} = (\cos\phi, \sin\phi, 0)$$
$$\mathbf{N} = (-\sin\phi, \cos\phi, 0)$$

where $\phi$ is the angle that $\mathbf{T}$ makes with the $x$-axis. On differentiating these equations with respect to $s$ using the chain rule we obtain:

$$\begin{aligned}\mathbf{T}' &= (-\sin\phi, \cos\phi, 0)\,\phi' = \phi'\mathbf{N} \\ \mathbf{N}' &= (-\cos\phi, -\sin\phi, 0)\,\phi' = -\phi'\mathbf{T}\end{aligned}$$

Now, since $\tau = 0$ then from Eqs. 6 and 7 we get:

$$\begin{aligned}\mathbf{T}' &= \kappa\mathbf{N} \\ \mathbf{N}' &= -\kappa\mathbf{T}\end{aligned}$$

On comparing the last two equations with the previous two equations we conclude:

$$\begin{aligned}\phi' &= \kappa \\ \phi &= \int \kappa ds \\ \phi &= \int \frac{ds}{3s+5} \\ \phi &= \frac{\ln(3s+5)}{3}\end{aligned}$$

where we assumed the constant of integration to be zero. Now, from Eq. 5 we get:

$$\begin{aligned}\mathbf{r}' &= \mathbf{T} \\ \mathbf{r} &= \int \mathbf{T} ds \\ \mathbf{r} &= \int \left(\cos\left(\frac{\ln(3s+5)}{3}\right), \sin\left(\frac{\ln(3s+5)}{3}\right), 0\right) ds \\ \mathbf{r} &= (f_x + c_1, f_y + c_2, c_3)\end{aligned}$$

where

$$\begin{aligned}f_x &= \frac{(3s+5)}{10}\left[\sin\left(\frac{\ln(3s+5)}{3}\right) + 3\cos\left(\frac{\ln(3s+5)}{3}\right)\right] \\ f_y &= \frac{(3s+5)}{10}\left[3\sin\left(\frac{\ln(3s+5)}{3}\right) - \cos\left(\frac{\ln(3s+5)}{3}\right)\right]\end{aligned}$$

and $c_1, c_2, c_3$ are real constants.

43. Write down the Frenet-Serret formulae of a space curve in a rectangular Cartesian coordinate system explaining all the symbols involved.
**Answer**:

$$\begin{aligned}\frac{dT^i}{ds} &= \kappa N^i \\ \frac{dN^i}{ds} &= \tau B^i - \kappa T^i\end{aligned}$$

$$\frac{dB^i}{ds} = -\tau N^i$$

where $T^i, N^i, B^i$ are the $i^{th}$ $(i = 1, 2, 3)$ components of the tangent, normal and binormal vectors $\mathbf{T}, \mathbf{N}, \mathbf{B}$ of a space curve, $s$ is a natural parameter and $\kappa$ and $\tau$ are the curvature and torsion of the curve. These equations may also be given in a more compact vector form by:

$$\mathbf{T}' = \kappa\mathbf{N} \tag{12}$$
$$\mathbf{N}' = \tau\mathbf{B} - \kappa\mathbf{T} \tag{13}$$
$$\mathbf{B}' = -\tau\mathbf{N} \tag{14}$$

where the prime symbolizes derivative with respect to $s$.

44. By integrating the Frenet-Serret formulae, obtain the solution of a space curve with $\kappa = A$ and $\tau = B$ where $A, B > 0$ are real constants.
**Answer**:

$$\begin{aligned}
\mathbf{T}' &= \kappa\mathbf{N} \\
\mathbf{T}' &= -\frac{\kappa}{\tau}\mathbf{B}' \\
\mathbf{T} &= -\frac{\kappa}{\tau}\mathbf{B} + \mathbf{C} \\
\mathbf{T} &= -\frac{\kappa}{\tau}\left(\frac{\mathbf{N}' + \kappa\mathbf{T}}{\tau}\right) + \mathbf{C} \\
\left(1 + \frac{\kappa^2}{\tau^2}\right)\mathbf{T} &= -\frac{\kappa}{\tau^2}\mathbf{N}' + \mathbf{C} \\
\left(1 + \frac{\kappa^2}{\tau^2}\right)\mathbf{r}' &= -\frac{\kappa}{\tau^2}\mathbf{N}' + \mathbf{C} \\
\left(1 + \frac{\kappa^2}{\tau^2}\right)\mathbf{r} &= -\frac{\kappa}{\tau^2}\mathbf{N} + s\mathbf{C} + \mathbf{D} \\
\mathbf{r} &= \left(1 + \frac{A^2}{B^2}\right)^{-1}\left(-\frac{A}{B^2}\mathbf{N} + s\mathbf{C} + \mathbf{D}\right)
\end{aligned}$$

where in line 1 we use Eq. 12, in line 2 we use Eq. 14, in line 3 we integrate the two sides with respect to $s$ and $\mathbf{C}$ is a constant, in line 4 we substitute for $\mathbf{B}$ using Eq. 13, line 5 is a simple algebraic manipulation, in line 6 we use Eq. 5, in line 7 we integrate the two sides with respect to $s$ and $\mathbf{D}$ is a constant, and in line 8 we give $\mathbf{r}$ (which is the spatial representation of the curve) in terms of $A$, $B$ and other curve parameters. Now, $\mathbf{D}$ can be dropped with no loss of generality because it is just a rigid translation in space. Also, from line 3 we can see that $\mathbf{C}$ has no component along $\mathbf{N}$ and hence it is perpendicular to $\mathbf{N}$. So, if $\mathbf{C}$ is oriented along the $z$ axis then $\mathbf{N}$ must be parallel to the $xy$ plane and hence it can be represented as $(\cos\theta, \sin\theta, 0)$ where $\theta$ is the angle between the **N** orientation and the $x$-axis. Accordingly, the equation of the last line becomes:

$$\begin{aligned}
\mathbf{r} &= \left(\frac{B^2}{A^2+B^2}\right)\left[-\frac{A}{B^2}(\cos\theta,\sin\theta,0)+s(0,0,c)\right] \\
&= \left(\frac{B^2}{A^2+B^2}\right)\left(\frac{-A}{B^2}\cos\theta,\frac{-A}{B^2}\sin\theta,sc\right) \\
&= \left(\frac{-A}{A^2+B^2}\cos\theta,\frac{-A}{A^2+B^2}\sin\theta,\frac{cB^2}{A^2+B^2}s\right)
\end{aligned}$$

where $c$ is a constant. On comparing this equation with Eq. 11 which we obtained in Exercise 31 we see that this is a helix with:

$$\begin{aligned}
a &= \frac{-A}{A^2+B^2} \\
b &= \frac{\pm acB^2}{\sqrt{\left(A^2+B^2\right)^2-c^2B^4}} \\
\theta &= \frac{s}{\sqrt{a^2+b^2}}
\end{aligned}$$

45. Identify the type of the curve in the previous question.
**Answer**: It is circular helix.

46. What are the "intrinsic" or "natural" equations of a curve?
**Answer**: They are the equations of the curvature $\kappa$ and torsion $\tau$ as functions of a natural parameter of the curve, that is:

$$\kappa=\kappa(s) \qquad \text{and} \qquad \tau=\tau(s)$$

where $s$ is a natural parameter of the curve.

47. Find the intrinsic equations of the catenary defined by the equations:

$$\begin{aligned}
x &= a\cosh\left(\frac{t}{a}\right) \\
z &= t
\end{aligned}$$

where $a$ is a positive constant and $t$ is a general parameter.
**Answer**: : We have:

$$\begin{aligned}
\dot{\mathbf{r}} &= \frac{d\mathbf{r}}{dt}=\frac{d}{dt}\left(a\cosh(t/a),0,t\right)=\left(\sinh(t/a),0,1\right) \\
\ddot{\mathbf{r}} &= \frac{d\dot{\mathbf{r}}}{dt}=\left(\frac{\cosh(t/a)}{a},0,0\right)
\end{aligned}$$

$$\begin{aligned}
\dot{\mathbf{r}} \times \ddot{\mathbf{r}} &= \begin{bmatrix} \mathbf{i} & \mathbf{j} & \mathbf{k} \\ \sinh(t/a) & 0 & 1 \\ \frac{\cosh(t/a)}{a} & 0 & 0 \end{bmatrix} = \left(0, \frac{\cosh(t/a)}{a}, 0\right) \\
|\dot{\mathbf{r}}| &= \sqrt{\sinh^2(t/a) + 1} = \sqrt{\cosh^2(t/a)} = \cosh(t/a) \\
|\dot{\mathbf{r}} \times \ddot{\mathbf{r}}| &= \sqrt{\frac{\cosh^2(t/a)}{a^2}} = \frac{\cosh(t/a)}{a} \\
\kappa &= \frac{|\dot{\mathbf{r}} \times \ddot{\mathbf{r}}|}{|\dot{\mathbf{r}}|^3} = \frac{\cosh(t/a)}{a\cosh^3(t/a)} = \frac{1}{a\cosh^2(t/a)}
\end{aligned}$$

Now, it was shown in Exercise 20 that for a curve represented spatially by $\mathbf{r}$ and parameterized naturally by $s$ and generally by $t$ the relation between $s$ and $t$ is given by $\left|\frac{ds}{dt}\right| = \left|\frac{d\mathbf{r}}{dt}\right|$. Hence:

$$\begin{aligned}
\left|\frac{ds}{dt}\right| &= \left|\frac{d\mathbf{r}}{dt}\right| \\
\left|\frac{ds}{dt}\right| &= |(\sinh(t/a), 0, 1)| \\
\left|\frac{ds}{dt}\right| &= \sqrt{\sinh^2(t/a) + 1} \\
\left|\frac{ds}{dt}\right| &= \cosh(t/a) \\
\frac{ds}{dt} &= \pm\cosh(t/a) \\
s &= \pm a\sinh(t/a) + c
\end{aligned}$$

where $c$ is the constant of integration. If we ignore the negative part and set $c = 0$ then we have:

$$\begin{aligned}
s &= a\sinh(t/a) \\
s^2 &= a^2\sinh^2(t/a) \\
s^2 + a^2 &= a^2\sinh^2(t/a) + a^2 \\
s^2 + a^2 &= a^2\left(\sinh^2(t/a) + 1\right) \\
s^2 + a^2 &= a^2\cosh^2(t/a)
\end{aligned}$$

Hence:

$$\kappa = \frac{1}{a\cosh^2(t/a)} = \frac{a}{a^2\cosh^2(t/a)} = \frac{a}{s^2 + a^2}$$

Regarding the torsion, the catenary is a plane curve and hence its torsion is zero. Therefore, the intrinsic equations of the given catenary are:

$$\kappa(s) = \frac{a}{s^2 + a^2} \qquad \text{and} \qquad \tau(s) = 0$$

48. Find the parametric representation of a curve with the following intrinsic equations: $\kappa = \sqrt{c/s}$ and $\tau = 0$ where $c > 0$ is a real constant and $s > 0$ is a natural parameter.
**Answer**: The curve is plane because $\tau = 0$. So, if we repeat the method that we used in Exercise 42 then we have:

$$\begin{aligned}
\phi' &= \kappa \\
\phi &= \int \kappa ds \\
\phi &= \int \sqrt{c/s}\, ds \\
\phi &= 2\sqrt{cs}
\end{aligned}$$

where we assumed the constant of integration to be zero. Hence, from Eq. 5 we get:

$$\begin{aligned}
\mathbf{r}' &= \mathbf{T} \\
\mathbf{r} &= \int \mathbf{T} ds \\
\mathbf{r} &= \int \left(\cos\left(2\sqrt{cs}\right), \sin\left(2\sqrt{cs}\right), 0\right) ds \\
\mathbf{r} &= (f_x + c_1, f_y + c_2, c_3)
\end{aligned}$$

where

$$\begin{aligned}
f_x &= \frac{2\sqrt{cs}\sin(2\sqrt{cs}) + \cos(2\sqrt{cs})}{2c} \\
f_y &= \frac{\sin(2\sqrt{cs}) - 2\sqrt{cs}\cos(2\sqrt{cs})}{2c}
\end{aligned}$$

and $c_1, c_2, c_3$ are real constants.

49. Prove that the curvature of a $t$-parameterized space curve is given by:

$$\kappa = \frac{|\dot{\mathbf{r}} \times \ddot{\mathbf{r}}|}{|\dot{\mathbf{r}}|^3}$$

**Answer**: We have:

$$\begin{aligned}
\dot{\mathbf{r}} &= \frac{d\mathbf{r}}{dt} = \frac{d\mathbf{r}}{ds}\frac{ds}{dt} = \mathbf{r}'\dot{s} \\
\ddot{\mathbf{r}} &= \frac{d\dot{\mathbf{r}}}{dt} = \frac{ds}{dt}\frac{d}{dt}\left(\frac{d\mathbf{r}}{ds}\right) + \frac{d\mathbf{r}}{ds}\frac{d^2s}{dt^2} = \frac{ds}{dt}\frac{d}{ds}\left(\frac{d\mathbf{r}}{ds}\right)\frac{ds}{dt} + \frac{d\mathbf{r}}{ds}\frac{d^2s}{dt^2} \\
&= \left(\frac{ds}{dt}\right)^2\frac{d^2\mathbf{r}}{ds^2} + \frac{d\mathbf{r}}{ds}\frac{d^2s}{dt^2} = (\dot{s})^2\mathbf{r}'' + \ddot{s}\mathbf{r}' \\
\dot{\mathbf{r}} \times \ddot{\mathbf{r}} &= \mathbf{r}'\dot{s} \times \left[(\dot{s})^2\mathbf{r}'' + \ddot{s}\mathbf{r}'\right] = (\dot{s})^3(\mathbf{r}' \times \mathbf{r}'') \\
|\dot{\mathbf{r}} \times \ddot{\mathbf{r}}| &= |\dot{s}|^3 |\mathbf{r}' \times \mathbf{r}''|
\end{aligned}$$

$$|\mathbf{r}' \times \mathbf{r}''| = \frac{|\dot{\mathbf{r}} \times \ddot{\mathbf{r}}|}{|\dot{s}|^3} \tag{15}$$

Now:

$$|\mathbf{r}' \times \mathbf{r}''| = |\mathbf{r}'|\,|\mathbf{r}''| \sin\theta$$

where $\theta$ is the angle between $\mathbf{r}'$ and $\mathbf{r}''$ and hence:

$$|\mathbf{r}' \times \mathbf{r}''| = |\mathbf{r}'|\,|\mathbf{r}''|$$

since $\mathbf{r}'$ $(= \mathbf{T})$ and $\mathbf{r}''$ $(= \mathbf{T}' = \kappa\mathbf{N})$ are orthogonal. Also:

$$\begin{aligned} |\mathbf{r}'| &= |\mathbf{T}| = 1 \\ |\mathbf{r}''| &= |\mathbf{T}'| = \kappa \end{aligned}$$

and hence:

$$|\mathbf{r}' \times \mathbf{r}''| = \kappa \tag{16}$$

Moreover:

$$|\dot{s}| \equiv \left|\frac{ds}{dt}\right| = \left|\frac{d\mathbf{r}}{dt}\right| \equiv |\dot{\mathbf{r}}| \tag{17}$$

as we found in Exercise 20. Hence:

$$\begin{aligned} |\mathbf{r}' \times \mathbf{r}''| &= \frac{|\dot{\mathbf{r}} \times \ddot{\mathbf{r}}|}{|\dot{s}|^3} \\ \kappa &= \frac{|\dot{\mathbf{r}} \times \ddot{\mathbf{r}}|}{|\dot{\mathbf{r}}|^3} \end{aligned}$$

where line 1 is from Eq. 15 while line 2 is from Eqs. 16 and 17.

50. Prove that the torsion of a $t$-parameterized space curve with non-vanishing curvature is given by:

$$\tau = \frac{\dot{\mathbf{r}} \cdot (\ddot{\mathbf{r}} \times \dddot{\mathbf{r}})}{|\dot{\mathbf{r}} \times \ddot{\mathbf{r}}|^2}$$

**Answer**: We have:

$$\begin{aligned} \mathbf{r}' &= \frac{d\mathbf{r}}{ds} = \frac{d\mathbf{r}}{dt}\frac{dt}{ds} = \dot{\mathbf{r}}t' \\ \mathbf{r}'' &= \frac{d^2\mathbf{r}}{ds^2} = \frac{dt}{ds}\frac{d}{ds}\left(\frac{d\mathbf{r}}{dt}\right) + \frac{d\mathbf{r}}{dt}\frac{d^2t}{ds^2} = \frac{dt}{ds}\frac{d}{dt}\left(\frac{d\mathbf{r}}{dt}\right)\frac{dt}{ds} + \frac{d\mathbf{r}}{dt}\frac{d^2t}{ds^2} \\ &= \left(\frac{dt}{ds}\right)^2\frac{d^2\mathbf{r}}{dt^2} + \frac{d\mathbf{r}}{dt}\frac{d^2t}{ds^2} = (t')^2\,\ddot{\mathbf{r}} + \dot{\mathbf{r}}t'' \\ \mathbf{r}''' &= \frac{d\mathbf{r}''}{ds} = \frac{d}{ds}\left[\left(\frac{dt}{ds}\right)^2\frac{d^2\mathbf{r}}{dt^2} + \frac{d\mathbf{r}}{dt}\frac{d^2t}{ds^2}\right] \end{aligned}$$

$$
\begin{aligned}
&= \frac{d^2\mathbf{r}}{dt^2}\frac{d}{ds}\left[\left(\frac{dt}{ds}\right)^2\right] + \left(\frac{dt}{ds}\right)^2 \frac{d}{ds}\left(\frac{d^2\mathbf{r}}{dt^2}\right) + \frac{d^2t}{ds^2}\frac{d}{ds}\left(\frac{d\mathbf{r}}{dt}\right) + \frac{d\mathbf{r}}{dt}\frac{d}{ds}\left(\frac{d^2t}{ds^2}\right) \\
&= \frac{d^2\mathbf{r}}{dt^2}\left[2\left(\frac{dt}{ds}\right)\left(\frac{d^2t}{ds^2}\right)\right] + \left(\frac{dt}{ds}\right)^2 \frac{d}{dt}\left(\frac{d^2\mathbf{r}}{dt^2}\right)\frac{dt}{ds} + \frac{d^2t}{ds^2}\frac{d}{dt}\left(\frac{d\mathbf{r}}{dt}\right)\frac{dt}{ds} + \frac{d\mathbf{r}}{dt}\frac{d^3t}{ds^3} \\
&= 2\left(\frac{dt}{ds}\right)\left(\frac{d^2t}{ds^2}\right)\frac{d^2\mathbf{r}}{dt^2} + \left(\frac{dt}{ds}\right)^3\frac{d^3\mathbf{r}}{dt^3} + \frac{d^2t}{ds^2}\left(\frac{d^2\mathbf{r}}{dt^2}\right)\frac{dt}{ds} + \frac{d\mathbf{r}}{dt}\frac{d^3t}{ds^3} \\
&= 3\left(\frac{dt}{ds}\right)\left(\frac{d^2t}{ds^2}\right)\frac{d^2\mathbf{r}}{dt^2} + \left(\frac{dt}{ds}\right)^3\frac{d^3\mathbf{r}}{dt^3} + \frac{d\mathbf{r}}{dt}\frac{d^3t}{ds^3} \\
&= 3t't''\ddot{\mathbf{r}} + (t')^3\dddot{\mathbf{r}} + t'''\dot{\mathbf{r}}
\end{aligned}
$$

Therefore:

$$
\begin{aligned}
& \mathbf{r}'\cdot(\mathbf{r}''\times\mathbf{r}''') \\
&= t'\dot{\mathbf{r}}\cdot\left[\left\{(t')^2\ddot{\mathbf{r}} + t''\dot{\mathbf{r}}\right\}\times\left\{3t't''\ddot{\mathbf{r}} + (t')^3\dddot{\mathbf{r}} + t'''\dot{\mathbf{r}}\right\}\right] \\
&= t'\dot{\mathbf{r}}\cdot\left[(t')^2\ddot{\mathbf{r}}\times\left\{3t't''\ddot{\mathbf{r}} + (t')^3\dddot{\mathbf{r}} + t'''\dot{\mathbf{r}}\right\} + t''\dot{\mathbf{r}}\times\left\{3t't''\ddot{\mathbf{r}} + (t')^3\dddot{\mathbf{r}} + t'''\dot{\mathbf{r}}\right\}\right] \\
&= t'\dot{\mathbf{r}}\cdot\left[\left\{\mathbf{0} + (t')^5(\ddot{\mathbf{r}}\times\dddot{\mathbf{r}}) + (t')^2 t'''(\ddot{\mathbf{r}}\times\dot{\mathbf{r}})\right\}\right. \\
&\quad \left. + \left\{3t'(t'')^2(\dot{\mathbf{r}}\times\ddot{\mathbf{r}}) + t''(t')^3(\dot{\mathbf{r}}\times\dddot{\mathbf{r}}) + \mathbf{0}\right\}\right] \\
&= t'\dot{\mathbf{r}}\cdot\left[(t')^5(\ddot{\mathbf{r}}\times\dddot{\mathbf{r}}) + (t')^2 t'''(\ddot{\mathbf{r}}\times\dot{\mathbf{r}}) + 3t'(t'')^2(\dot{\mathbf{r}}\times\ddot{\mathbf{r}}) + t''(t')^3(\dot{\mathbf{r}}\times\dddot{\mathbf{r}})\right] \\
&= (t')^6\dot{\mathbf{r}}\cdot(\ddot{\mathbf{r}}\times\dddot{\mathbf{r}}) + 0 + 0 + 0 \\
&= (t')^6\dot{\mathbf{r}}\cdot(\ddot{\mathbf{r}}\times\dddot{\mathbf{r}})
\end{aligned}
$$

that is:

$$\mathbf{r}'\cdot(\mathbf{r}''\times\mathbf{r}''') = (t')^6\dot{\mathbf{r}}\cdot(\ddot{\mathbf{r}}\times\dddot{\mathbf{r}}) \tag{18}$$

Now, $\mathbf{r}' = \mathbf{T}$ (Eq. 5), $\mathbf{r}'' = \mathbf{T}' = \kappa\mathbf{N}$ (Eq. 6) and

$$
\begin{aligned}
\mathbf{r}''' &= \mathbf{T}'' \\
&= \kappa'\mathbf{N} + \kappa\mathbf{N}' \\
&= \kappa'\mathbf{N} + \kappa(\mathbf{B}\times\mathbf{T})' \\
&= \kappa'\mathbf{N} + \kappa(\mathbf{B}'\times\mathbf{T}) + \kappa(\mathbf{B}\times\mathbf{T}') \\
&= \kappa'\mathbf{N} + \kappa(-\tau\mathbf{N}\times\mathbf{T}) + \kappa(\mathbf{B}\times\kappa\mathbf{N}) \\
&= \kappa'\mathbf{N} - \kappa\tau(\mathbf{N}\times\mathbf{T}) + \kappa^2(\mathbf{B}\times\mathbf{N}) \\
&= \kappa'\mathbf{N} + \kappa\tau\mathbf{B} - \kappa^2\mathbf{T}
\end{aligned}
$$

where Eqs. 6 and 8 are used plus the fact that $\mathbf{T}, \mathbf{N}, \mathbf{B}$ are right handed vector set. Hence:

$$\mathbf{r}'\cdot(\mathbf{r}''\times\mathbf{r}''') = \mathbf{T}\cdot\left(\kappa\mathbf{N}\times\left[\kappa'\mathbf{N} + \kappa\tau\mathbf{B} - \kappa^2\mathbf{T}\right]\right) \tag{19}$$

$$\begin{aligned}
&= \mathbf{T}\cdot\left(\mathbf{0}+\kappa^2\tau\{\mathbf{N}\times\mathbf{B}\}-\kappa^3\{\mathbf{N}\times\mathbf{T}\}\right)\\
&= \mathbf{T}\cdot\left(\kappa^2\tau\mathbf{T}+\kappa^3\mathbf{B}\right)\\
&= \kappa^2\tau\left(\mathbf{T}\cdot\mathbf{T}\right)+\kappa^3\left(\mathbf{T}\cdot\mathbf{B}\right)\\
&= \kappa^2\tau+0\\
&= \kappa^2\tau
\end{aligned}$$

Moreover:

$$(t')^6 \equiv \left(\frac{dt}{ds}\right)^6 = \left(\frac{1}{ds/dt}\right)^6 = \left(\frac{1}{|ds/dt|}\right)^6 = \left(\frac{1}{|d\mathbf{r}/dt|}\right)^6 = \left(\frac{1}{|\dot{\mathbf{r}}|}\right)^6 = \frac{1}{|\dot{\mathbf{r}}|^6} \tag{20}$$

where we used the result of Exercise 20.
On substituting from Eqs. 19 and 20 into Eq. 18 we obtain:

$$\begin{aligned}
\mathbf{r}'\cdot\left(\mathbf{r}''\times\mathbf{r}'''\right) &= (t')^6\,\dot{\mathbf{r}}\cdot\left(\ddot{\mathbf{r}}\times\dddot{\mathbf{r}}\right)\\
\kappa^2\tau &= \frac{\dot{\mathbf{r}}\cdot\left(\ddot{\mathbf{r}}\times\dddot{\mathbf{r}}\right)}{|\dot{\mathbf{r}}|^6}\\
\tau &= \frac{\dot{\mathbf{r}}\cdot\left(\ddot{\mathbf{r}}\times\dddot{\mathbf{r}}\right)}{\kappa^2\,|\dot{\mathbf{r}}|^6}\\
\tau &= \frac{\dot{\mathbf{r}}\cdot\left(\ddot{\mathbf{r}}\times\dddot{\mathbf{r}}\right)}{\left(\kappa\,|\dot{\mathbf{r}}|^3\right)^2}\\
\tau &= \frac{\dot{\mathbf{r}}\cdot\left(\ddot{\mathbf{r}}\times\dddot{\mathbf{r}}\right)}{|\dot{\mathbf{r}}\times\ddot{\mathbf{r}}|^2}
\end{aligned}$$

where in the last line we used the equation $\kappa = \frac{|\dot{\mathbf{r}}\times\ddot{\mathbf{r}}|}{|\dot{\mathbf{r}}|^3}$ which we obtained in the previous exercise.

51. Which of the two main curve parameters, $\kappa$ and $\tau$, is necessarily non-negative and why?
**Answer**: It is the curvature $\kappa$ because it is the magnitude of a vector (i.e. $\kappa = |\mathbf{r}''|$) according to our convention.

52. Show that along an $s$-parameterized curve $\mathbf{r}(s)$, the following relation holds true:

$$\mathbf{r}'\cdot\left(\mathbf{r}''\times\mathbf{r}'''\right)=\kappa^2\tau$$

**Answer**: This was done in Exercise 50 (see Eq. 19).

53. Can we obtain the curvature and torsion of circle as a special case of the curvature and torsion of helix? If yes, how? Is this consistent with the definition of helix as given in the book?
**Answer**: Yes. As stated in the book, the circle can be seen as a degenerate form of helix corresponding to $b = 0$. Hence the parametric equations of helix become:

$$x = a\cos\theta$$

$$\begin{aligned} y &= a\sin\theta \\ z &= 0 \end{aligned}$$

which are the parametric equations of a circle confined to the $xy$ plane with radius $a$. Now, the curvature and torsion of helix are given by (see Eqs. 9 and 10 in Exercise 28):

$$\kappa = \frac{a}{a^2+b^2} \qquad \text{and} \qquad \tau = \frac{b}{a^2+b^2}$$

Hence, when $b = 0$ they become:

$$\kappa = \frac{1}{a} \qquad \text{and} \qquad \tau = 0$$

which are the curvature and torsion of a circle with radius $a$. So, it is consistent with the definition of helix as given in the book.

54. Find the curvature and torsion of a space curve represented by: $\mathbf{r}(t) = (t^3, t+1, -t^2)$ at the point with $t = 2.4$.
**Answer**: We have:

$$\begin{aligned} \dot{\mathbf{r}} &= \frac{d\mathbf{r}}{dt} = \left(3t^2, 1, -2t\right) \\ \ddot{\mathbf{r}} &= \frac{d\dot{\mathbf{r}}}{dt} = (6t, 0, -2) \\ \dddot{\mathbf{r}} &= \frac{d\ddot{\mathbf{r}}}{dt} = (6, 0, 0) \\ \dot{\mathbf{r}} \times \ddot{\mathbf{r}} &= \begin{bmatrix} \mathbf{i} & \mathbf{j} & \mathbf{k} \\ 3t^2 & 1 & -2t \\ 6t & 0 & -2 \end{bmatrix} = \left(-2, -6t^2, -6t\right) \\ \ddot{\mathbf{r}} \times \dddot{\mathbf{r}} &= \begin{bmatrix} \mathbf{i} & \mathbf{j} & \mathbf{k} \\ 6t & 0 & -2 \\ 6 & 0 & 0 \end{bmatrix} = (0, -12, 0) \end{aligned}$$

Hence:

$$\begin{aligned} \kappa &= \frac{|\dot{\mathbf{r}} \times \ddot{\mathbf{r}}|}{|\dot{\mathbf{r}}|^3} \\ &= \frac{\sqrt{4 + 36t^4 + 36t^2}}{(9t^4 + 1 + 4t^2)^{3/2}} \\ &= \frac{\sqrt{36t^4 + 36t^2 + 4}}{(9t^4 + 4t^2 + 1)^{3/2}} \\ \tau &= \frac{\dot{\mathbf{r}} \cdot (\ddot{\mathbf{r}} \times \dddot{\mathbf{r}})}{|\dot{\mathbf{r}} \times \ddot{\mathbf{r}}|^2} \end{aligned}$$

$$
\begin{aligned}
&= \frac{(3t^2, 1, -2t) \cdot (0, -12, 0)}{36t^4 + 36t^2 + 4} \\
&= \frac{-12}{36t^4 + 36t^2 + 4} \\
&= \frac{-3}{9t^4 + 9t^2 + 1}
\end{aligned}
$$

At the point with $t = 2.4$ we have:

$$
\begin{aligned}
\kappa &= \frac{\sqrt{36\,(2.4)^4 + 36\,(2.4)^2 + 4}}{\left(9\,(2.4)^4 + 4\,(2.4)^2 + 1\right)^{3/2}} \simeq 0.00647 \\
\tau &= \frac{-3}{9\,(2.4)^4 + 9\,(2.4)^2 + 1} \simeq -0.00854
\end{aligned}
$$

55. Give an example of a space curve whose curvature and torsion are variables.
**Answer**: The curve of the previous exercise is a good example. The curves of Exercise 26 are also good examples.

56. Investigate the relation between the curvature and torsion at corresponding points of two space curves which are mirror-reflection of each other with respect to a given plane.
**Answer**: It is obvious that the curvature and torsion at the corresponding points of the two curves are equal in magnitude. Moreover, since the curvature is non-negative (see Exercise 51) then the two curvatures are equal. However, the torsions should be of opposite sign on the two curves. This is because according to Eq. 19 we have: $\tau = \mathbf{r}' \cdot (\mathbf{r}'' \times \mathbf{r}''')/\kappa^2$ and since the handedness of the $\mathbf{r}', \mathbf{r}'', \mathbf{r}'''$ vectors will change by reflection then the sign of $\mathbf{r}' \cdot (\mathbf{r}'' \times \mathbf{r}''')$ will change and hence the sign of $\tau$ will also change.

57. Investigate the relation between the curvature and torsion at corresponding points of two space curves which are symmetric with respect to a given point.
**Answer**: This is the same as the previous exercise apart from the fact that the $\mathbf{r}', \mathbf{r}'', \mathbf{r}'''$ vectors will not change their handedness and hence the two curves should have equal curvatures and equal torsions.

58. Discuss, in detail, the concept of "1D inhabitant" in the context of classifying the properties of space curves.
**Answer**: The concept of "1D inhabitant" in its relation to curves is the same as the concept of "2D inhabitant" in its relation to surfaces. So, the idea is that such a creature is confined to his 1D world (or 1D space which is the curve) and hence his perception is restricted to the intrinsic properties of this 1D space since it has no access to an embedding space of higher dimensionality (such as 2D surface or 3D ordinary space) to have an external view to this 1D space or have any notion about its extrinsic properties.

59. Establish a correspondence between the two main parameters of space curve (i.e. curvature and torsion) and the first and second fundamental forms of space surface.

**Answer**: The curvature and torsion uniquely characterize space curve according to the fundamental theorem of curves. Similarly, the first and second fundamental forms of space surface characterize space surface (although within certain restrictions) according to the fundamental theorem of surfaces. Hence, the curvature and torsion in curves play a similar role to the role of the first and second fundamental forms in surfaces.
Also, the curvature and torsion play in the Frenet-Serret formulae for space curves a similar role to the role of the coefficients of the first and second fundamental forms in the Gauss-Weingarten equations for space surfaces.

60. Discuss the resemblance between $\kappa$ and $\tau$ of space curve and the curvature tensor of space surface.
**Answer**: It may be claimed that $\kappa$ and $\tau$ of space curve characterize and quantify (from an external perspective) the "distortion" that is suffered by the space curve from its "ideal" straight and plane state situation since $\kappa = 0$ for straight line and $\tau = 0$ for plane curve. Similarly, the curvature tensor of space surface characterizes and quantifies (from an external perspective) the "distortion" that is suffered by the space surface from its "ideal" plane state situation since $b_{\alpha\beta} = 0$ for plane surface.

61. State, in words, the mathematical relation: $\tau = \underline{\epsilon}^{ijk} T_i N_j \frac{\delta N_k}{\delta s}$ using technical terms for all the notations and symbols used in this equation.
**Answer**: In general curvilinear coordinate systems, the torsion of an $s$-parameterized curve is equal to the scalar triple product of the tangent vector, the normal vector and the absolute derivative of the normal vector with respect to $s$. In the above equation, $\tau$ is the torsion, $\underline{\epsilon}^{ijk}$ is the 3D contravariant absolute permutation tensor, $T_i$ is the tangent vector, $N_j$ is the normal vector, $\delta N_k/\delta s$ is the absolute derivative of the normal vector with respect to $s$, and $s$ is a natural parameter.

62. What is the significance of the curvature and torsion of space curves as measures of their variation in the embedding space?
**Answer**: The curvature $\kappa$ of space curve is a measure of its deviation from its straight state and hence $\kappa = 0$ for straight line. The torsion $\tau$ of space curve is a measure of its deviation from its plane state and hence $\tau = 0$ for plane curve.

63. What is the relation between the curvature and the radius of curvature of a space curve? Is there an advantage in using one of these or the other as the main concept?
**Answer**: They are the reciprocal of each other, that is:

$$R_\kappa = \frac{1}{\kappa}$$

where $R_\kappa$ is the radius of curvature and $\kappa$ ($\neq 0$) is the curvature.
It may be claimed that there is an advantage in using the concept of "curvature" as the principal concept instead of "radius of curvature" since the curvature can be defined at all points of a smooth curve where a tangent vector is defined, including those with vanishing curvature, while the radius of curvature is defined only at those points with non-vanishing curvature.

64. Outline the physical significance of the relation: $\kappa = |\mathbf{r}''|$.
**Answer**: It means that the curvature of a space curve is equal to the magnitude of the acceleration along the curve when the time is represented by a natural parameter.

65. Express $\kappa$ in terms of $\mathbf{r}$ and its first and second derivatives where $\mathbf{r}$ is a spatial representation of a curve parameterized by a general parameter $t$.
**Answer**:

$$\kappa = \frac{|\dot{\mathbf{r}} \times \ddot{\mathbf{r}}|}{|\dot{\mathbf{r}}|^3}$$

66. Express $\tau$ in terms of $\mathbf{N}'$ and $\mathbf{B}$ of a naturally parameterized curve.
**Answer**:

$$\tau = \mathbf{N}' \cdot \mathbf{B}$$

67. Express $\tau$ in terms of $\mathbf{r}$ and its first, second and third derivatives where $\mathbf{r}$ is a spatial representation of a curve parameterized by a general parameter $t$.
**Answer**:

$$\tau = \frac{\dot{\mathbf{r}} \cdot (\ddot{\mathbf{r}} \times \dddot{\mathbf{r}})}{|\dot{\mathbf{r}} \times \ddot{\mathbf{r}}|^2}$$

68. What is the relation between the torsion and the radius of torsion of a space curve?
**Answer**: The radius of torsion is the absolute value of the reciprocal of torsion, that is:

$$R_\tau = \left|\frac{1}{\tau}\right|$$

where $R_\tau$ is the radius of torsion and $\tau$ ($\neq 0$) is the torsion.

69. Define geodesic torsion in words and state its mathematical relation to $\mathbf{r}$ and $\mathbf{n}$ and their derivatives.
**Answer**: The geodesic torsion of a surface curve $C$ at a given point $P$ is the torsion of the geodesic curve that passes through $P$ in the tangent direction of $C$ at $P$. The geodesic torsion $\tau_g$ of a surface curve represented spatially by $\mathbf{r}(s)$ is given by the following scalar triple product:

$$\tau_g = \mathbf{n} \cdot (\mathbf{n}' \times \mathbf{r}')$$

where $\mathbf{n}$ is the unit normal vector to the surface and the primes represent differentiation with respect to the natural parameter $s$.

70. What is the significance of the relation: $\tau_g = \tau - \frac{d\phi}{ds}$ and what the symbols in this relation mean?
**Answer**: This relation, which is known as the Bonnet formula, expresses the geodesic torsion $\tau_g$ in terms of the torsion $\tau$ of the curve and the derivative of $\phi$ with respect to the natural parameter $s$ where $\phi$ is the angle between the unit normal vector $\mathbf{n}$ to the surface and the principal normal vector $\mathbf{N}$ of the curve. So, when this angle is constant (e.g. when $\mathbf{n}$ and $\mathbf{N}$ are collinear along the curve as it is the case in geodesic curves) the geodesic torsion becomes equal to the torsion of the curve.

71. What is the condition for the torsion and the geodesic torsion of a space curve to be equal?
**Answer**: The condition (according to the relation: $\tau_g = \tau - \frac{d\phi}{ds}$ which we discussed in the previous exercise) is $\frac{d\phi}{ds} = 0$.

72. Prove that along a sufficiently smooth curve represented by $\mathbf{r}(t)$, the vector $\ddot{\mathbf{r}}$ at a given point $P$ on the curve is parallel to the osculating plane at $P$.
**Answer**: We have:

$$\begin{aligned}
\dot{\mathbf{r}} &= \frac{d\mathbf{r}}{dt} = \frac{d\mathbf{r}}{ds}\frac{ds}{dt} = \mathbf{r}'\dot{s} \\
\ddot{\mathbf{r}} &= \frac{d\dot{\mathbf{r}}}{dt} = \frac{ds}{dt}\frac{d}{dt}\left(\frac{d\mathbf{r}}{ds}\right) + \frac{d\mathbf{r}}{ds}\frac{d^2s}{dt^2} = \frac{ds}{dt}\frac{d}{ds}\left(\frac{d\mathbf{r}}{ds}\right)\frac{ds}{dt} + \frac{d\mathbf{r}}{ds}\frac{d^2s}{dt^2} \\
&= \left(\frac{ds}{dt}\right)^2\frac{d^2\mathbf{r}}{ds^2} + \frac{d\mathbf{r}}{ds}\frac{d^2s}{dt^2} = (\dot{s})^2\,\mathbf{r}'' + \ddot{s}\mathbf{r}'
\end{aligned}$$

Now, $\mathbf{r}' = \mathbf{T}$ (Eq. 5) and $\mathbf{r}'' = \mathbf{T}' = \kappa\mathbf{N}$ (Eq. 6). Hence:

$$\ddot{\mathbf{r}} = (\dot{s})^2\,\mathbf{r}'' + \ddot{s}\mathbf{r}' = (\dot{s})^2\,\kappa\mathbf{N} + \ddot{s}\mathbf{T}$$

i.e. $\ddot{\mathbf{r}}$ is a linear combination of $\mathbf{T}$ and $\mathbf{N}$ and hence it is parallel to the osculating plane since the osculating plane is the span of $\mathbf{T}$ and $\mathbf{N}$ (or in other words the osculating plane is perpendicular to $\mathbf{B}$).

73. Obtain the equation of the osculating plane of a curve represented parametrically by: $\mathbf{r}(t) = (3\cos t, 2\sin t, \cos t + 5\sin t)$ at a general point on the curve.
**Answer**: We have:

$$\begin{aligned}
\dot{\mathbf{r}} &= \frac{d\mathbf{r}}{dt} = (-3\sin t, 2\cos t, -\sin t + 5\cos t) \\
\ddot{\mathbf{r}} &= \frac{d\dot{\mathbf{r}}}{dt} = (-3\cos t, -2\sin t, -\cos t - 5\sin t) \\
\dot{\mathbf{r}} \times \ddot{\mathbf{r}} &= \begin{bmatrix} \mathbf{i} & \mathbf{j} & \mathbf{k} \\ -3\sin t & 2\cos t & -\sin t + 5\cos t \\ -3\cos t & -2\sin t & -\cos t - 5\sin t \end{bmatrix} = (-2, -15, 6) \\
\mathbf{B} &= \frac{\dot{\mathbf{r}} \times \ddot{\mathbf{r}}}{|\dot{\mathbf{r}} \times \ddot{\mathbf{r}}|} = \frac{(-2, -15, 6)}{\sqrt{265}}
\end{aligned}$$

The osculating plane is the plane passing through the point $(3\cos t, 2\sin t, \cos t + 5\sin t)$ $\left[\text{i.e. the point corresponding to a given } t\right]$ with normal vector $\mathbf{B}$, that is:

$$\begin{aligned}
\frac{(-2, -15, 6)}{\sqrt{265}} \cdot (x - 3\cos t, y - 2\sin t, z - \cos t - 5\sin t) &= 0 \\
-2\,(x - 3\cos t) - 15\,(y - 2\sin t) + 6\,(z - \cos t - 5\sin t) &= 0 \\
-2x + 6\cos t - 15y + 30\sin t + 6z - 6\cos t - 30\sin t &= 0 \\
-2x - 15y + 6z &= 0
\end{aligned}$$

74. Derive the second of the Frenet-Serret formulae using the other two formulae (see Exercise 43).
**Answer**:

$$\begin{aligned}
\mathbf{N} &= \mathbf{B} \times \mathbf{T} \\
\mathbf{N}' &= (\mathbf{B}' \times \mathbf{T}) + (\mathbf{B} \times \mathbf{T}') \\
\mathbf{N}' &= (-\tau\mathbf{N} \times \mathbf{T}) + (\mathbf{B} \times \kappa\mathbf{N}) \\
\mathbf{N}' &= -\tau\,(\mathbf{N} \times \mathbf{T}) + \kappa\,(\mathbf{B} \times \mathbf{N}) \\
\mathbf{N}' &= -\tau\,(-\mathbf{B}) + \kappa\,(-\mathbf{T}) \\
\mathbf{N}' &= \tau\mathbf{B} - \kappa\mathbf{T}
\end{aligned}$$

where line 1 is because $\mathbf{T}, \mathbf{N}, \mathbf{B}$ are orthonormal right handed vector set, line 2 is the product rule of differentiation, line 3 is based on using the other two of Frenet-Serret formulae, and line 5 is because $\mathbf{T}, \mathbf{N}, \mathbf{B}$ are orthonormal right handed vector set.

75. Define Darboux vector $\mathbf{d}$ and hence verify that the following relations are valid:

$$\mathbf{T}' = \mathbf{d} \times \mathbf{T} \qquad \mathbf{N}' = \mathbf{d} \times \mathbf{N} \qquad \mathbf{B}' = \mathbf{d} \times \mathbf{B}$$

**Answer**: Darboux vector $\mathbf{d}$ is defined by the equation:

$$\mathbf{d} = \tau\mathbf{T} + \kappa\mathbf{B}$$

where $\tau$ and $\kappa$ are the torsion and curvature while $\mathbf{T}$ and $\mathbf{B}$ are the tangent and binormal unit vectors. Hence:

$$\begin{aligned}
\mathbf{T}' &= \mathbf{d} \times \mathbf{T} \\
&= (\tau\mathbf{T} + \kappa\mathbf{B}) \times \mathbf{T} \\
&= (\tau\mathbf{T} \times \mathbf{T}) + (\kappa\mathbf{B} \times \mathbf{T}) \\
&= \mathbf{0} + (\kappa\mathbf{N}) \\
&= \kappa\mathbf{N}
\end{aligned}$$

which is the first of Frenet-Serret formulae (Eq. 12).

$$\begin{aligned}
\mathbf{N}' &= \mathbf{d} \times \mathbf{N} \\
&= (\tau\mathbf{T} + \kappa\mathbf{B}) \times \mathbf{N} \\
&= (\tau\mathbf{T} \times \mathbf{N}) + (\kappa\mathbf{B} \times \mathbf{N}) \\
&= \tau\mathbf{B} - \kappa\mathbf{T}
\end{aligned}$$

which is the second of Frenet-Serret formulae (Eq. 13).

$$\begin{aligned}
\mathbf{B}' &= \mathbf{d} \times \mathbf{B} \\
&= (\tau\mathbf{T} + \kappa\mathbf{B}) \times \mathbf{B}
\end{aligned}$$

$$\begin{aligned} &= (\tau\mathbf{T}\times\mathbf{B}) + (\kappa\mathbf{B}\times\mathbf{B}) \\ &= (-\tau\mathbf{N}) + \mathbf{0} \\ &= -\tau\mathbf{N} \end{aligned}$$

which is the third of Frenet-Serret formulae (Eq. 14).

76. Explain all the symbols and notations used in the following relation:

$$\tau = \frac{\dot{\mathbf{r}}\cdot(\ddot{\mathbf{r}}\times\dddot{\mathbf{r}})}{(\dot{\mathbf{r}}\cdot\dot{\mathbf{r}})(\ddot{\mathbf{r}}\cdot\ddot{\mathbf{r}}) - (\dot{\mathbf{r}}\cdot\ddot{\mathbf{r}})^2}$$

**Answer**: $\tau$ is the curve torsion, $\mathbf{r}$ is the curve spatial representation, the single, double and triple overdots represent the first, second and third order derivative of $\mathbf{r}$ with respect to $t$ where $t$ is a general parameter of the curve in its $\mathbf{r}(t)$ representation, the dot (i.e. $\cdot$) is the dot product operator and the cross (i.e. $\times$) is the cross product operator.

77. Prove that for a curve with helical shape, the Darboux vector is constant along the curve.

**Answer**: Circular helix is generally parameterized by:

$$\mathbf{r}(t) = (a\cos t, a\sin t, bt)$$

where $a$ and $b$ are real constants and $t$ is a general parameter. Accordingly:

$$\begin{aligned} \dot{\mathbf{r}} &= \frac{d\mathbf{r}}{dt} = (-a\sin t, a\cos t, b) \\ \ddot{\mathbf{r}} &= \frac{d\dot{\mathbf{r}}}{dt} = (-a\cos t, -a\sin t, 0) \\ \dot{\mathbf{r}}\times\ddot{\mathbf{r}} &= \begin{bmatrix} \mathbf{i} & \mathbf{j} & \mathbf{k} \\ -a\sin t & a\cos t & b \\ -a\cos t & -a\sin t & 0 \end{bmatrix} = \left(ab\sin t, -ab\cos t, a^2\right) \\ \mathbf{T} &= \frac{\dot{\mathbf{r}}}{|\dot{\mathbf{r}}|} = \frac{(-a\sin t, a\cos t, b)}{\sqrt{a^2+b^2}} \\ \mathbf{B} &= \frac{\dot{\mathbf{r}}\times\ddot{\mathbf{r}}}{|\dot{\mathbf{r}}\times\ddot{\mathbf{r}}|} = \frac{(ab\sin t, -ab\cos t, a^2)}{a\sqrt{a^2+b^2}} = \frac{(b\sin t, -b\cos t, a)}{\sqrt{a^2+b^2}} \end{aligned}$$

Also, for helix $\kappa = \frac{a}{a^2+b^2}$ and $\tau = \frac{b}{a^2+b^2}$ (see Eqs. 9 and 10 in Exercise 28). Hence:

$$\begin{aligned} \mathbf{d} &= \tau\mathbf{T} + \kappa\mathbf{B} \\ &= \frac{b}{a^2+b^2}\frac{(-a\sin t, a\cos t, b)}{\sqrt{a^2+b^2}} + \frac{a}{a^2+b^2}\frac{(b\sin t, -b\cos t, a)}{\sqrt{a^2+b^2}} \\ &= \frac{(-ab\sin t, ab\cos t, b^2)}{(a^2+b^2)^{3/2}} + \frac{(ab\sin t, -ab\cos t, a^2)}{(a^2+b^2)^{3/2}} \\ &= \frac{(0, 0, a^2+b^2)}{(a^2+b^2)^{3/2}} \end{aligned}$$

$$= \frac{(0, 0, 1)}{\sqrt{a^2 + b^2}}$$

Now, since $a$ and $b$ are constants then the Darboux vector is constant along the curve, as required.

78. State the Frenet-Serret formulae in a single equation using the Darboux vector.
**Answer**:

$$(\mathbf{T}', \mathbf{N}', \mathbf{B}') = \mathbf{d} \times (\mathbf{T}, \mathbf{N}, \mathbf{B})$$

79. Write down the Frenet-Serret formulae assuming a general curvilinear coordinate system.
**Answer**:

$$\begin{aligned}
\frac{\delta T^i}{\delta s} &= \frac{dT^i}{ds} + \Gamma^i_{jk} T^j \frac{dx^k}{ds} = \kappa N^i \\
\frac{\delta N^i}{\delta s} &= \frac{dN^i}{ds} + \Gamma^i_{jk} N^j \frac{dx^k}{ds} = \tau B^i - \kappa T^i \\
\frac{\delta B^i}{\delta s} &= \frac{dB^i}{ds} + \Gamma^i_{jk} B^j \frac{dx^k}{ds} = -\tau N^i
\end{aligned}$$

80. Give a brief explanation of why the solution of a space curve cannot be obtained in general by a direct integration of the Frenet-Serret equations.
**Answer**: Direct integration is a very primitive method for obtaining a solution for a system of differential equations, and hence it can be useful only for special cases. So, in general we need more sophisticated mathematical methods to obtain a solution for a system of differential equations like the Frenet-Serret equations.

81. Give a brief definition for the osculating circle and the osculating sphere of a space curve.
**Answer**: The osculating circle of a smooth space curve $C$ at a given point $P$ on the curve is defined as the circle whose radius is the radius of curvature of $C$ at $P$, it is tangent to $C$ at $P$, and it is confined to the osculating plane of $C$ at $P$.
The osculating sphere of a smooth space curve $C$ at a given point $P$ on the curve is defined as the limit of a sphere passing through $P$ and three neighboring points on the curve as these three points converge to $P$.

82. How can the osculating circle and the osculating sphere of a space curve be defined using the concept of limit?
**Answer**: Following the manner of defining the tangent line to a curve as a limit of the secant line, the osculating circle to a curve at a given point $P$ may be defined geometrically as the limit of a circle passing through $P$ and two other points on the curve as these two points converge to $P$ while staying on the curve. The definition of the osculating sphere as a limit was given in the answer of the previous question.

83. Prove that for a given space curve $C$, the binormal lines of $C$ and the tangent lines to the locus of the centers of spherical curvature of $C$ are parallel at their corresponding points.
**Answer**: The equation of the center of spherical curvature of a curve $C$ at a given point $P$ is:

$$\mathbf{r}_S = \mathbf{r}_P + \frac{1}{\kappa}\mathbf{N} - \frac{\kappa'}{\tau\kappa^2}\mathbf{B}$$

where $\mathbf{r}_S$ is the position of the center of spherical curvature of $C$ at $P$, $\mathbf{r}_P$ is the position of $P$, $\mathbf{B}$ and $\mathbf{N}$ are the binormal and principal normal vectors of $C$ at $P$, $\kappa$ and $\tau$ are the curvature and torsion of $C$ at $P$, and the prime symbolizes derivative with respect to a natural parameter $s$ of $C$. On differentiating this equation with respect to $s$ we get:

$$\begin{aligned}
\frac{d\mathbf{r}_S}{ds} &= \mathbf{r}'_P + \left(\frac{1}{\kappa}\mathbf{N}\right)' - \left(\frac{\kappa'}{\tau\kappa^2}\mathbf{B}\right)' \\
&= \mathbf{r}'_P + \left(-\frac{\kappa'}{\kappa^2}\mathbf{N} + \frac{1}{\kappa}\mathbf{N}'\right) - \left(\left[\frac{\kappa'}{\tau\kappa^2}\right]'\mathbf{B} + \frac{\kappa'}{\tau\kappa^2}\mathbf{B}'\right) \\
&= \mathbf{r}'_P - \frac{\kappa'}{\kappa^2}\mathbf{N} + \frac{1}{\kappa}\mathbf{N}' - \left[\frac{\kappa'}{\tau\kappa^2}\right]'\mathbf{B} - \frac{\kappa'}{\tau\kappa^2}\mathbf{B}' \\
&= \mathbf{T}_P - \frac{\kappa'}{\kappa^2}\mathbf{N} + \frac{1}{\kappa}(\tau\mathbf{B} - \kappa\mathbf{T}_P) - \left[\frac{\kappa'}{\tau\kappa^2}\right]'\mathbf{B} - \frac{\kappa'}{\tau\kappa^2}(-\tau\mathbf{N}) \\
&= \mathbf{T}_P - \frac{\kappa'}{\kappa^2}\mathbf{N} + \frac{\tau}{\kappa}\mathbf{B} - \mathbf{T}_P - \left[\frac{\kappa'}{\tau\kappa^2}\right]'\mathbf{B} + \frac{\kappa'}{\kappa^2}\mathbf{N} \\
&= \frac{\tau}{\kappa}\mathbf{B} - \left[\frac{\kappa'}{\tau\kappa^2}\right]'\mathbf{B} \\
&= \left(\frac{\tau}{\kappa} - \left[\frac{\kappa'}{\tau\kappa^2}\right]'\right)\mathbf{B}
\end{aligned}$$

where in line 2 we use the product rule of differentiation, while in line 4 we use Eqs. 5, 13 and 14.
The last line means that the tangent vector $\frac{d\mathbf{r}_S}{ds}$ to the locus of the centers of spherical curvature of $C$ is along the orientation of $\mathbf{B}$ and hence the binormal lines of $C$ (which are along $\mathbf{B}$) and the tangent lines to the locus of the centers of spherical curvature of $C$ (which are along $\frac{d\mathbf{r}_S}{ds}$) are parallel at their corresponding points, as required.

84. Derive an expression for the position of the center of curvature of a $t$-parameterized twisted curve represented spatially by $\mathbf{r}(t)$ in terms of $\mathbf{r}$ and its derivatives.
**Answer**: The equation of the center of curvature of a curve $C$ at a given point $P$ is:

$$\mathbf{r}_C = \mathbf{r}_P + \frac{1}{\kappa}\mathbf{N}$$

where $\mathbf{r}_C$ is the position of the center of curvature of $C$ at $P$, $\mathbf{r}_P$ is the position of $P$, $\kappa$ is the curvature of $C$ at $P$, and $\mathbf{N}$ is the principal normal vector of $C$ at $P$. Now, for a

$t$-parameterized twisted curve represented spatially by $\mathbf{r}(t)$ we have:

$$\mathbf{N} = \frac{\dot{\mathbf{r}} \times (\ddot{\mathbf{r}} \times \dot{\mathbf{r}})}{|\dot{\mathbf{r}}|\,|\ddot{\mathbf{r}} \times \dot{\mathbf{r}}|} \qquad \text{and} \qquad \kappa = \frac{|\dot{\mathbf{r}} \times \ddot{\mathbf{r}}|}{|\dot{\mathbf{r}}|^3}$$

Hence, the position of the center of curvature $\mathbf{r}_C$ is given in terms of $\mathbf{r}$ (which is subscripted with $P$ to indicate its association with a given point on $C$) and its derivatives by:

$$\begin{aligned}\mathbf{r}_C &= \mathbf{r}_P + \left(\frac{|\dot{\mathbf{r}}_P|^3}{|\dot{\mathbf{r}}_P \times \ddot{\mathbf{r}}_P|}\right)\frac{\dot{\mathbf{r}}_P \times (\ddot{\mathbf{r}}_P \times \dot{\mathbf{r}}_P)}{|\dot{\mathbf{r}}_P|\,|\ddot{\mathbf{r}}_P \times \dot{\mathbf{r}}_P|} \\ &= \mathbf{r}_P + \left(\frac{|\dot{\mathbf{r}}_P|^2}{|\ddot{\mathbf{r}}_P \times \dot{\mathbf{r}}_P|^2}\right)\dot{\mathbf{r}}_P \times (\ddot{\mathbf{r}}_P \times \dot{\mathbf{r}}_P)\end{aligned}$$

85. Find the spatial coordinates of the center of the osculating circle of a space curve represented by: $\mathbf{r}(t) = (t, \sqrt{t}, t^2)$ at the point with $t = 2.6$.
**Answer**: We have:

$$\begin{aligned}\dot{\mathbf{r}}_P &= \left(1, \frac{1}{2\sqrt{t}}, 2t\right) \\ \ddot{\mathbf{r}}_P &= \left(0, \frac{-1}{4t^{3/2}}, 2\right) \\ \ddot{\mathbf{r}}_P \times \dot{\mathbf{r}}_P &= \begin{bmatrix} \mathbf{i} & \mathbf{j} & \mathbf{k} \\ 0 & \frac{-1}{4t^{3/2}} & 2 \\ 1 & \frac{1}{2\sqrt{t}} & 2t \end{bmatrix} = \left(\frac{-3}{2\sqrt{t}}, 2, \frac{1}{4t^{3/2}}\right) \\ \dot{\mathbf{r}}_P \times (\ddot{\mathbf{r}}_P \times \dot{\mathbf{r}}_P) &= \begin{bmatrix} \mathbf{i} & \mathbf{j} & \mathbf{k} \\ 1 & \frac{1}{2\sqrt{t}} & 2t \\ \frac{-3}{2\sqrt{t}} & 2 & \frac{1}{4t^{3/2}} \end{bmatrix} = \left(\frac{1}{8t^2} - 4t, -\frac{1}{4t^{3/2}} - 3\sqrt{t}, 2 + \frac{3}{4t}\right)\end{aligned}$$

Using the formula that we obtained in the previous exercise with $\mathbf{r}_P$ being the point on the curve corresponding to $t = 2.6$, we get:

$$\begin{aligned}\mathbf{r}_C &= \mathbf{r}_P + \left(\frac{|\dot{\mathbf{r}}_P|^2}{|\ddot{\mathbf{r}}_P \times \dot{\mathbf{r}}_P|^2}\right)\dot{\mathbf{r}}_P \times (\ddot{\mathbf{r}}_P \times \dot{\mathbf{r}}_P) \\ &= (2.6, \sqrt{2.6}, 2.6^2) + \\ &\quad \frac{1 + \frac{1}{4(2.6)} + 4\,(2.6)^2}{\frac{9}{4(2.6)} + 4 + \frac{1}{16(2.6)^3}}\left(\frac{1}{8\,(2.6)^2} - 4\,(2.6), -\frac{1}{4\,(2.6)^{3/2}} - 3\sqrt{2.6}, 2 + \frac{3}{4\,(2.6)}\right) \\ &\simeq (-57.39, -26.69, 19.98)\end{aligned}$$

86. What is the relation between the osculating circle and the osculating plane at a given point of a space curve?
**Answer**: The osculating circle lies in the osculating plane.

87. Explain all the symbols involved in the equation: $\mathbf{r}_S = \mathbf{r}_P + R_\kappa \mathbf{N} + \text{sgn}(\tau) R_\tau R'_\kappa \mathbf{B}$ which is related to the center of the osculating sphere.
**Answer**: $\mathbf{r}_S$ is the position vector of the center of the osculating sphere, $\mathbf{r}_P$ is the position vector of the point of the curve to which the osculating sphere belongs, $R_\kappa$ and $R_\tau$ are the radius of curvature and the radius of torsion of the curve at the given point, $\mathbf{N}$ and $\mathbf{B}$ are the principal normal and the binormal vectors of the curve at the given point, the prime symbolizes differentiation with respect to a natural parameter of the curve, and $\text{sgn}(\tau)$ is the sign function of the torsion of the curve.

88. Write down the formula for the radius of the osculating sphere explaining all the symbols used.
**Answer**:
$$|\mathbf{r}_S - \mathbf{r}_P| = \sqrt{R_\kappa^2 + (R_\tau R'_\kappa)^2}$$
The symbols are explained in the answer of the previous exercise.

89. What is the difference between the concept of parallelism in Euclidean and non-Euclidean spaces?
**Answer**: In Euclidean spaces parallelism is an absolute property as it is defined without reference to a peripheral object, while in non-Euclidean spaces parallelism is defined in reference to a given curve.

90. State the mathematical condition for a vector field to be parallel along a given surface curve in terms of its absolute derivative.
**Answer**: The necessary and sufficient condition for a vector field $A^\alpha$ to be parallel along a surface curve $u^\beta = u^\beta(t)$ is that its absolute derivative along the curve vanishes identically, that is:
$$\frac{\delta A^\alpha}{\delta t} \equiv A^\alpha{}_{;\beta} \frac{du^\beta}{dt} \equiv \frac{dA^\alpha}{dt} + \Gamma^\alpha_{\gamma\beta} A^\gamma \frac{du^\beta}{dt} = 0$$

91. What is the meaning of describing parallel propagation as path dependent? What are the consequences of this dependency?
**Answer**: It means that parallel propagation and its implications are dependent on the specific curve (or path) along which the propagation takes place. This path dependency is more clarified in the following.
Some consequences of this dependency are:
• Given two points $P_1$ and $P_2$ on a surface, the vector obtained at $P_2$ by parallel propagation of a vector from $P_1$ along a given surface curve $C$ connecting $P_1$ to $P_2$ depends on the curve $C$.
• Starting from a given point $P$ on a closed surface curve $C$ enclosing a simply connected region on the surface, parallel propagation of a vector field around $C$ starting from $P$ does not necessarily result in the same vector field when arriving at $P$.

# Chapter 3
# Surfaces in Space

1. Give the mathematical definition of space surface and explain the difference between a surface and its trace according to this definition.
   **Answer**: A surface in a 3D manifold is a mapping from a subset of the parameters plane to a 3D space, that is $S : \Omega \rightarrow \mathbb{R}^3$, where $\Omega$ is a subset of $\mathbb{R}^2$ plane and $S$ is a sufficiently smooth injective function. The image of this mapping in $\mathbb{R}^3$ is known as the trace of the surface. So, the surface technically is a mathematical function while its trace is the image of this function in the target space.

2. What is the mathematical condition for a surface to be regular at a particular point in terms of its basis vectors?
   **Answer**: The condition is that its basis vectors, $\mathbf{E}_1$ and $\mathbf{E}_2$, at that point are not collinear, i.e. $\mathbf{E}_1 \times \mathbf{E}_2 \neq \mathbf{0}$.

3. State the three main mathematical methods for defining a space surface and compare them explaining any advantages or disadvantages in using one of these methods or the others in various contexts.
   **Answer**: A surface embedded in a 3D space can be defined explicitly as $z = f(x, y)$, or implicitly as $F(x, y, z) = 0$, or parametrically as $x(u^1, u^2)$, $y(u^1, u^2)$, $z(u^1, u^2)$ where $u^1$ and $u^2$ are the coordinates on the parameters plane.
   The first method is the most explicit, intuitive and straightforward and hence it should be preferred as a first choice when there is no other factor dictating or favoring the use of the other methods. The main limitation of this method is that it is rather simple and technically primitive and hence it does not lend itself to sophisticated mathematical manipulation and management. Moreover, it is not available always since the mathematical relation that represents the surface may not lend itself to this explicit form.
   The second method is second in the aforementioned factors and hence it may be preferred as a second choice when the first method is not available.
   The third method is the most sophisticated and general method and hence it lends itself to mathematical manipulation and management more easily and naturally and can deal with more complex mathematical situations that the other two methods may not be capable to do.

4. Classify the three methods of the last question into two main categories and discuss these categories.
   **Answer**: The three methods can be classified as direct (which includes the first and second methods) and indirect (which includes the third method). In the first category the space surface is described directly by the three spatial coordinates of the surface points according to a given 3D coordinate system where these coordinates are linked by

a single functional relation, while in the second category the space surface is described parametrically using two independent parameters to determine the three spatial coordinates of the surface points according to a given 3D coordinate system and hence three functional relations are needed, i.e. one parametric relation for each coordinate.

5. What "coordinate patch of class $C^n$" means? What are the mathematical conditions that should be satisfied by such a patch?
**Answer**: A coordinate patch of a surface is an injective, bicontinuous, regular, parametric representation of a part of the surface.[14] Technically, a coordinate patch of class $C^n$ $(n > 0)$ on a space surface $S$ is a functional mapping of an open set $\Omega$ in the $uv$ parameters plane onto $S$ that satisfies the following conditions:
• The functional mapping relation is of class $C^n$ over $\Omega$.
• The mapping is one-to-one and bicontinuous over $\Omega$.
• $\mathbf{E}_1 \times \mathbf{E}_2 \neq \mathbf{0}$ at any point in $\Omega$ where $\mathbf{E}_1$ and $\mathbf{E}_2$ are the surface basis vectors.

6. Show that a Monge patch of the form $\mathbf{r}(u, v) = (u, v, f(u, v))$ is regular of class $C^n$ if $f$ is of this class.
**Answer**: By definition, $\mathbf{r}(u, v) = (u, v, f(u, v))$ is of class $C^n$ since $f$ is of this class. So, what we need is to show that it is regular. Now:

$$\begin{aligned} \mathbf{E}_1 &= \partial_u \mathbf{r} = (1, 0, \partial_u f) \\ \mathbf{E}_2 &= \partial_v \mathbf{r} = (0, 1, \partial_v f) \\ \mathbf{E}_1 \times \mathbf{E}_2 &= (-\partial_u f, -\partial_v f, 1) \neq \mathbf{0} \end{aligned}$$

and hence it is regular.

7. Give a rigorous definition of "tangent vector" of a surface curve at a particular point on the surface.
**Answer**: A tangent vector $\mathbf{v}$ of a surface curve $C$ at a particular point $P$ on the surface $S$ is given mathematically by:

$$\mathbf{v} = c \frac{d\mathbf{r}_P}{dt}$$

where $c$ is a real constant $(c \neq 0)$, $\mathbf{r} = \mathbf{r}(t)$ is the spatial parametric representation of the curve with $t$ being a general parameter, and the subscript $P$ indicates that the derivative belongs to the point $P$ where the curve is assumed regular there. In terms of the surface coordinates, the above equation can be expressed as:

$$\mathbf{v} = c \left( \frac{\partial \mathbf{r}_P}{\partial u} \frac{du}{dt} + \frac{\partial \mathbf{r}_P}{\partial v} \frac{dv}{dt} \right) = c \left( \mathbf{E}_1 \frac{du}{dt} + \mathbf{E}_2 \frac{dv}{dt} \right)$$

where the spatial representation of $C$ is parameterized as $\mathbf{r}(u(t), v(t))$ and $\mathbf{E}_1$ and $\mathbf{E}_2$ are the surface basis vectors at $P$. Accordingly, $\mathbf{v}$ is a linear combination of the surface basis vectors at $P$ and hence it belongs to the tangent space of $S$ at $P$.

[14] The patch may include the entire surface as a special case when the whole surface can be represented by a single patch.

8. How is the tangent plane of a surface at a particular point related to the basis vectors of the surface at that point?
**Answer**: The tangent plane is a linear combination of the surface basis vectors, $\mathbf{E}_1$ and $\mathbf{E}_2$, at that point. In other words, the tangent plane is the span of $\mathbf{E}_1$ and $\mathbf{E}_2$.

9. Find the equation of the tangent plane to the ellipsoid that is represented parametrically by: $\mathbf{r}(\theta, \phi) = (2\sin\theta\,\cos\phi,\ 1.5\sin\theta\,\sin\phi,\ 0.5\cos\theta)$ at the point with $\theta = 1.3$ and $\phi = 0.72$.
**Answer**: We have:

$$\begin{aligned}
\mathbf{E}_1 &= \partial_\theta \mathbf{r} = (2\cos\theta\cos\phi,\ 1.5\cos\theta\sin\phi,\ -0.5\sin\theta) \\
\mathbf{E}_2 &= \partial_\phi \mathbf{r} = (-2\sin\theta\sin\phi,\ 1.5\sin\theta\cos\phi,\ 0) \\
\mathbf{E}_1 \times \mathbf{E}_2 &= \left(0.75\sin^2\theta\cos\phi, \sin^2\theta\sin\phi, 3\cos\theta\sin\theta\right)
\end{aligned}$$

The tangent plane is the plane passing through the given point (i.e. the point with $\theta = 1.3$ and $\phi = 0.72$ which we can label as $P$) and having a normal vector $\mathbf{E}_1 \times \mathbf{E}_2$ corresponding to $\theta = 1.3$ and $\phi = 0.72$, that is:

$$\begin{aligned}
(\mathbf{r} - \mathbf{r}_P) \cdot (\mathbf{E}_1 \times \mathbf{E}_2) &= 0 \\
\mathbf{r} \cdot (\mathbf{E}_1 \times \mathbf{E}_2) - \mathbf{r}_P \cdot (\mathbf{E}_1 \times \mathbf{E}_2) &= 0 \\
\mathbf{r} \cdot (\mathbf{E}_1 \times \mathbf{E}_2) &= \mathbf{r}_P \cdot (\mathbf{E}_1 \times \mathbf{E}_2) \\
x0.75\sin^2\theta\cos\phi + y\sin^2\theta\sin\phi + z3\cos\theta\sin\theta &= 1.5\sin\theta \\
x0.75\sin^2 1.3\cos 0.72 + y\sin^2 1.3\sin 0.72 + z3\cos 1.3\sin 1.3 &= 1.5\sin 1.3 \\
0.5235x + 0.6122y + 0.7733z &\simeq 1.4453
\end{aligned}$$

10. Find the equation of the tangent plane of a surface represented parametrically by: $\mathbf{r}(u, v) = (6v, 2u, 1.4u^2 + 6)$ at the point with $u = 1.2$ and $v = 3.6$.
**Answer**: We have:

$$\begin{aligned}
\mathbf{E}_1 &= \partial_u \mathbf{r} = (0, 2, 2.8u) \\
\mathbf{E}_2 &= \partial_v \mathbf{r} = (6, 0, 0) \\
\mathbf{E}_1 \times \mathbf{E}_2 &= (0, 16.8u, -12)
\end{aligned}$$

The tangent plane is the plane passing through the given point (i.e. the point with $u = 1.2$ and $v = 3.6$ which we can label as $P$) and having a normal vector $\mathbf{E}_1 \times \mathbf{E}_2$ corresponding to $u = 1.2$ and $v = 3.6$, that is:

$$\begin{aligned}
(\mathbf{r} - \mathbf{r}_P) \cdot (\mathbf{E}_1 \times \mathbf{E}_2) &= 0 \\
\mathbf{r} \cdot (\mathbf{E}_1 \times \mathbf{E}_2) - \mathbf{r}_P \cdot (\mathbf{E}_1 \times \mathbf{E}_2) &= 0 \\
\mathbf{r} \cdot (\mathbf{E}_1 \times \mathbf{E}_2) &= \mathbf{r}_P \cdot (\mathbf{E}_1 \times \mathbf{E}_2) \\
y16.8u - 12z &= 16.8u^2 - 72 \\
y16.8\,(1.2) - 12z &= 16.8\,(1.2)^2 - 72 \\
20.16y - 12z &= -47.808
\end{aligned}$$

11. Show that for a circular cone the tangent plane at all points of any one of its generators is the same.
**Answer**: A cone is represented parametrically in a 3D space by:

$$\mathbf{r}(\rho,\phi) = (\rho\cos\phi,\ \rho\sin\phi,\ c\rho)$$

where $\rho, \phi$ are polar coordinates ($\rho \geq 0$ and $0 \leq \phi < 2\pi$) and $c$ is a positive constant. Hence, we have:

$$\begin{aligned}
\mathbf{E}_1 &= \partial_\rho \mathbf{r} = (\cos\phi,\ \sin\phi,\ c) \\
\mathbf{E}_2 &= \partial_\phi \mathbf{r} = (-\rho\sin\phi,\ \rho\cos\phi,\ 0) \\
\mathbf{E}_1 \times \mathbf{E}_2 &= (-c\rho\cos\phi, -c\rho\sin\phi, \rho) \\
|\mathbf{E}_1 \times \mathbf{E}_2| &= \sqrt{c^2\rho^2\cos^2\phi + c^2\rho^2\sin^2\phi + \rho^2} = \rho\sqrt{c^2+1} \\
\mathbf{n} &= \frac{\mathbf{E}_1 \times \mathbf{E}_2}{|\mathbf{E}_1 \times \mathbf{E}_2|} = \frac{(-c\cos\phi, -c\sin\phi, 1)}{\sqrt{c^2+1}}
\end{aligned}$$

Now, along a generator of a circular cone only $\rho$ varies. Moreover, the above equation shows that $\mathbf{n}$ is independent of $\rho$ which means that $\mathbf{n}$ is constant along the generator. In more technical terms, along the generator $\partial_\phi \mathbf{n} = \mathbf{0}$ (since $\phi$ is constant) and $\partial_\rho \mathbf{n} = \mathbf{0}$ (since $\mathbf{n}$ is independent of $\rho$) and hence $\mathbf{n}$ is constant. Therefore, the tangent plane (which is determined by $\mathbf{n}$) along any generator of cone is the same, as required.

12. Discuss the following statement explaining its meaning in simple words: "The tangent space at a specific point $P$ of a surface is a property of the surface at $P$ and hence it is independent of the patch that contains $P$".
**Answer**: It simply means that the tangent plane is determined by the local properties of the surface, namely the properties of the point to which the tangent plane belongs, and hence it does not depend on the patch to which the point belongs, i.e. it is independent of any local or global extension over the surface.

13. Does the tangent space of a surface at a given point depend on the particular parameterization of the surface?
**Answer**: No, it does not depend on the particular parameterization of the surface because it is a property of the surface at the specific point and not a property of the particular parameterization.

14. Find the equation of the plane that passes through the point $(-1, 3, -9)$ and spanned by the two vectors $(3, 0.5, 1.2)$ and $(0.9, 3, 6.8)$.
**Answer**: The normal vector to the plane is given by the cross product of the two vectors, that is:

$$(3, 0.5, 1.2) \times (0.9, 3, 6.8) = (-0.2, -19.32, 8.55)$$

Hence, the equation of the plane is given by:

$$[(x, y, z) - (-1, 3, -9)] \cdot (-0.2, -19.32, 8.55) = 0$$

$$\begin{aligned}(x,y,z)\cdot(-0.2,-19.32,8.55)-(-1,3,-9)\cdot(-0.2,-19.32,8.55) &= 0\\ -0.2x-19.32y+8.55z+134.71 &= 0\end{aligned}$$

15. Define, mathematically, the normal unit vector $\mathbf{n}$ of a surface at a given point in terms of the two basis vectors of the surface at that point.
**Answer**: It is given by:

$$\mathbf{n}=\frac{\mathbf{E}_1\times\mathbf{E}_2}{|\mathbf{E}_1\times\mathbf{E}_2|}$$

where $\mathbf{E}_1$ and $\mathbf{E}_2$ are the surface basis vectors at the given point.

16. Calculate, symbolically, the normal unit vector $\mathbf{n}$ at a general point of a surface defined by: $\mathbf{S}(x,y)=(x,y,f)$ where $f=f(x,y)$ is a differentiable function.
**Answer**: We have:

$$\begin{aligned}\mathbf{E}_1 &= \partial_x\mathbf{S}=(1,0,\partial_x f)\\ \mathbf{E}_2 &= \partial_y\mathbf{S}=(0,1,\partial_y f)\\ \mathbf{E}_1\times\mathbf{E}_2 &= (-\partial_x f,-\partial_y f,1)\\ \mathbf{n} &= \frac{\mathbf{E}_1\times\mathbf{E}_2}{|\mathbf{E}_1\times\mathbf{E}_2|}=\frac{(-\partial_x f,-\partial_y f,1)}{\sqrt{(\partial_x f)^2+(\partial_y f)^2+1}}\end{aligned}$$

17. Find the equation of the normal line of a hyperboloid of one sheet given by: $\mathbf{r}(\xi,\theta)=(1.6\cosh\xi\,\cos\theta,\,2.1\cosh\xi\,\sin\theta,\,0.4\sinh\xi)$ at the point with $\xi=3.2$ and $\theta=1.5$.
**Answer**: We have:

$$\begin{aligned}\mathbf{E}_1 &= \partial_\xi\mathbf{r}=(1.6\sinh\xi\cos\theta,\,2.1\sinh\xi\sin\theta,\,0.4\cosh\xi)\\ \mathbf{E}_2 &= \partial_\theta\mathbf{r}=(-1.6\cosh\xi\sin\theta,\,2.1\cosh\xi\cos\theta,\,0)\\ \mathbf{E}_1\times\mathbf{E}_2 &= \left(-0.84\cosh^2\xi\cos\theta,-0.64\cosh^2\xi\sin\theta,3.36\sinh\xi\cosh\xi\right)\end{aligned}$$

The equation of the normal line is given by:

$$\begin{aligned}\mathbf{r} &= \mathbf{r}_P+k\,(\mathbf{E}_1\times\mathbf{E}_2)_P\\ (x,y,z) &= (1.6\cosh 3.2\,\cos 1.5,\,2.1\cosh 3.2\,\sin 1.5,\,0.4\sinh 3.2)+\\ &\quad k\left(-0.84\cosh^2 3.2\,\cos 1.5,-0.64\cosh^2 3.2\,\sin 1.5,3.36\sinh 3.2\,\cosh 3.2\right)\\ (x,y,z) &\simeq (1.3906-8.9700k,\,25.7373-96.3734k,\,4.8984+505.5484k)\end{aligned}$$

where $\mathbf{r}$ is the position of an arbitrary point on the normal line, $\mathbf{r}_P$ is the position of the given point on the surface (i.e. the point with $\xi=3.2$ and $\theta=1.5$), $k$ is a real parameter ($-\infty<k<\infty$), and the subscript $P$ refers to the given point.

18. Find the equation of the normal line of a surface represented by: $\mathbf{r}(u,v)=(3u,u^2+v,5v)$ at the point with $u=2.5$ and $v=-1.8$.
**Answer**: We have:

$$\mathbf{E}_1 = \partial_u\mathbf{r}=(3,2u,0)$$

$$\begin{aligned} \mathbf{E}_2 &= \partial_v \mathbf{r} = (0, 1, 5) \\ \mathbf{E}_1 \times \mathbf{E}_2 &= (10u, -15, 3) \end{aligned}$$

The equation of the normal line is given by:

$$\begin{aligned} \mathbf{r} &= \mathbf{r}_P + k\left(\mathbf{E}_1 \times \mathbf{E}_2\right)_P \\ (x, y, z) &= \left(3 \times 2.5, 2.5^2 - 1.8, -5 \times 1.8\right) + k\left(10 \times 2.5, -15, 3\right) \\ (x, y, z) &= (7.5, 4.45, -9) + k\left(25, -15, 3\right) \\ (x, y, z) &= (7.5 + 25k, 4.45 - 15k, -9 + 3k) \end{aligned}$$

where $\mathbf{r}$ is the position of an arbitrary point on the normal line, $\mathbf{r}_P$ is the position of the given point on the surface (i.e. the point with $u = 2.5$ and $v = -1.8$), $k$ is a real parameter ($-\infty < k < \infty$), and the subscript $P$ refers to the given point.

19. Define "Monge patch" giving its three forms. Which of these forms is the most common in use?
**Answer**: Monge patch is a coordinate patch in a 3D space defined parametrically by one of the following three forms:

$$\begin{aligned} \mathbf{r}(u, v) &= (f(u, v), u, v) \\ \mathbf{r}(u, v) &= (u, f(u, v), v) \\ \mathbf{r}(u, v) &= (u, v, f(u, v)) \end{aligned}$$

where $f$ is a differentiable function of $u$ and $v$. The most common form in use is the third form.

20. Define, briefly, the following terms with some examples representing these concepts: simple surface, simply connected region on a surface, closed surface, compact surface, elementary surface, oriented surface and developable surface.
**Answer**:[15]
Simple surface is a continuously deformed plane by compression, stretching and bending such as cylinders and cones.
Simply connected region on a surface is a region with no holes or gaps that separate its parts such as the region identified by $-3 \le x \le 9$ and $5 \le y \le 11$ on the $xy$ plane.
Closed surface is a simple surface with no open edges such as spheres and ellipsoids.
Compact surface is a surface which is bounded and closed such as tori and ellipsoids.
Elementary surface is a simple surface that possesses a single coordinate patch basis, and hence it is an orientable surface which can be mapped bicontinuously to an open set in the plane. Examples of elementary surface are planes, cones and elliptic paraboloids.
Oriented surface is an orientable surface over which the direction of the normal vector is determined such as spheres or cylinders with outward normal vectors.
Developable surface is a surface that can be flattened into a plane by unfolding without local distortion by compression or stretching such as cylinders and cones.

[15] The purpose of these definitions is to provide a general idea about these concepts. For rigorous and technical definitions the reader should consult the specialized literature.

21. Which of the following is a simple surface and which is not: cylinder, hyperboloid of two sheets, torus, Klein bottle, and elliptic paraboloid?
**Answer**:
Cylinder, hyperboloid of two sheets, torus and elliptic paraboloid are simple. Klein bottle is not simple.

22. Is the surface represented by the equation: $x^2 + y^2 + z^2 = 9$ compact? What about the surface represented by: $x^2 - y^2 - z^2 = 4$?
**Answer**: The first surface is a sphere and hence it is compact. The second surface is a hyperboloid of two sheets and hence it is not compact.

23. Why the Mobius strip is not an orientable surface? Give another example of a non-orientable surface.
**Answer**: Because a normal vector moved continuously around the Mobius strip from a given point will return, following a complete round, to that point in the opposite direction. Another example of a non-orientable surface is Klein bottle.

24. What is conformal mapping? What is direct and inverse conformal mapping?
**Answer**: Conformal mapping is a differentiable regular mapping between two surfaces that preserves the size of the angles between oriented intersecting curves on the surfaces. Conformal mapping is described as direct if it preserves the sense of the angles and inverse if it reverses it.

25. Describe stereographic mapping making a simple sketch representing this type of mapping.
**Answer**: Stereographic mapping is a projection of the unit sphere onto a plane where each point of the sphere is projected, through the line connecting this point to the north pole of the sphere, onto the point of intersection of that line with the plane. This plane is the tangent plane to the sphere at its south pole. A simple sketch of stereographic mapping is shown in Figure 14.

26. Prove that stereographic mapping is conformal.
**Answer**:[16] The essence of the following proof is to show that if $C_1$ and $C_2$ are two arbitrary curves on the stereographic sphere $S_s$ where $C_1$ and $C_2$ intersect at a given point $P$ on $S_s$ then the angle $\alpha$ between their tangents $T_1$ and $T_2$ at $P$ is equal to the angle $\bar{\alpha}$ of intersection of the stereographic images of these tangents on the stereographic plane $\Pi_s$. We rely in this on the intuitive premise that the tangent to a given curve $C$ on $S_s$ at a given point $P$ is stereographically projected on the tangent of the stereographic image of $C$ at the stereographic image $\bar{P}$ of $P$. Accordingly, if the angle between the two tangents on $S_s$ is equal to the angle between their images on $\Pi_s$ then the angle between the two curves on $S_s$ should also be equal to the angle between their images

[16] There are different ways for proving the conformality of stereographic mapping. The proof that we present here is rather simple. We also rely in some steps on intuition or known geometric facts whose proof can be found in the literature.

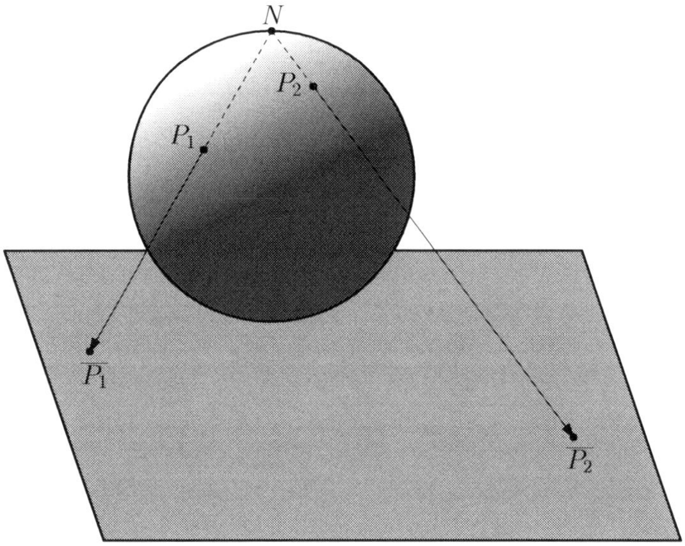


Figure 14: A simple sketch of stereographic mapping where the points $P_1$ and $P_2$ on the unit sphere are projected respectively on the points $\overline{P_1}$ and $\overline{P_2}$ on the plane with $N$ representing the north pole of the sphere.

on $\Pi_s$.
So, let start by claiming that if $PA$ is a straight line segment that is tangent to $S_s$ at $P$ and meets $\Pi_s$ at point $A$ then the length of $\bar{P}A$ (where $\bar{P}$ is the stereographic image of $P$) is equal to the length of $PA$ itself (this claim is justified in the upcoming note). Also let assume that $T_1$ and $T_2$ meet $\Pi_s$ at points $A$ and $B$. Now, the tangent triangle $PAB$ and the image triangle $\bar{P}AB$ have a common side $AB$; moreover, $PA$ and $\bar{P}A$ are equal in length and $PB$ and $\bar{P}B$ are also equal in length. Therefore, the two triangles are identical due to the equality of their three sides and hence the angle $APB$ and the angle $A\bar{P}B$ are equal (i.e. $\alpha = \bar{\alpha}$), as required.
Note: in this note we justify the claim that if $PA$ is a straight line segment that is tangent to $S_s$ at $P$ and meets $\Pi_s$ at point $A$ then the length of $\bar{P}A$ is equal to the length of $PA$. First, let stat from a simple case where the tangent is along a meridian of the sphere and for the sake of simplicity (with no loss of generality) we can assume that this meridian is contained in the $xz$ plane. Referring to Figure 15 (left frame) we note the following:[17]
$\angle NPS = \pi/2$ because it subtends the diameter. Hence, $\angle SP\bar{P} = \pi/2$.
$\triangle SAP$ is isosceles because $AP$ and $AS$ are tangents to $S_s$ and hence $AS = AP$. Therefore, $\angle PSA = \angle SPA$.
$\angle P\bar{P}A + \angle PSA = \pi/2$ because $\triangle SP\bar{P}$ is a right triangle. Also, $\angle \bar{P}PA + \angle SPA = \pi/2$ because $\angle SP\bar{P} = \pi/2$. Hence, $\angle P\bar{P}A = \angle \bar{P}PA$ because $\angle PSA = \angle SPA$. Therefore, $\triangle PA\bar{P}$ is isosceles and $\bar{P}A = PA$. Accordingly, the above claim (i.e. "if $PA$ is a

[17] Here, we use $\angle$ for angle and $\triangle$ for triangle. Also, for simplicity we use equalities like $AS = AP$ to mean the equality of the lengths of these segments.

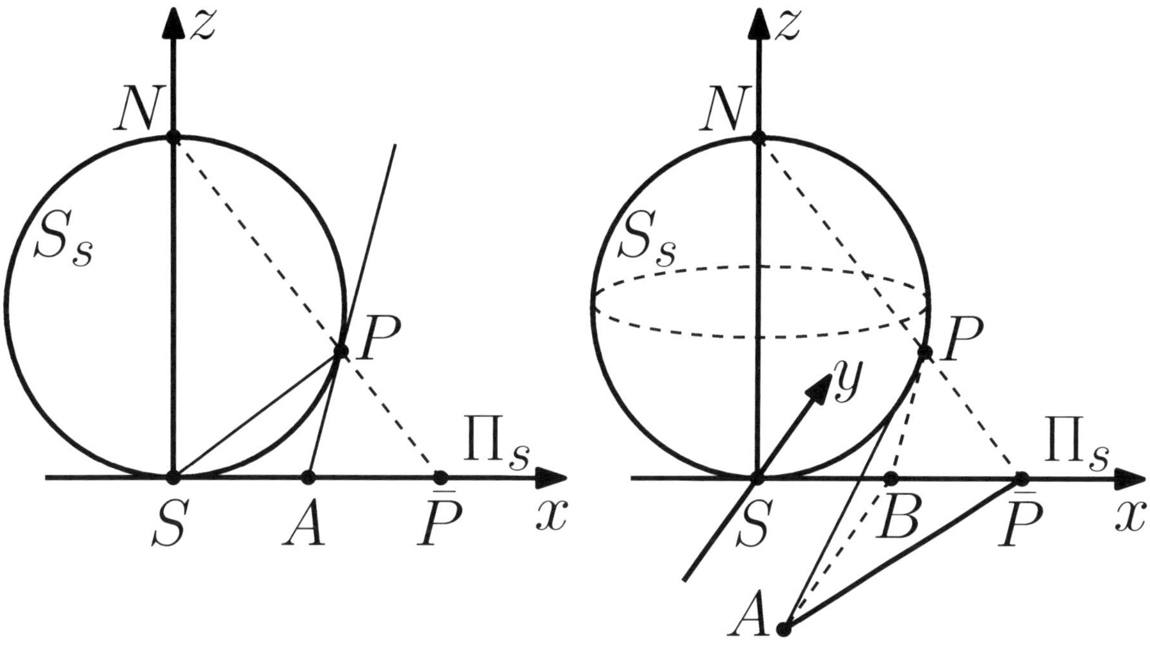


Figure 15: The left frame is a 2D plot of the stereographic device where the stereographic sphere $S_s$ is represented by a great circle (with $N$ and $S$ representing its north and south poles), the stereographic plane $\Pi_s$ is represented by the horizontal line (i.e. $x$-axis), and $PA$ is a tangent segment to $S_s$ at $P$ (along the meridian that passes through $P$) with $\bar{P}$ being the stereographic image of $P$. The right frame is a 3D plot of the stereographic device where $PA$ is a tangent segment to $S_s$ at $P$ with $\bar{P}$ being the stereographic image of $P$ while point $B$ is the midpoint of the segment $S\bar{P}$ and hence point $B$ in the right frame corresponds to point $A$ in the left frame.

straight line segment ... etc.") is justified in this simple case. We should also note that since $AS = AP$ and $AP = A\bar{P}$ then $AS = A\bar{P}$ and hence $A$ is the midpoint of the segment $S\bar{P}$.

Now, let generalize this simple case by having a tangent to $S_s$ at $P$ that is not along the meridian. Such a tangent should lie in the tangent plane of $S_s$ at $P$ and for the sake of simplicity (with no loss of generality) we can assume that $P$ is in the $xz$ plane and hence the tangent plane of $S_s$ at $P$ is parallel to the $y$-axis. Referring to Figure 15 (right frame noting that $B$ in this frame corresponds to $A$ in the left frame) we note the following:

$AB$ is a common side to $\triangle ABP$ and $\triangle AB\bar{P}$.

$PB = \bar{P}B$ according to our finding earlier in the simple case where $B$ here corresponds to $A$ there.

$\angle ABP = \pi/2$ because $AB$ is the intersection of the $xy$ plane with the tangent plane (which is parallel to the $y$-axis) and $BP$ is the intersection of the $xz$ plane with the tangent plane (which is parallel to the $y$-axis).

$\angle AB\bar{P} = \pi/2$ because $AB$ is the intersection of the $xy$ plane with the tangent plane

(which is parallel to the $y$-axis) and $B\bar{P}$ is the intersection of the $xz$ plane with the $xy$ plane (which contains the $y$-axis). Hence, $\angle ABP = \angle AB\bar{P}$.
Therefore, the two triangles (i.e. $\triangle ABP$ and $\triangle AB\bar{P}$) are identical due to the equality of two corresponding sides and the angle between and hence $\bar{P}A = PA$, as required.

27. Define isometric mapping giving an example of such mapping between two types of surface.
**Answer**: Isometric mapping is a one-to-one mapping between two surfaces that preserves distances. An example of isometric mapping is the projection of a rectangular plane sheet onto a cylinder.

28. What are the mathematical conditions for two surfaces to be isometric? Does isometry relate to the intrinsic or extrinsic properties of the surface and why?
**Answer**: The necessary and sufficient condition for a mapping from surface $S$ to surface $\bar{S}$ to be isometric is that the coefficients of the first fundamental form for any patch on $S$ are identical to the corresponding coefficients of the first fundamental form of its image on $\bar{S}$, that is:

$$E = \bar{E} \qquad F = \bar{F} \qquad G = \bar{G} \tag{21}$$

where $E, F, G$ and $\bar{E}, \bar{F}, \bar{G}$ are the coefficients of the first fundamental form of $S$ and $\bar{S}$ respectively at their corresponding points. Isometry relates to the intrinsic properties of the surface because the defining conditions of isometry are based on the coefficients of the first fundamental form of the surface.

29. What is the distinctive property of an isometric relation between two surfaces in terms of angles, arc lengths and areas defined on the two surfaces?
**Answer**: Isometric mapping preserves angles, arc lengths and areas.

30. What is the relation between conformal mapping and isometric mapping?
**Answer**: Isometric mapping is a subset of conformal mapping because every isometric mapping is conformal (since it preserves angles) but not every conformal mapping is isometric (since preserving angles does not necessarily mean preserving distances).

31. What is local isometry? What is the difference between local isometry and global isometry?
**Answer**: Local isometry is a mapping that preserves distances but it is not injective. Global isometry is injective but local isometry is not.

32. Show that Eq. 21 applies to local isometric mapping.
**Answer**: Global and local isometry are identical apart from being injective or not (which is an extensive attribute since it is based on the relation between the two surfaces on a large scale). Now, the conditions in Eq. 21 are entirely related to the intrinsic properties of the surfaces at the individual points with no indication to any extensive attribute over the surface (such as being injective or not) and hence if they apply to

global isometry (as they do) they should also apply to local isometry. Accordingly, Eq. 21 applies to local isometry as to global isometry. The reader is also referred to Exercise 81.

33. A surface $S_1$ is mapped isometrically onto another surface $S_2$. How will the intrinsic properties of $S_1$ be affected by this mapping?
**Answer**: The intrinsic properties are invariant under isometric mapping and hence the intrinsic properties of $S_1$ will not be affected by this mapping.

34. Make a clear distinction between the tangent surface of a curve and the tangent plane of a surface describing each of these briefly.
**Answer**: The tangent surface of a space curve is a surface generated by the assembly of all the tangent lines to the curve while the tangent plane is a plane made of all the tangent vectors to the surface at a given point on the surface. We note the following differences between the two:
• Tangent surface belongs to a curve while tangent plane belongs to a surface.
• Tangent surface is a surface and hence it is twisted in general while tangent plane is a plane.
• Tangent surface belongs to the curve as a whole and hence there is a single tangent surface to a given curve while tangent plane belongs to a point on a surface and hence there are infinitely many tangent planes to a given surface.

35. What is the meaning of "branch of the tangent surface of a curve $C$ at a given point $P$ on the curve"?
**Answer**: It is the tangent line of $C$ at $P$.

36. Give a brief definition of involute and evolute.
**Answer**: Involute is a space curve $C_i$ embedded in the tangent surface of another space curve $C_e$ and it is orthogonal to all the tangent lines of $C_e$ at their intersection points. The curve $C_e$ is described as the evolute of the curve $C_i$, i.e. $C_i$ is an involute of $C_e$ and $C_e$ is an evolute of $C_i$.

37. Write down a general mathematical relation representing the position vector of a point $P$ on a space surface as a function of the surface coordinates, $u^1$ and $u^2$, where the surface is embedded in a 3D Euclidean space coordinated by a rectangular Cartesian system.
**Answer**:
$$\mathbf{r}(u^1, u^2) = x(u^1, u^2)\mathbf{e}_1 + y(u^1, u^2)\mathbf{e}_2 + z(u^1, u^2)\mathbf{e}_3$$
where $\mathbf{r}$ is the position vector of a point on the surface as a function of the surface coordinates $u^1$ and $u^2$, $x, y, z$ are the Cartesian coordinates of the point which are also functions of the surface coordinates, and $\mathbf{e}_1, \mathbf{e}_2, \mathbf{e}_3$ are the Cartesian basis vectors.

38. Describe, in detail, how a coordinate grid is constructed on a space surface with a clear definition of the coordinate curves used to build this grid.

**Answer**: We start by adopting a coordinate plane identified by two mutually independent coordinates, say $u^1$ and $u^2$. This coordinate plane corresponds to the surface by a given mapping and hence any point on the surface can be identified by two coordinates from the coordinate plane. We then construct a surface grid serving as a curvilinear positioning system for the surface by constructing coordinate curves where one of the surface coordinate variables is held fixed in turn while the other is varied. Hence, each one of the following two surface functions:

$$\mathbf{r}(u^1, c_2) \qquad \text{and} \qquad \mathbf{r}(c_1, u^2)$$

defines a coordinate curve for the surface, where $c_1$ and $c_2$ are given real constants. These two coordinate curves meet at the common surface point $(c_1, c_2)$. The grid is then generated by varying $c_1$ and $c_2$ uniformly to obtain coordinate curves at regular intervals in its domain.

39. Make a simple and fully labeled sketch of a space surface coordinated by a curvilinear grid with the two covariant surface basis vectors and the unit normal vector to the surface at one point.
**Answer**: The sketch should be similar to Figure 16.

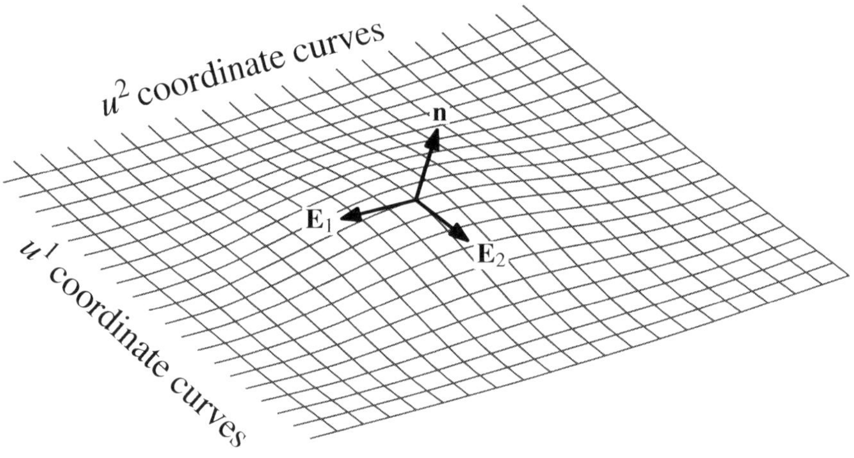


Figure 16: A space surface identified by a coordinate grid with the surface covariant basis vectors, $\mathbf{E}_1$ and $\mathbf{E}_2$, and the normal vector to the surface $\mathbf{n}$ at a particular point on the surface.

40. A surface is represented spatially by: $\mathbf{r}(u, v) = (u, 5, 2v)$. Discuss the type and the main properties of this surface.
**Answer**: This is obviously a plane as can be easily verified by noting that:

$$\begin{aligned}
\mathbf{E}_1 &= \partial_u \mathbf{r} = (1, 0, 0) \\
\mathbf{E}_2 &= \partial_v \mathbf{r} = (0, 0, 2) \\
\mathbf{E}_1 \times \mathbf{E}_2 &= (0, -2, 0)
\end{aligned}$$

i.e. the normal vector to the surface is constant over the entire surface and hence the surface must be a plane. We also note that since the $y$ coordinate of $\mathbf{r}$ is constant then the surface is parallel to the $xz$ plane. This can also be easily concluded from the fact that the normal vector to the surface has only a $y$ component. So in brief, this surface is a plane parallel to the $xz$ plane.

41. Define, symbolically, the covariant basis vectors of a space surface in terms of the coordinates of the ambient space $x^i$ and the coordinates of the surface $u^\alpha$.
**Answer**:
$$\mathbf{E}_\alpha = \frac{\partial x^i}{\partial u^\alpha}\mathbf{e}_i$$
where $\mathbf{E}_\alpha$ represents covariant basis vectors of the surface, $\mathbf{e}_i$ represents basis vectors of the ambient space, and $i = 1, 2, 3$ and $\alpha = 1, 2$.

42. Give the symbol used in tensor notation to represent the surface basis vectors $\mathbf{E}_\alpha$. From analyzing this symbol, describe how the surface basis vectors can be regarded as covariant and contravariant vectors at the same time.
**Answer**: The symbol is $x^i_\alpha$ which can be defined as:
$$x^i_\alpha \equiv \frac{\partial x^i}{\partial u^\alpha} = [\mathbf{E}_\alpha]^i$$
where the symbols are explained in the previous question. Accordingly, the surface basis vectors, as defined by this symbol, can be seen as 2D covariant surface vectors (by noting their $\alpha$ index) or as 3D contravariant space vectors (by noting their $i$ index).

43. Write, in full tensor notation, the equation representing the covariant form of the unit normal vector to a space surface.
**Answer**:
$$n_i = \frac{1}{2}\underline{\epsilon}^{\alpha\beta}\underline{\epsilon}_{ijk}x^j_\alpha x^k_\beta$$
where $n_i$ is the covariant form of the unit normal vector $\mathbf{n}$, $\underline{\epsilon}^{\alpha\beta}$ is the absolute contravariant permutation tensor for the surface, $\underline{\epsilon}_{ijk}$ is the absolute covariant permutation tensor for the space, and $x^j_\alpha \equiv \partial x^j/\partial u^\alpha$ and $x^k_\beta \equiv \partial x^k/\partial u^\beta$ are surface basis vectors. The Latin indices range over 1,2,3 while the Greek indices range over 1,2.

44. Although $\mathbf{E}_1$ and $\mathbf{E}_2$ are linearly independent at the regular points of the surface they are not necessarily orthogonal or of unit length. Is it possible to construct an orthonormal set of basis vectors from $\mathbf{E}_1$ and $\mathbf{E}_2$? If so, how?
**Answer**: Yes, it is possible to construct an orthonormal set of basis vectors from $\mathbf{E}_1$ and $\mathbf{E}_2$ by orthonormalization, that is:
$$\underline{\mathbf{E}}_1 = \frac{\mathbf{E}_1}{|\mathbf{E}_1|} = \frac{\mathbf{E}_1}{\sqrt{a_{11}}} \qquad \text{and} \qquad \underline{\mathbf{E}}_2 = \frac{a_{11}\mathbf{E}_2 - a_{12}\mathbf{E}_1}{\sqrt{a_{11}a}}$$
where $a$ is the determinant of the surface covariant metric tensor, the indexed $a$ are the coefficients of this tensor, and $\underline{\mathbf{E}}_1$ and $\underline{\mathbf{E}}_2$ are orthonormal surface basis vectors.

45. Following a transformation from an unbarred surface coordinate system to a barred surface coordinate system, what are the mathematical expressions representing the barred basis set, $\bar{\mathbf{E}}_1$ and $\bar{\mathbf{E}}_2$, in terms of the unbarred basis set, $\mathbf{E}_1$ and $\mathbf{E}_2$? Give these expressions in vector and tensor notations.
**Answer**: They are:

$$\begin{aligned} \bar{\mathbf{E}}_\alpha &= \frac{\partial \mathbf{r}}{\partial \bar{u}^\alpha} = \mathbf{E}_\beta \frac{\partial u^\beta}{\partial \bar{u}^\alpha} \\ \frac{\partial x^i}{\partial \bar{u}^\alpha} &= \frac{\partial x^i}{\partial u^\beta}\frac{\partial u^\beta}{\partial \bar{u}^\alpha} \end{aligned}$$

where $\mathbf{r}$ is the spatial representation of the surface, the indexed $u$ are the surface coordinates, $x^i$ is the $i^{th}$ spatial coordinate, and $\alpha, \beta = 1, 2$ and $i = 1, 2, 3$ with summation over $\beta$.

46. Define, descriptively and mathematically, the set of contravariant basis vectors for a space surface discussing how these vectors can be regarded as covariant and contravariant vectors simultaneously.
**Answer**: The contravariant basis vectors $\mathbf{E}^\alpha$ for a space surface are defined as the gradient of the surface coordinate curves with respect to the spatial coordinates, that is:

$$\mathbf{E}^\alpha = \nabla u^\alpha$$

where $\nabla \equiv \mathbf{e}_i \frac{\partial}{\partial x^i}$ (with $x^i$ being the $i^{th}$ spatial coordinate) and $u^\alpha$ is the $\alpha^{th}$ surface coordinate. From the tensor notation of these vectors which is given by:

$$[\mathbf{E}^\alpha]_i \equiv \frac{\partial u^\alpha}{\partial x^i} \qquad (\alpha = 1, 2 \text{ and } i = 1, 2, 3)$$

we can see that these vectors are covariant space vectors considering their $i$ index and contravariant surface vectors considering their $\alpha$ index.

47. How can the covariant and contravariant basis sets of a space surface be obtained from each other? State this in words and mathematically defining all the symbols involved.
**Answer**: The surface covariant basis vectors are obtained from the surface contravariant basis vectors by using the index lowering operator (which is the surface covariant metric tensor) while the surface contravariant basis vectors are obtained from the surface covariant basis vectors by using the index raising operator (which is the surface contravariant metric tensor), that is:

$$\mathbf{E}_\alpha = a_{\alpha\beta}\mathbf{E}^\beta \qquad \text{and} \qquad \mathbf{E}^\alpha = a^{\alpha\beta}\mathbf{E}_\beta$$

where $\mathbf{E}_\alpha$ and $\mathbf{E}_\beta$ are the surface covariant basis vectors, $\mathbf{E}^\alpha$ and $\mathbf{E}^\beta$ are the surface contravariant basis vectors, and $a_{\alpha\beta}$ and $a^{\alpha\beta}$ are the covariant and contravariant forms of the surface metric tensor.

48. What is the significance of the following relation involving the covariant and contravariant surface basis vectors and the Kronecker delta: $\mathbf{E}_\alpha \cdot \mathbf{E}^\beta = \delta^\beta_\alpha$?
**Answer**: It represents the fact that the covariant and contravariant forms of the surface basis vectors are reciprocal vector sets, i.e the dot product of their corresponding vectors is unity while the dot product of their non-corresponding vectors is zero which means they are orthogonal.

49. Define, mathematically, the coefficients of the surface covariant metric tensor in terms of the surface covariant basis vectors in Euclidean and Riemannian spaces.
**Answer**: They are given in Euclidean and Riemannian spaces respectively by:

$$\begin{aligned} a_{\alpha\beta} &= \mathbf{E}_\alpha \cdot \mathbf{E}_\beta = \frac{\partial \mathbf{r}}{\partial u^\alpha} \cdot \frac{\partial \mathbf{r}}{\partial u^\beta} = \frac{\partial x^i}{\partial u^\alpha} \frac{\partial x^i}{\partial u^\beta} \\ a_{\alpha\beta} &= \mathbf{E}_\alpha \cdot \mathbf{E}_\beta = \frac{\partial \mathbf{r}}{\partial u^\alpha} \cdot \frac{\partial \mathbf{r}}{\partial u^\beta} = g_{ij} \frac{\partial x^i}{\partial u^\alpha} \frac{\partial x^j}{\partial u^\beta} \end{aligned}$$

where $a_{\alpha\beta}$ are the coefficients of the surface covariant metric tensor, $\mathbf{E}_\alpha$ and $\mathbf{E}_\beta$ are the surface covariant basis vectors, $\mathbf{r}$ is the spatial representation of the surface, $u^\alpha$ and $u^\beta$ are the surface coordinates, $x^i$ and $x^j$ are the space coordinates, $g_{ij}$ is the space covariant metric tensor, and $\alpha, \beta = 1, 2$ while $i, j = 1, 2, 3$.

50. Give the fundamental relation that provides the important link between the metric tensor of a surface and the metric tensor of its enveloping space.
**Answer**: It is:

$$a_{\alpha\beta} = g_{ij} x^i_\alpha x^j_\beta \tag{22}$$

where $a_{\alpha\beta}$ is the surface covariant metric tensor, $g_{ij}$ is the space covariant metric tensor, $x^i$ and $x^j$ are the coordinates of the enveloping space, the subscripts $\alpha$ and $\beta$ symbolize partial derivatives with respect to the surface coordinates $u^\alpha$ and $u^\beta$, and $\alpha, \beta = 1, 2$ while $i, j = 1, 2, 3$. As explained earlier, $x^i_\alpha$ and $x^j_\beta$ represent surface basis vectors in tensor notation. It should be obvious that this equation is the same as the second equation in the previous question with minor notational difference.

51. Find the surface basis vectors, $\mathbf{E}_1$ and $\mathbf{E}_2$, and the coefficients of the first fundamental form of a surface parameterized by: $\mathbf{r}(u, v) = (au \cos v, bu \sin v, cu^2)$ where $a, b, c$ are constants.
**Answer**: We have:

$$\begin{aligned} \mathbf{E}_1 &= \partial_u \mathbf{r} = (a \cos v, b \sin v, 2cu) \\ \mathbf{E}_2 &= \partial_v \mathbf{r} = (-au \sin v, bu \cos v, 0) \\ E &= \mathbf{E}_1 \cdot \mathbf{E}_1 = a^2 \cos^2 v + b^2 \sin^2 v + 4c^2 u^2 \\ F &= \mathbf{E}_1 \cdot \mathbf{E}_2 = -a^2 u \cos v \sin v + b^2 u \sin v \cos v \\ G &= \mathbf{E}_2 \cdot \mathbf{E}_2 = a^2 u^2 \sin^2 v + b^2 u^2 \cos^2 v \end{aligned}$$

52. Find, symbolically, the first fundamental form of a surface represented by: $\mathbf{r}(u,v) = (f_1(u), f_2(u), v)$ where $f_1$ and $f_2$ are continuous functions of the given coordinate.
**Answer**: We have:

$$\begin{aligned}
\mathbf{E}_1 &= \partial_u \mathbf{r} = (\partial_u f_1, \partial_u f_2, 0) \\
\mathbf{E}_2 &= \partial_v \mathbf{r} = (0, 0, 1) \\
E &= \mathbf{E}_1 \cdot \mathbf{E}_1 = (\partial_u f_1)^2 + (\partial_u f_2)^2 \\
F &= \mathbf{E}_1 \cdot \mathbf{E}_2 = 0 \\
G &= \mathbf{E}_2 \cdot \mathbf{E}_2 = 1
\end{aligned}$$

Hence, the first fundamental form of the surface is:

$$\begin{aligned}
I_S &= E(du)^2 + 2F\,dudv + G(dv)^2 \\
&= \left[(\partial_u f_1)^2 + (\partial_u f_2)^2\right](du)^2 + 0 + (dv)^2 \\
&= (du\,\partial_u f_1)^2 + (du\,\partial_u f_2)^2 + (dv)^2
\end{aligned}$$

53. Given the fact that for 3D Cartesian systems: $I_S = (dx^1)^2 + (dx^2)^2 + (dx^3)^2$ plus the transformation equations from spherical to Cartesian coordinates in 3D, derive $I_S$ for spherical coordinate systems.
**Answer**: The transformation equations from spherical to Cartesian are:

$$x^1 = r\sin\theta\cos\phi \qquad x^2 = r\sin\theta\sin\phi \qquad x^3 = r\cos\theta$$

Hence:

$$\begin{aligned}
dx^1 &= \left(\frac{\partial}{\partial r}dr + \frac{\partial}{\partial\theta}d\theta + \frac{\partial}{\partial\phi}d\phi\right) r\sin\theta\cos\phi \\
&= \sin\theta\cos\phi dr + r\cos\theta\cos\phi d\theta - r\sin\theta\sin\phi d\phi \\
dx^2 &= \left(\frac{\partial}{\partial r}dr + \frac{\partial}{\partial\theta}d\theta + \frac{\partial}{\partial\phi}d\phi\right) r\sin\theta\sin\phi \\
&= \sin\theta\sin\phi dr + r\cos\theta\sin\phi d\theta + r\sin\theta\cos\phi d\phi \\
dx^3 &= \left(\frac{\partial}{\partial r}dr + \frac{\partial}{\partial\theta}d\theta + \frac{\partial}{\partial\phi}d\phi\right) r\cos\theta \\
&= \cos\theta dr - r\sin\theta d\theta
\end{aligned}$$

Hence, $I_S$ for spherical coordinate systems is:

$$\begin{aligned}
I_S &= \left(dx^1\right)^2 + \left(dx^2\right)^2 + \left(dx^3\right)^2 \\
&= (\sin\theta\cos\phi dr + r\cos\theta\cos\phi d\theta - r\sin\theta\sin\phi d\phi)^2 + \\
&\quad (\sin\theta\sin\phi dr + r\cos\theta\sin\phi d\theta + r\sin\theta\cos\phi d\phi)^2 + \\
&\quad (\cos\theta dr - r\sin\theta d\theta)^2 \\
&= \sin^2\theta\cos^2\phi\,(dr)^2 + 2r\cos\theta\sin\theta\cos^2\phi dr d\theta - 2r\sin^2\theta\sin\phi\cos\phi dr d\phi +
\end{aligned}$$

$$
\begin{aligned}
& r^2 \cos^2\theta \cos^2\phi \,(d\theta)^2 - 2r^2 \sin\theta \cos\theta \sin\phi \cos\phi d\theta d\phi + \\
& r^2 \sin^2\theta \sin^2\phi \,(d\phi)^2 + \\
& \sin^2\theta \sin^2\phi \,(dr)^2 + 2r \sin\theta \cos\theta \sin^2\phi dr d\theta + 2r \sin^2\theta \cos\phi \sin\phi dr d\phi + \\
& r^2 \cos^2\theta \sin^2\phi \,(d\theta)^2 + 2r^2 \sin\theta \cos\theta \cos\phi \sin\phi d\theta d\phi + \\
& r^2 \sin^2\theta \cos^2\phi \,(d\phi)^2 + \\
& \cos^2\theta \,(dr)^2 - 2r \cos\theta \sin\theta dr d\theta + r^2 \sin^2\theta \,(d\theta)^2 \\
= \; & (dr)^2 + r^2 (d\theta)^2 + r^2 \sin^2\theta \,(d\phi)^2
\end{aligned}
$$

which is the well known expression for the square of line element $\left[\text{i.e. } (ds)^2\right]$ in spherical coordinate systems.

54. Given the fact that for 3D Cartesian systems: $I_S = (dx^1)^2 + (dx^2)^2 + (dx^3)^2$ plus the transformation equations between general curvilinear and Cartesian coordinate systems in 3D, prove that for general curvilinear systems: $I_S = a_{\alpha\beta} du^\alpha du^\beta$.
**Answer**: In this answer, we label the general curvilinear coordinates with $q^1, q^2, q^3$ while the Cartesian coordinates are already labeled with $x^1, x^2, x^3$. We also use $\mathbf{r} \equiv (x^1, x^2, x^3)$ and hence $d\mathbf{r} \equiv (dx^1, dx^2, dx^3)$. The transformation equations between general curvilinear and Cartesian coordinate systems in 3D are:

$$
dx^i = \frac{\partial x^i}{\partial q^j} dq^j \qquad \text{or} \qquad d\mathbf{r} = \frac{\partial \mathbf{r}}{\partial q^j} dq^j
$$

where $i, j = 1, 2, 3$. Now, from the definition of the surface covariant metric tensor $a_{\alpha\beta}$ we have:

$$
\begin{aligned}
a_{\alpha\beta} &= \mathbf{E}_\alpha \cdot \mathbf{E}_\beta \\
a_{\alpha\beta} du^\alpha du^\beta &= (\mathbf{E}_\alpha \cdot \mathbf{E}_\beta)\, du^\alpha du^\beta \\
&= \left( \frac{\partial \mathbf{r}}{\partial u^\alpha} \cdot \frac{\partial \mathbf{r}}{\partial u^\beta} \right) du^\alpha du^\beta \\
&= \left( \frac{\partial \mathbf{r}}{\partial u^\alpha} du^\alpha \right) \cdot \left( \frac{\partial \mathbf{r}}{\partial u^\beta} du^\beta \right) \\
&= \left( \frac{\partial \mathbf{r}}{\partial q^i} \frac{\partial q^i}{\partial u^\alpha} du^\alpha \right) \cdot \left( \frac{\partial \mathbf{r}}{\partial q^j} \frac{\partial q^j}{\partial u^\beta} du^\beta \right) \\
&= \left( \frac{\partial \mathbf{r}}{\partial q^i} dq^i \right) \cdot \left( \frac{\partial \mathbf{r}}{\partial q^j} dq^j \right) \\
&= d\mathbf{r} \cdot d\mathbf{r} \\
&= \left(dx^1, dx^2, dx^3\right) \cdot \left(dx^1, dx^2, dx^3\right) \\
&= \left(dx^1\right)^2 + \left(dx^2\right)^2 + \left(dx^3\right)^2 \\
&= I_S
\end{aligned}
$$

55. Define, mathematically, each of the following vectors: $x^1_\alpha$, $x^2_\alpha$, $x^3_\alpha$, $x^i_1$, $x^i_2$. Also, discuss their attributes as space and surface vectors and their variance type.

**Answer**: We have:[18]

$$x^1_\alpha = \left(\frac{\partial x^1}{\partial u^1}, \frac{\partial x^1}{\partial u^2}\right) \qquad x^2_\alpha = \left(\frac{\partial x^2}{\partial u^1}, \frac{\partial x^2}{\partial u^2}\right) \qquad x^3_\alpha = \left(\frac{\partial x^3}{\partial u^1}, \frac{\partial x^3}{\partial u^2}\right)$$

$$x^i_1 = \left(\frac{\partial x^1}{\partial u^1}, \frac{\partial x^2}{\partial u^1}, \frac{\partial x^3}{\partial u^1}\right) \qquad x^i_2 = \left(\frac{\partial x^1}{\partial u^2}, \frac{\partial x^2}{\partial u^2}, \frac{\partial x^3}{\partial u^2}\right)$$

Hence, $x^1_\alpha$, $x^2_\alpha$, $x^3_\alpha$ can be seen as three space vectors which are covariantly-transformed with respect to the two surface coordinates $u^\alpha$ while $x^i_1$, $x^i_2$ can be seen as two surface vectors which are contravariantly-transformed with respect to the three space coordinates $x^i$.

56. Discuss how a surface vector can also be considered as a space vector stating the mathematical link between the surface and space representations. Are these representations equivalent? If so, how?
**Answer**: A surface vector $A^\alpha$ ($\alpha = 1, 2$), defined as a linear combination of the surface basis vectors $\mathbf{E}_1$ and $\mathbf{E}_2$, can also be considered as a space vector $A^i$ ($i = 1, 2, 3$) where the two representations are linked through the relation:

$$A^i = \frac{\partial x^i}{\partial u^\alpha} A^\alpha = x^i_\alpha A^\alpha \tag{23}$$

The two representations are equivalent because:

$$a_{\alpha\beta} A^\alpha A^\beta = g_{ij} x^i_\alpha x^j_\beta A^\alpha A^\beta = g_{ij} x^i_\alpha A^\alpha x^j_\beta A^\beta = g_{ij} A^i A^j$$

where Eq. 22 is used in the first equality while Eq. 23 is used in the last equality.

57. Write down, using tensor notation, the mathematical relation that correlates the surface basis vectors to the unit normal vector to the surface.
**Answer**:

$$x^i_\alpha = \underline{\epsilon}^{ijk} \underline{\epsilon}_{\alpha\beta} x^\beta_j n_k$$

where $x^i_\alpha$ and $x^\beta_j$ represent the covariant and contravariant surface basis vectors, $\underline{\epsilon}^{ijk}$ is the space contravariant absolute permutation tensor, $\underline{\epsilon}_{\alpha\beta}$ is the surface covariant absolute permutation tensor, and $n_k$ is the unit normal vector to the surface.

58. What is the significance of the following relation which involves the surface contravariant and covariant metric tensors and the Kronecker delta: $a^{\alpha\gamma} a_{\gamma\beta} = \delta^\alpha_\beta$?
**Answer**: It means that the contravariant surface metric tensor and the covariant surface metric tensor are inverses of each other.

[18] We note that there may be an abuse of notation in the following equalities. However, our objective is to interpret these symbols rather than setting equalities. An equivalence symbol, i.e. $\equiv$, may be more appropriate than an equality symbol.

59. Give the matrix $\left[a^{\alpha\beta}\right]$ that represents the contravariant form of the surface metric tensor in terms of the coefficients of the first fundamental form.
**Answer**:

$$\left[a^{\alpha\beta}\right] = \frac{1}{EG - F^2}\begin{bmatrix} G & -F \\ -F & E \end{bmatrix}$$

where $E, F, G$ are the coefficients of the first fundamental form.

60. Give the matrix $\left[a^{\alpha}_{\beta}\right]$ that represents the mixed form of the surface metric tensor.
**Answer**:

$$\left[a^{\alpha}_{\beta}\right] = \begin{bmatrix} 1 & 0 \\ 0 & 1 \end{bmatrix}$$

61. Write down the relation that represents the transformation of the covariant metric tensor between unbarred and barred surface coordinate systems.
**Answer**:

$$\bar{a}_{\alpha\beta} = a_{\gamma\delta}\frac{\partial u^{\gamma}}{\partial \bar{u}^{\alpha}}\frac{\partial u^{\delta}}{\partial \bar{u}^{\beta}}$$

where $\bar{a}_{\alpha\beta}$ and $a_{\gamma\delta}$ are the surface covariant metric tensor in the barred and unbarred systems while the indexed $\bar{u}$ and $u$ are the surface coordinates in these systems.

62. How are the determinants of the surface covariant metric tensor of two transformed coordinate systems of a given surface linked?
**Answer**: They are linked by the following relation:

$$\bar{a} = J^2 a$$

where $\bar{a}$ and $a$ are the determinants of the covariant form of the surface metric tensor in the barred and unbarred systems and $J$ is the Jacobian of the transformation between the two surface systems.

63. Express the Christoffel symbols of the first kind $[\alpha\beta, \gamma]$ for a surface in terms of the surface covariant basis vectors and their partial derivatives.
**Answer**:

$$[\alpha\beta, \gamma] = \frac{\partial \mathbf{E}_{\alpha}}{\partial u^{\beta}} \cdot \mathbf{E}_{\gamma}$$

where the indexed $\mathbf{E}$ are surface covariant basis vectors and $u^{\beta}$ is surface coordinate.

64. Derive the mathematical relation between the partial derivative of the surface metric tensor $\partial_{\beta} a_{\alpha\gamma}$ and the Christoffel symbols of the first kind for the surface.
**Answer**: From the definition of the Christoffel symbols of the first kind we have:

$$\begin{aligned} [\alpha\beta, \gamma] &= \frac{1}{2}\left(\partial_{\beta} a_{\alpha\gamma} + \partial_{\alpha} a_{\beta\gamma} - \partial_{\gamma} a_{\alpha\beta}\right) \\ [\gamma\beta, \alpha] &= \frac{1}{2}\left(\partial_{\beta} a_{\gamma\alpha} + \partial_{\gamma} a_{\beta\alpha} - \partial_{\alpha} a_{\gamma\beta}\right) \end{aligned}$$

where the second equation is obtained from the first by exchanging $\alpha$ and $\gamma$. On adding the two sides of these equations we obtain:

$$\begin{aligned}
[\alpha\beta,\gamma]+[\gamma\beta,\alpha] &= \frac{1}{2}\left(\partial_\beta a_{\alpha\gamma}+\partial_\alpha a_{\beta\gamma}-\partial_\gamma a_{\alpha\beta}+\partial_\beta a_{\gamma\alpha}+\partial_\gamma a_{\beta\alpha}-\partial_\alpha a_{\gamma\beta}\right)\\
&= \frac{1}{2}\left(\partial_\beta a_{\alpha\gamma}+\partial_\alpha a_{\beta\gamma}-\partial_\gamma a_{\alpha\beta}+\partial_\beta a_{\alpha\gamma}+\partial_\gamma a_{\alpha\beta}-\partial_\alpha a_{\beta\gamma}\right)\\
&= \frac{1}{2}\left(\partial_\beta a_{\alpha\gamma}+\partial_\beta a_{\alpha\gamma}\right)\\
&= \partial_\beta a_{\alpha\gamma}
\end{aligned}$$

where the second line is justified by the symmetry of the metric tensor.

65. How will the coefficients of the surface metric tensor be affected by scaling the surface up or down by a constant positive scalar factor?
**Answer**: The coefficients of the surface metric tensor will be scaled by the square of the scaling factor.

66. Give the covariant and contravariant types of the surface metric tensor for a Monge patch of the form $\mathbf{r}(u,v)=(u,v,f(u,v))$.
**Answer**:

$$[a_{\alpha\beta}] = \begin{bmatrix} 1+f_u^2 & f_uf_v \\ f_uf_v & 1+f_v^2 \end{bmatrix} \tag{24}$$

$$[a^{\alpha\beta}] = \frac{1}{1+f_u^2+f_v^2}\begin{bmatrix} 1+f_v^2 & -f_uf_v \\ -f_uf_v & 1+f_u^2 \end{bmatrix} \tag{25}$$

where the subscripts $u$ and $v$ stand for partial derivatives of $f$ with respect to these surface coordinates.

67. Discuss how the concept of "length of straight segment" is extended to the length of a polygonal arc. Also discuss how the concept of "length of polygonal arc" is extended to the length of an arc of a twisted space curve.
**Answer**: The concept of length of straight segment is extended to the length of a polygonal arc by taking the sum of the lengths of the segments that make the polygonal arc. The concept of length of polygonal arc is extended to the length of an arc of a twisted space curve by taking the limit of the length of an asymptotic polygonal arc as the length of the longest straight line segment of the asymptotic polygonal arc tends to zero.

68. Derive the following relation which links the length $ds$ of an element of arc of a curve residing on a 2D surface to the covariant metric tensor $a_{\alpha\beta}$ of the surface: $(ds)^2 = a_{\alpha\beta}du^\alpha du^\beta$.
**Answer**: We have:

$$(ds)^2 = d\mathbf{r}\cdot d\mathbf{r}$$

$$
\begin{aligned}
&= \left(\frac{\partial \mathbf{r}}{\partial u^\alpha} du^\alpha\right) \cdot \left(\frac{\partial \mathbf{r}}{\partial u^\beta} du^\beta\right) \\
&= \frac{\partial \mathbf{r}}{\partial u^\alpha} \cdot \frac{\partial \mathbf{r}}{\partial u^\beta} du^\alpha du^\beta \\
&= \mathbf{E}_\alpha \cdot \mathbf{E}_\beta \, du^\alpha du^\beta \\
&= a_{\alpha\beta} du^\alpha du^\beta
\end{aligned}
$$

where line 1 is based on the definition of $ds$ from first principles, line 2 is the chain rule in multi-variable differentiation, line 3 is based on the definition of dot product, line 4 is based on the definition of the surface covariant basis vectors, and line 5 is based on the relation between the covariant basis vectors and the coefficients of the covariant metric tensor.

69. Is the length of a surface curve an intrinsic or extrinsic property and why?
**Answer**: It is intrinsic because it depends only on the coefficients of the metric tensor (or the coefficients of the first fundamental form) as can be seen from the relation that we derived in the previous exercise.

70. Give the formula for the length, $L$, of a segment of a $t$-parameterized surface curve in terms of the coefficients of the first fundamental form of the surface.
**Answer**:

$$
L = \int_{t_1}^{t_2} \sqrt{E\left(\frac{du^1}{dt}\right)^2 + 2F\frac{du^1}{dt}\frac{du^2}{dt} + G\left(\frac{du^2}{dt}\right)^2} \, dt
$$

where $E, F, G$ are the coefficients of the first fundamental form, $u^1$ and $u^2$ are the surface coordinates, and $t_1$ and $t_2$ are the values of the parameter $t$ that correspond to the start and end points of the segment.

71. Using the metric tensor, verify the following relation where $f$ represents a Monge patch of the form $\mathbf{r}(u,v) = (u, v, f(u,v))$:

$$
ds = \sqrt{(1+f_u^2)\, dudu + 2f_u f_v dudv + (1+f_v^2)\, dvdv}
$$

**Answer**: The metric tensor of a Monge patch of the form $\mathbf{r}(u,v) = (u, v, f(u,v))$ is:

$$
[a_{\alpha\beta}] \equiv \begin{bmatrix} a_{11} & a_{12} \\ a_{21} & a_{22} \end{bmatrix} = \begin{bmatrix} 1+f_1^2 & f_1 f_2 \\ f_1 f_2 & 1+f_2^2 \end{bmatrix} \equiv \begin{bmatrix} 1+f_u^2 & f_u f_v \\ f_u f_v & 1+f_v^2 \end{bmatrix}
$$

where the subscripts $1, 2, u, v$ mean partial derivative with respect to $u^1, u^2, u, v$ and $(u^1, u^2) \equiv (u, v)$. Hence:

$$
\begin{aligned}
(ds)^2 &= a_{\alpha\beta} du^\alpha du^\beta \\
&= a_{11} du^1 du^1 + a_{12} du^1 du^2 + a_{21} du^2 du^1 + a_{22} du^2 du^2 \\
&= \left(1+f_1^2\right) du^1 du^1 + f_1 f_2 du^1 du^2 + f_1 f_2 du^2 du^1 + \left(1+f_2^2\right) du^2 du^2 \\
&= \left(1+f_1^2\right) du^1 du^1 + 2 f_1 f_2 du^1 du^2 + \left(1+f_2^2\right) du^2 du^2
\end{aligned}
$$

$$
\begin{aligned}
&= \left(1+f_u^2\right) dudu + 2f_u f_v dudv + \left(1+f_v^2\right) dvdv \\
ds &= \sqrt{\left(1+f_u^2\right) dudu + 2f_u f_v dudv + \left(1+f_v^2\right) dvdv}
\end{aligned}
$$

72. Develop an analytical expression for the length of an element of arc, $ds$, of the catenary parameterized by the equations: $x = a \cosh\left(\frac{\xi}{a}\right)$ and $z = \xi$.
**Answer**: We have:

$$
\begin{aligned}
ds &= \sqrt{(dx)^2 + (dz)^2} \\
&= \sqrt{\left(\frac{dx}{d\xi}\right)^2 + \left(\frac{dz}{d\xi}\right)^2}\, d\xi \\
&= \sqrt{\sinh^2\left(\frac{\xi}{a}\right) + 1}\, d\xi \\
&= \sqrt{\cosh^2\left(\frac{\xi}{a}\right)}\, d\xi \\
&= \cosh\left(\frac{\xi}{a}\right) d\xi
\end{aligned}
$$

73. Discuss how the concept of "area of polygonal plane fragment" is extended to the area of a surface consisting of polygonal plane fragments. Also discuss how the concept of "area of surface made of polygonal plane fragments" is extended to the area of a generalized twisted space surface.
**Answer**: The concept of area of polygonal plane fragment is extended to the area of a surface consisting of polygonal plane fragments by taking the sum of the areas of the fragments that make the surface. The concept of area of surface made of polygonal plane fragments is extended to the area of a generalized twisted space surface by taking the limit of the area of its asymptotic polygonal surface as the area of the largest of its polygonal fragments tends to zero.

74. Derive the mathematical expression for the area $d\sigma$ of an infinitesimal element of a surface and the expression for the area $\sigma$ of a surface patch.
**Answer**: Regarding the first part, we have:

$$
\begin{aligned}
d\sigma &= \left|d\mathbf{r}_1 \times d\mathbf{r}_2\right| \\
&= \left|\left(\frac{\partial \mathbf{r}}{\partial u^1} du^1\right) \times \left(\frac{\partial \mathbf{r}}{\partial u^2} du^2\right)\right| \\
&= \left|\frac{\partial \mathbf{r}}{\partial u^1} \times \frac{\partial \mathbf{r}}{\partial u^2}\right| du^1 du^2 \\
&= \left|\mathbf{E}_1 \times \mathbf{E}_2\right| du^1 du^2 \\
&= \sqrt{\left|\mathbf{E}_1\right|^2 \left|\mathbf{E}_2\right|^2 - \left(\mathbf{E}_1 \cdot \mathbf{E}_2\right)^2}\, du^1 du^2
\end{aligned}
$$

$$= \sqrt{a_{11}a_{22} - (a_{12})^2}\, du^1 du^2$$
$$= \sqrt{a}\, du^1 du^2$$

where line 1 is based on the definition of vector cross product and its relation to the area of the parallelogram that is defined by the vectors, line 2 is the chain rule in multi-variable differentiation, line 3 is based on the definition of cross product plus the assumption that $du^1$ and $du^2$ are positive, line 4 is based on the definition of the surface covariant basis vectors, line 5 is a well known identity,[19] line 6 is based on the relation between the covariant basis vectors and the coefficients of the covariant metric tensor (i.e. $a_{\alpha\beta} = \mathbf{E}_\alpha \cdot \mathbf{E}_\beta$), and line 7 is based on the definition of determinant plus the fact that the metric tensor is symmetric (i.e. $a_{21} = a_{12}$).
Regarding the second part, we have:

$$\begin{aligned} \sigma &= \int_\Omega d\sigma \\ &= \iint_\Omega \sqrt{a_{11}a_{22} - (a_{12})^2}\, du^1 du^2 \\ &= \iint_\Omega \sqrt{a}\, du^1 du^2 \end{aligned}$$

where line 1 is a definition based on the integral as a limit of sum of infinitesimal area elements, while line 2 and line 3 are from the result of the first part.

75. Derive the following equation $\sigma = \iint_\Omega \sqrt{1 + f_u^2 + f_v^2}\, dudv$ for the surface area of a Monge patch of the form $\mathbf{r}(u, v) = (u, v, f(u, v))$.
**Answer**: The covariant metric tensor for this form of Monge patch is given by:

$$[a_{\alpha\beta}] = \begin{bmatrix} 1 + f_u^2 & f_u f_v \\ f_u f_v & 1 + f_v^2 \end{bmatrix}$$

Hence, its determinant is:

$$a = \left(1 + f_u^2\right)\left(1 + f_v^2\right) - \left(f_u f_v\right)^2 = 1 + f_u^2 + f_v^2$$

Using the result of the previous exercise, we have:

$$\sigma = \iint_\Omega \sqrt{a}\, du^1 du^2$$

[19] This identity can be easily shown as follows:

$$\begin{aligned} |\mathbf{E}_1|^2 |\mathbf{E}_2|^2 &= |\mathbf{E}_1|^2 |\mathbf{E}_2|^2 \times 1 \\ &= |\mathbf{E}_1|^2 |\mathbf{E}_2|^2 \left(\sin^2\theta + \cos^2\theta\right) \\ &= |\mathbf{E}_1|^2 |\mathbf{E}_2|^2 \sin^2\theta + |\mathbf{E}_1|^2 |\mathbf{E}_2|^2 \cos^2\theta \\ &= |\mathbf{E}_1 \times \mathbf{E}_2|^2 + |\mathbf{E}_1 \cdot \mathbf{E}_2|^2 \\ |\mathbf{E}_1 \times \mathbf{E}_2|^2 &= |\mathbf{E}_1|^2 |\mathbf{E}_2|^2 - (\mathbf{E}_1 \cdot \mathbf{E}_2)^2 \end{aligned}$$

$$\begin{aligned} &= \iint_\Omega \sqrt{a}\,dudv \\ &= \iint_\Omega \sqrt{1+f_u^2+f_v^2}\,dudv \end{aligned}$$

where line 2 is justified by the fact $(u^1, u^2) \equiv (u, v)$.

76. A cone is represented parametrically in a 3D space by: $\mathbf{r}(\rho,\phi) = (\rho\cos\phi,\ \rho\sin\phi,\ c\rho)$ where $\rho, \phi$ are polar coordinates ($\rho \ge 0$ and $0 \le \phi < 2\pi$) and $c$ is a positive constant. Find the area[20] of the part of the cone corresponding to $0 \le \rho \le A$ where $A$ is a given positive constant.
**Answer**: We have:

$$\begin{aligned} \mathbf{E}_1 &= \partial_\rho \mathbf{r} = (\cos\phi,\ \sin\phi,\ c) \\ \mathbf{E}_2 &= \partial_\phi \mathbf{r} = (-\rho\sin\phi,\ \rho\cos\phi,\ 0) \\ \mathbf{E}_1 \times \mathbf{E}_2 &= (-c\rho\cos\phi, -c\rho\sin\phi, \rho) \\ |\mathbf{E}_1 \times \mathbf{E}_2| &= \sqrt{c^2\rho^2\cos^2\phi + c^2\rho^2\sin^2\phi + \rho^2} \\ &= \rho\sqrt{c^2+1} \end{aligned}$$

Hence:

$$\begin{aligned} \sigma &= \int_\Omega d\sigma \\ &= \iint_\Omega |\mathbf{E}_1 \times \mathbf{E}_2|\,du^1du^2 \\ &= \int_{\phi=0}^{\phi=2\pi}\int_{\rho=0}^{\rho=A} \rho\sqrt{c^2+1}d\rho d\phi \\ &= \sqrt{c^2+1}\int_{\phi=0}^{\phi=2\pi}\left[\frac{\rho^2}{2}\right]_0^A d\phi \\ &= \sqrt{c^2+1}\int_{\phi=0}^{\phi=2\pi}\frac{A^2}{2}d\phi \\ &= \frac{A^2\sqrt{c^2+1}}{2}\int_{\phi=0}^{\phi=2\pi} d\phi \\ &= \frac{A^2\sqrt{c^2+1}}{2}\left[\phi\right]_0^{2\pi} \\ &= \frac{A^2\sqrt{c^2+1}}{2}2\pi \\ &= \pi A^2\sqrt{c^2+1} \end{aligned}$$

[20] We mean the curved part only (i.e. without the area of the base).

77. Derive the mathematical expression: $\cos\theta = g_{ij}A^iB^j$ for the angle $\theta$ between two unit surface vectors, **A** and **B**, where $g_{ij}$ is the space covariant metric tensor. Also give the mathematical expression of $\sin\theta$ for the angle between **A** and **B**.
**Answer**: We have:

$$\begin{aligned}
\cos\theta &= \frac{\mathbf{A}\cdot\mathbf{B}}{|\mathbf{A}|\,|\mathbf{B}|} \\
&= \mathbf{A}\cdot\mathbf{B} \\
&= A^\alpha\mathbf{E}_\alpha\cdot B^\beta\mathbf{E}_\beta \\
&= (\mathbf{E}_\alpha\cdot\mathbf{E}_\beta)\,A^\alpha B^\beta \\
&= a_{\alpha\beta}A^\alpha B^\beta \\
&= g_{ij}x^i_\alpha x^j_\beta A^\alpha B^\beta \\
&= g_{ij}A^iB^j
\end{aligned}$$

where line 1 is based on the definition of the dot product, line 2 is justified by the fact that **A** and **B** are unit vectors, line 3 is the definition of surface vector, line 5 is based on the relation between the covariant basis vectors and the covariant metric tensor, line 6 is based on the equation $a_{\alpha\beta} = g_{ij}x^i_\alpha x^j_\beta$ which is given and verified in the book, and line 7 is based on the equation $A^i = x^i_\alpha A^\alpha$ which is given and verified in the book (also see Exercise 56).
The mathematical expression of $\sin\theta$ for the angle between **A** and **B** is:

$$\sin\theta = \underline{\epsilon}_{\alpha\beta}A^\alpha B^\beta$$

where $\underline{\epsilon}_{\alpha\beta}$ is the surface covariant absolute permutation tensor.

78. Give two mathematical expressions for the coefficients of the surface covariant curvature tensor $b_{\alpha\beta}$.
**Answer**:

$$\begin{aligned}
b_{\alpha\beta} &= -\mathbf{E}_\alpha\cdot\frac{\partial\mathbf{n}}{\partial u^\beta} \\
b_{\alpha\beta} &= \frac{\partial\mathbf{E}_\alpha}{\partial u^\beta}\cdot\mathbf{n}
\end{aligned}$$

where $\mathbf{E}_\alpha$ is surface basis vector, **n** is normal unit vector to the surface and $u^\beta$ is surface coordinate.

79. Using a spherical coordinates representation, find the determinant of the surface covariant curvature tensor of a sphere.
**Answer**: A sphere centered at the origin of coordinates is represented parametrically in spherical coordinates by: $\mathbf{r}(\theta,\phi) = a(\sin\theta\cos\phi, \sin\theta\sin\phi, \cos\theta)$ where $a > 0$ is a constant (which is its radius), $0 \le \theta \le \pi$ and $0 \le \phi < 2\pi$. Hence:

$$\mathbf{E}_1 = \partial_\theta\mathbf{r} = a\,(\cos\theta\cos\phi, \cos\theta\sin\phi, -\sin\theta)$$

$$
\begin{aligned}
\mathbf{E}_2 &= \partial_\phi \mathbf{r} = a\left(-\sin\theta\sin\phi, \sin\theta\cos\phi, 0\right) \\
\mathbf{E}_1 \times \mathbf{E}_2 &= \begin{vmatrix} \mathbf{i} & \mathbf{j} & \mathbf{k} \\ a\cos\theta\cos\phi & a\cos\theta\sin\phi & -a\sin\theta \\ -a\sin\theta\sin\phi & a\sin\theta\cos\phi & 0 \end{vmatrix} \\
&= a^2\left(\sin^2\theta\cos\phi, \sin^2\theta\sin\phi, \cos\theta\sin\theta\right) \\
|\mathbf{E}_1 \times \mathbf{E}_2| &= a^2\sqrt{\sin^4\theta\cos^2\phi + \sin^4\theta\sin^2\phi + \cos^2\theta\sin^2\theta} = a^2\sin\theta \\
\mathbf{n} &= \frac{\mathbf{E}_1 \times \mathbf{E}_2}{|\mathbf{E}_1 \times \mathbf{E}_2|} = \left(\sin\theta\cos\phi, \sin\theta\sin\phi, \cos\theta\right)
\end{aligned}
$$

Using the equation $b_{\alpha\beta} = -\mathbf{E}_\alpha \cdot \frac{\partial \mathbf{n}}{\partial u^\beta}$ where $(u^1, u^2) \equiv (\theta, \phi)$ we get:

$$
\begin{aligned}
b_{11} &= -\mathbf{E}_1 \cdot \frac{\partial \mathbf{n}}{\partial u^1} \\
&= -a\left(\cos\theta\cos\phi, \cos\theta\sin\phi, -\sin\theta\right) \cdot \partial_\theta\left(\sin\theta\cos\phi, \sin\theta\sin\phi, \cos\theta\right) \\
&= -a\left(\cos\theta\cos\phi, \cos\theta\sin\phi, -\sin\theta\right) \cdot \left(\cos\theta\cos\phi, \cos\theta\sin\phi, -\sin\theta\right) \\
&= -a\left(\cos^2\theta\cos^2\phi + \cos^2\theta\sin^2\phi + \sin^2\theta\right) \\
&= -a \\
b_{12} &= -\mathbf{E}_1 \cdot \frac{\partial \mathbf{n}}{\partial u^2} \\
&= -a\left(\cos\theta\cos\phi, \cos\theta\sin\phi, -\sin\theta\right) \cdot \partial_\phi\left(\sin\theta\cos\phi, \sin\theta\sin\phi, \cos\theta\right) \\
&= -a\left(\cos\theta\cos\phi, \cos\theta\sin\phi, -\sin\theta\right) \cdot \left(-\sin\theta\sin\phi, \sin\theta\cos\phi, 0\right) \\
&= -a\left(-\sin\theta\cos\theta\sin\phi\cos\phi + \sin\theta\cos\theta\sin\phi\cos\phi - 0\right) \\
&= 0 \\
b_{21} &= -\mathbf{E}_2 \cdot \frac{\partial \mathbf{n}}{\partial u^1} \\
&= -a\left(-\sin\theta\sin\phi, \sin\theta\cos\phi, 0\right) \cdot \partial_\theta\left(\sin\theta\cos\phi, \sin\theta\sin\phi, \cos\theta\right) \\
&= -a\left(-\sin\theta\sin\phi, \sin\theta\cos\phi, 0\right) \cdot \left(\cos\theta\cos\phi, \cos\theta\sin\phi, -\sin\theta\right) \\
&= -a\left(-\sin\theta\cos\theta\sin\phi\cos\phi + \sin\theta\cos\theta\sin\phi\cos\phi - 0\right) \\
&= 0 \\
b_{22} &= -\mathbf{E}_2 \cdot \frac{\partial \mathbf{n}}{\partial u^2} \\
&= -a\left(-\sin\theta\sin\phi, \sin\theta\cos\phi, 0\right) \cdot \partial_\phi\left(\sin\theta\cos\phi, \sin\theta\sin\phi, \cos\theta\right) \\
&= -a\left(-\sin\theta\sin\phi, \sin\theta\cos\phi, 0\right) \cdot \left(-\sin\theta\sin\phi, \sin\theta\cos\phi, 0\right) \\
&= -a\left(\sin^2\theta\sin^2\phi + \sin^2\theta\cos^2\phi + 0\right) \\
&= -a\sin^2\theta
\end{aligned}
$$

Hence, the determinant of the surface covariant curvature tensor of a sphere is:

$$b = b_{11}b_{22} - b_{12}b_{21} = a^2\sin^2\theta$$

80. Show that $\partial_\beta \mathbf{E}_\alpha = \partial_\alpha \mathbf{E}_\beta$.

**Answer**: We have:

$$\begin{aligned}
\frac{\partial \mathbf{E}_\alpha}{\partial u^\beta} &= \frac{\partial}{\partial u^\beta}\left(\frac{\partial \mathbf{r}}{\partial u^\alpha}\right) \\
&= \frac{\partial^2 \mathbf{r}}{\partial u^\beta \partial u^\alpha} \\
&= \frac{\partial^2 \mathbf{r}}{\partial u^\alpha \partial u^\beta} \\
&= \frac{\partial}{\partial u^\alpha}\left(\frac{\partial \mathbf{r}}{\partial u^\beta}\right) \\
&= \frac{\partial \mathbf{E}_\beta}{\partial u^\alpha}
\end{aligned}$$

where line 1 and line 5 are based on the definition of the surface basis vectors while line 3 is based on the commutativity of the partial differential operators.

81. Prove that if two surfaces, $S_1$ and $S_2$, are mapped isometrically one on the other then they have identical first fundamental form coefficients at their corresponding points, i.e. $E_1 = E_2$, $F_1 = F_2$ and $G_1 = G_2$.
**Answer**: Assume that $S_1$ is mapped isometrically onto $S_2$ and we have an arbitrary curve $C_1$ on $S_1$ with an arbitrary start point $A$ and an arbitrary end point $B$ and $C_1$ passes through an arbitrary point $P_1$. We also assume that the coordinate curves are $t$-parameterized, i.e. $u(t)$ and $v(t)$. So, the curve $C_1$ will be mapped onto a curve $C_2$ on $S_2$ and the point $P_1$ will be mapped on point $P_2$ on $C_2$. Since the mapping is isometric then the length $L_1$ of $C_1$ is equal to the length $L_2$ of $C_2$, that is:

$$\begin{aligned}
L_1 &= \int_{t_A}^{t_B} \sqrt{E_1\left(\frac{du}{dt}\right)^2 + 2F_1\frac{du}{dt}\frac{dv}{dt} + G_1\left(\frac{dv}{dt}\right)^2}\, dt = \\
L_2 &= \int_{t_A}^{t_B} \sqrt{E_2\left(\frac{du}{dt}\right)^2 + 2F_2\frac{du}{dt}\frac{dv}{dt} + G_2\left(\frac{dv}{dt}\right)^2}\, dt
\end{aligned}$$

Now, since everything is arbitrary (i.e. the curve $C_1$, its start and end points, the $t$-parameterization of the surface coordinates) then we should have:

$$E_1 = E_2 \qquad F_1 = F_2 \qquad G_1 = G_2$$

at any corresponding points on the curves including the point $P_1$ (which is also an arbitrary point and hence it can be any point on $S_1$) and its image $P_2$ on $C_2$. In other words, the two surfaces have identical first fundamental form coefficients at their corresponding points, as required.

82. Give the matrix $\left[b^{\alpha\beta}\right]$ which represents the contravariant form of the surface curvature tensor in terms of the coefficients of the first and second fundamental forms.
**Answer**:

$$\left[b^{\alpha\beta}\right] = \frac{1}{a^2}\begin{bmatrix} eG^2 - 2fFG + gF^2 & fEG - eFG - gEF + fF^2 \\ fEG - eFG - gEF + fF^2 & gE^2 - 2fEF + eF^2 \end{bmatrix}$$

where $E, F, G$ and $e, f, g$ are the coefficients of the first and second fundamental forms and $a$ $(= EG - F^2)$ is the determinant of the surface covariant metric tensor.

83. Discuss and justify the relation $\bar{b} = J^2 b$ explaining all the symbols involved.
**Answer**: The transformation equation of the surface curvature tensor from unbarred to barred coordinate systems is given in tensor notation by:

$$\bar{b}_{\alpha\beta} = b_{\gamma\delta} \frac{\partial u^\gamma}{\partial \bar{u}^\alpha} \frac{\partial u^\delta}{\partial \bar{u}^\beta}$$

where $\bar{b}_{\alpha\beta}$ and $b_{\gamma\delta}$ are the surface covariant curvature tensor in the barred and unbarred systems while the indexed $\bar{u}$ and $u$ are the surface coordinates in the two systems. On taking the determinant of the two sides of this equation we obtain:

$$\bar{b} = J^2 b$$

where $\bar{b}$ and $b$ are the determinants of the surface covariant curvature tensor in the barred and unbarred coordinate systems and $J$ is the Jacobian of the transformation between the two systems.

84. What are the other symbols used by some authors to label the coefficients of the second fundamental form $e, f, g$? Discuss the advantages and disadvantages of using each one of these sets of symbols. Also, write the mathematical expression for the second fundamental form using the alternative symbols.
**Answer**: They are $L, M, N$.
The advantage of using $e, f, g$ is that they correspond intuitively to the coefficients of the first fundamental form $E, F, G$ making the formulae involving the coefficients of the first and second fundamental forms more symmetric and memorable. However, this notation may cause confusion in writing and reciting since the corresponding coefficients pronounce identically unless certain precautions are taken. On the other side, the use of $L, M, N$ is advantageous when reciting formulae especially if the formulae contain the coefficients of both fundamental forms; moreover, it is less susceptible to errors in the writing and typing of these formulae. However, the symmetry and correspondence between these coefficients in the formulae that involve them will become less obvious.
The mathematical expression for the second fundamental form using the alternative symbols is:

$$II_S = L(du^1)^2 + 2M\, du^1 du^2 + N(du^2)^2$$

85. How can we obtain the mixed form of the surface curvature tensor $b^\alpha{}_\beta$ from the covariant form of this tensor $b_{\alpha\beta}$?
**Answer**: By raising the first index of $b_{\alpha\beta}$ using the surface contravariant metric tensor, that is:

$$b^\alpha{}_\beta = a^{\alpha\gamma} b_{\gamma\beta}$$

86. Express the mean curvature $H$ and the Gaussian curvature $K$ of a surface in terms of the mixed form of the surface curvature tensor $b^{\alpha}{}_{\beta}$.
**Answer**:
$$H = \frac{\mathrm{tr}(b^{\alpha}{}_{\beta})}{2} \qquad\qquad K = \det(b^{\alpha}{}_{\beta})$$
where tr stands for trace and det stands for determinant.

87. Explain, in details, all the symbols and notations involved in the following relation:
$$b^{\alpha}{}_{\gamma} b_{\beta\delta} - b^{\alpha}{}_{\delta} b_{\beta\gamma} = \partial_{\gamma}\Gamma^{\alpha}_{\beta\delta} - \partial_{\delta}\Gamma^{\alpha}_{\beta\gamma} + \Gamma^{\omega}_{\beta\delta}\Gamma^{\alpha}_{\omega\gamma} - \Gamma^{\omega}_{\beta\gamma}\Gamma^{\alpha}_{\omega\delta}$$
**Answer**: $b^{\alpha}{}_{\gamma}$ and $b^{\alpha}{}_{\delta}$ symbolize mixed curvature tensor, $b_{\beta\delta}$ and $b_{\beta\gamma}$ symbolize covariant curvature tensor, the indexed $\Gamma$ symbolize the Christoffel symbols of the second kind, and all these symbols belong to 2D surface. All the indices represent surface coordinates and hence they range over 1,2. The symbols $\partial_{\gamma}$ and $\partial_{\delta}$ represent partial differential operators with respect to the $\gamma^{th}$ and $\delta^{th}$ surface coordinates, and the summation convention applies to the dummy index $\omega$.

88. Write down the matrix form of the surface covariant curvature tensor for a Monge patch of the form $\mathbf{r}(u, v) = (u, v, f(u, v))$.
**Answer**:
$$[b_{\alpha\beta}] = \frac{1}{\sqrt{1 + f_u^2 + f_v^2}} \begin{bmatrix} f_{uu} & f_{uv} \\ f_{vu} & f_{vv} \end{bmatrix} \tag{26}$$
where the subscripts $u$ and $v$ stand for partial derivatives of $f$ with respect to these surface coordinates.

89. What is the relation between the Riemann-Christoffel curvature tensor of the second kind and the curvature tensor of a surface?
**Answer**: It is:
$$R^{\alpha}{}_{\beta\gamma\delta} = b^{\alpha}{}_{\gamma} b_{\beta\delta} - b^{\alpha}{}_{\delta} b_{\beta\gamma}$$
where $R^{\alpha}{}_{\beta\gamma\delta}$ is the Riemann-Christoffel curvature tensor of the second kind and the indexed $b$ are the mixed and covariant forms of the surface curvature tensor.

90. Give all the coefficients of the Riemann-Christoffel curvature tensor and the coefficients of the curvature tensor for a plane surface.
**Answer**: All these coefficients are identically zero.

91. On a space surface, how many independent non-vanishing components the Riemann-Christoffel curvature tensor possesses?
**Answer**: It possesses only one independent non-vanishing component.

92. Explain the relation between the coefficients of the covariant metric tensor of a surface and the coefficients of its first fundamental form stating the necessary equations.
**Answer**: These coefficients are correspondingly identical, that is:
$$E = a_{11} \qquad\qquad F = a_{12} = a_{21} \qquad\qquad G = a_{22}$$

where $E, F, G$ are the coefficients of the first fundamental form while $a_{11}, a_{12}, a_{21}, a_{22}$ are the coefficients of the surface covariant metric tensor.

93. Derive the mathematical formula for $I_S$ in terms of the coefficients of the first fundamental form.
**Answer**: In Exercise 68 we derived $I_S$ in terms of the surface covariant metric tensor, that is:

$$\begin{aligned} I_S &\equiv (ds)^2 \\ &= a_{\alpha\beta} du^\alpha du^\beta \\ &= a_{11} du^1 du^1 + a_{12} du^1 du^2 + a_{21} du^2 du^1 + a_{22} du^2 du^2 \\ &= a_{11} \left(du^1\right)^2 + 2a_{12} du^1 du^2 + a_{22} \left(du^2\right)^2 \end{aligned}$$

where the last line is justified by the symmetry of the metric tensor. Now, since we have $E = a_{11}$, $F = a_{12}$ and $G = a_{22}$ (see the previous exercise), then we should have:

$$I_S = E \left(du^1\right)^2 + 2F du^1 du^2 + G \left(du^2\right)^2$$

94. Express the determinant of the surface covariant metric tensor as a function of the coefficients of the first fundamental form $E, F, G$.
**Answer**:

$$a = EG - F^2$$

95. Express $E, F, G$ as dot products of the covariant basis vectors of the surface and relate this to the space metric tensor.
**Answer**:

$$\begin{aligned} E &= \mathbf{E}_1 \cdot \mathbf{E}_1 = \frac{\partial \mathbf{r}}{\partial u^1} \cdot \frac{\partial \mathbf{r}}{\partial u^1} = g_{ij} \frac{\partial x^i}{\partial u^1} \frac{\partial x^j}{\partial u^1} \\ F &= \mathbf{E}_1 \cdot \mathbf{E}_2 = \frac{\partial \mathbf{r}}{\partial u^1} \cdot \frac{\partial \mathbf{r}}{\partial u^2} = g_{ij} \frac{\partial x^i}{\partial u^1} \frac{\partial x^j}{\partial u^2} \\ G &= \mathbf{E}_2 \cdot \mathbf{E}_2 = \frac{\partial \mathbf{r}}{\partial u^2} \cdot \frac{\partial \mathbf{r}}{\partial u^2} = g_{ij} \frac{\partial x^i}{\partial u^2} \frac{\partial x^j}{\partial u^2} \end{aligned}$$

where $g_{ij}$ is the space covariant metric tensor and the other symbols are as explained earlier.

96. Does the first fundamental form provide a unique characterization of the space surface as seen internally by a 2D inhabitant? Explain why.
**Answer**: Yes, because all the intrinsic properties are represented by the first fundamental form and all the extrinsic properties are represented by something else. Hence, the first fundamental form (which encompasses all the properties of the surface that can be detected by a 2D inhabitant) provides a unique characterization of the surface in the eye of a 2D inhabitant.
We note that "unique characterization" means that all the properties that can be seen from the given perspective are included while all the other properties are excluded.

97. Does the first fundamental form provide a unique characterization of the space surface as seen from the external ambient space? Explain why.
**Answer**: No, because the extrinsic properties of the surface (which provide extra characterization to the surface that can be seen from an external perspective in addition to the characterization provided by the first fundamental form) are not represented by the first fundamental form.

98. State the mathematical conditions for the provision of positive definiteness of the first fundamental form.
**Answer**: They are:

$$E > 0 \qquad \text{and} \qquad EG - F^2 > 0$$

99. Give the mathematical conditions that apply to the coefficients of the covariant metric tensor of two isometric surfaces at their corresponding points.
**Answer**: If we label the two surfaces as starred and non-starred (i.e. $S^*$ and $S$) then we should have:

$$a_{11}^* = a_{11} \qquad a_{12}^* = a_{21}^* = a_{12} = a_{21} \qquad a_{22}^* = a_{22}$$

where the starred and non-starred coefficients of the metric tensor belong to $S^*$ and $S$ respectively.

100. Explain how the second fundamental form characterizes the surface from the ambient space perspective and how the unit normal vector to the surface is employed in this characterization.
**Answer**: The second fundamental form characterizes the surface from the ambient space perspective by embedding in its definition the variation of an entity that belongs to the ambient space and hence it can characterize the surface from an external view. This entity is the unit normal vector to the surface which can only be observed externally from outside the surface by an observer that resides in the space that envelops the surface. Accordingly, the variation of the unit normal vector as it moves around the surface is used as an indicator to characterize the surface shape from an external point of view and that is how this indicator is exploited in the second fundamental form to represent the extrinsic geometry of the surface.

101. Express the determinant of the surface covariant curvature tensor as a function of the coefficients of the second fundamental form $e, f, g$.
**Answer**:

$$b = eg - f^2$$

102. Derive the mathematical relation of the second fundamental form $II_S$ in terms of the coefficients $e, f, g$. Also provide the main mathematical definitions for the coefficients $e, f, g$ in terms of the surface basis vectors and the unit normal vector to the surface.
**Answer**: Regarding the first part we have:

$$II_S = -d\mathbf{r} \cdot d\mathbf{n}$$

$$\begin{aligned}
&= -\left(\frac{\partial \mathbf{r}}{\partial u^\alpha} du^\alpha\right) \cdot \left(\frac{\partial \mathbf{n}}{\partial u^\beta} du^\beta\right) \\
&= -\left(\frac{\partial \mathbf{r}}{\partial u^\alpha} \cdot \frac{\partial \mathbf{n}}{\partial u^\beta}\right) du^\alpha du^\beta \\
&= b_{\alpha\beta} du^\alpha du^\beta \\
&= b_{11} du^1 du^1 + b_{12} du^1 du^2 + b_{21} du^2 du^1 + b_{22} du^2 du^2 \\
&= b_{11} \left(du^1\right)^2 + 2b_{12} du^1 du^2 + b_{22} \left(du^2\right)^2 \\
&= e(du^1)^2 + 2f\, du^1 du^2 + g(du^2)^2
\end{aligned}$$

where line 1 is a basic definition of the second fundamental form, line 2 is the chain rule in multi-variable differentiation, line 3 is based on the definition of dot product, line 4 is based on the definition of the covariant curvature tensor, line 6 is based on the symmetry of the covariant curvature tensor, and line 7 is based on the identities: $b_{11} \equiv e$, $b_{12} \equiv f$ and $b_{22} \equiv g$.
Regarding the second part of the question, the answer can be found in the first part, that is:

$$\begin{aligned}
e &= -\frac{\partial \mathbf{r}}{\partial u^1} \cdot \frac{\partial \mathbf{n}}{\partial u^1} = -\mathbf{E}_1 \cdot \frac{\partial \mathbf{n}}{\partial u^1} \\
f &= -\frac{\partial \mathbf{r}}{\partial u^1} \cdot \frac{\partial \mathbf{n}}{\partial u^2} = -\mathbf{E}_1 \cdot \frac{\partial \mathbf{n}}{\partial u^2} \\
g &= -\frac{\partial \mathbf{r}}{\partial u^2} \cdot \frac{\partial \mathbf{n}}{\partial u^2} = -\mathbf{E}_2 \cdot \frac{\partial \mathbf{n}}{\partial u^2}
\end{aligned}$$

103. What is the relation between the second fundamental form $II_S$ and the second order differential of the position vector $d^2\mathbf{r}$ of a surface?
**Answer**: The relation is $II_S = d^2\mathbf{r} \cdot \mathbf{n}$ which means that the second fundamental form is the projection of $d^2\mathbf{r}$ in the direction of the normal unit vector to the surface.

104. State the mathematical relations between the coefficients of the second fundamental form and the coefficients of the surface covariant curvature tensor.
**Answer**: They are:

$$e = b_{11} \qquad f = b_{12} = b_{21} \qquad g = b_{22}$$

where $e, f, g$ are the coefficients of the second fundamental form while $b_{11}, b_{12}, b_{21}, b_{22}$ are the coefficients of the surface covariant curvature tensor.

105. Express, in full tensor notation, the second fundamental form in terms of the coefficients of the surface covariant curvature tensor and link this to the expression involving the coefficients $e, f, g$.
**Answer**: It is:

$$II_S = b_{\alpha\beta} du^\alpha du^\beta$$

i.e. $II_S = b_{11}du^1du^1 + b_{12}du^1du^2 + b_{21}du^2du^1 + b_{22}du^2du^2$.
The expression involving the coefficients $e, f, g$ is:

$$II_S = e(du^1)^2 + 2f\, du^1du^2 + g(du^2)^2$$

The two expressions are identical noting that $e = b_{11}$, $f = b_{12} = b_{21}$ and $g = b_{22}$.

106. Explain the following relation which provides a bridge between the first and second fundamental forms: $II_S = \kappa_n I_S$.
**Answer**: This relation means that the second fundamental form $II_S$ of the surface at a given point and in a given direction is equal to the first fundamental form $I_S$ at that point and in that direction times the normal curvature $\kappa_n$ of the surface at that point and in that direction.

107. Derive the following relation where $f$ is a functional representation of a Monge patch of the form $\mathbf{r}(u, v) = (u, v, f(u, v))$:

$$II_S = \frac{f_{uu}dudu + 2f_{uv}dudv + f_{vv}dvdv}{\sqrt{1 + f_u^2 + f_v^2}}$$

**Answer**: We have (refer to Exercise 88):

$$[b_{\alpha\beta}] \equiv \begin{bmatrix} b_{11} & b_{12} \\ b_{21} & b_{22} \end{bmatrix} = \frac{1}{\sqrt{1 + f_u^2 + f_v^2}} \begin{bmatrix} f_{uu} & f_{uv} \\ f_{vu} & f_{vv} \end{bmatrix}$$

Hence:

$$\begin{aligned} II_S &= b_{11}du^1du^1 + 2b_{12}du^1du^2 + b_{22}du^2du^2 \\ &= \frac{f_{uu}dudu + 2f_{uv}dudv + f_{vv}dvdv}{\sqrt{1 + f_u^2 + f_v^2}} \end{aligned}$$

108. Discuss how the first and second fundamental forms represent the intrinsic and extrinsic geometry of the surface.
**Answer**: The first fundamental form is based on the metric of the surface and hence it represents the characterization of the surface geometry from an interior perspective. On the other hand, the second fundamental form is based on the curvature of the surface and hence it represents the characterization of the surface geometry from an exterior perspective.

109. If two surfaces have identical first and second fundamental forms, should they be congruent?
**Answer**: Yes. The existence of these surfaces should ensure that the first and second fundamental forms satisfy the required extra compatibility conditions.

110. What are the compatibility conditions linking the first and second fundamental forms which are needed to fully identify a surface associated with specific first and second fundamental forms and secure its existence?
**Answer**: The compatibility conditions are given by the Codazzi-Mainardi equations plus the following equation:

$$\begin{aligned} eg - f^2 &= F\left[\frac{\partial\Gamma^2_{22}}{\partial u} - \frac{\partial\Gamma^2_{12}}{\partial v} + \Gamma^1_{22}\Gamma^2_{11} - \Gamma^1_{12}\Gamma^2_{12}\right] + \\ & E\left[\frac{\partial\Gamma^1_{22}}{\partial u} - \frac{\partial\Gamma^1_{12}}{\partial v} + \Gamma^1_{22}\Gamma^1_{11} + \Gamma^2_{22}\Gamma^1_{12} - \Gamma^1_{12}\Gamma^1_{12} - \Gamma^2_{12}\Gamma^1_{22}\right] \end{aligned}$$

111. State, using mathematical technical terms, the fundamental theorem of space surfaces.
**Answer**: Given six sufficiently smooth functions $E, F, G, e, f, g$ on a subset of $\mathbb{R}^2$ satisfying the following conditions:
• $E, G > 0$ and $EG - F^2 > 0$, and
• $E, F, G, e, f, g$ satisfy the compatibility conditions of the previous question,
then there is a unique surface with $E, F, G$ as its first fundamental form coefficients and $e, f, g$ as its second fundamental form coefficients.

112. Give a brief definition of Dupin indicatrix and state its significance and usage in differential geometry.
**Answer**: Dupin indicatrix is an indicator of the departure of the surface from the tangent plane in the close proximity of the point of tangency. In quantitative terms, Dupin indicatrix is the family of conic sections given by the following quadratic equation:

$$eu^2 + 2fuv + gv^2 = \pm 1$$

where $e, f, g$ are the coefficients of the second fundamental form at the point with the coordinates $u$ and $v$. The significance of the Dupin indicatrix is that it characterizes the local shape of the surface in the neighborhood of a point and hence it is used to classify the surface points with respect to the local shape as elliptic, parabolic, hyperbolic or flat.

113. What is the shape of Dupin indicatrix at elliptic, parabolic and hyperbolic points on a smooth surface? What is the shape of Dupin indicatrix at flat points?
**Answer**: It is ellipse or circle at elliptic points. It is two parallel lines at parabolic points. It is two conjugate hyperbolas at hyperbolic points. It is not defined at flat points.

114. Make a simple sketch to illustrate Dupin indicatrix at an elliptic point, a parabolic point and a hyperbolic point on a surface marking the two principal directions in each case.
**Answer**: The sketch should be similar to Figure 17.

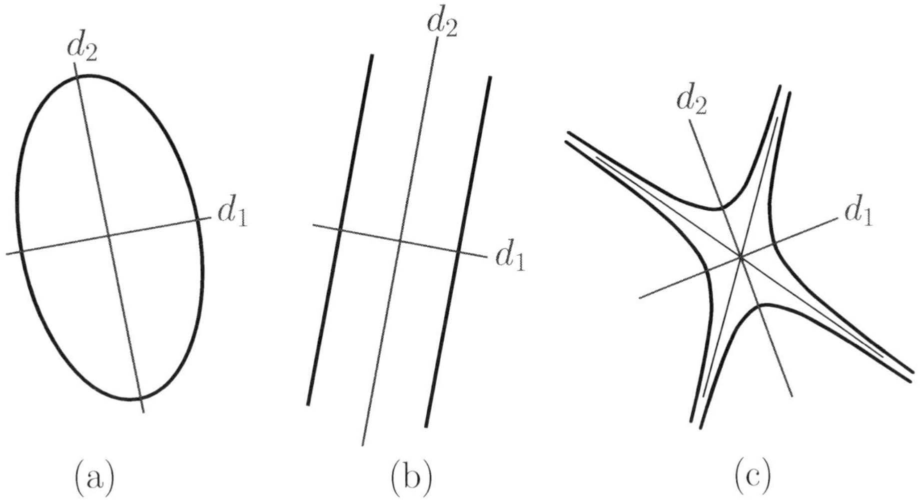


Figure 17: Dupin indicatrix at (a) elliptic point, (b) parabolic point and (c) hyperbolic point. The principal directions are labeled as $d_1$ and $d_2$.

115. Write down the mathematical expression for the third fundamental form $III_S$ in terms of the unit normal vector to the surface $\mathbf{n}$ and in terms of the coefficients $c_{\alpha\beta}$.
**Answer**:

$$III_S = d\mathbf{n} \cdot d\mathbf{n} = c_{\alpha\beta} du^\alpha du^\beta$$

116. Express the coefficients of the third fundamental form as a function of the coefficients of the surface metric and curvature tensors.
**Answer**:

$$c_{\alpha\beta} = a^{\gamma\delta} b_{\alpha\gamma} b_{\beta\delta}$$

where $c_{\alpha\beta}$ are the coefficients of the third fundamental form, $a^{\gamma\delta}$ is the surface contravariant metric tensor and the indexed $b$ are the coefficients of the surface covariant curvature tensor.

117. Explain all the symbols involved in the following equation: $KI_S - 2H\,II_S + III_S = 0$.
**Answer**: $I_S$, $II_S$ and $III_S$ are the first, second and third fundamental forms, $K$ is the Gaussian curvature and $H$ is the mean curvature.

118. Derive the equation in the last question using the Weingarten equations.
**Answer**: According to the definition of the third fundamental form, we have:

$$\begin{aligned} III_S &= d\mathbf{n} \cdot d\mathbf{n} \\ &= (\mathbf{n}_u du + \mathbf{n}_v dv) \cdot (\mathbf{n}_u du + \mathbf{n}_v dv) \\ &= (\mathbf{n}_u \cdot \mathbf{n}_u)\, dudu + 2\,(\mathbf{n}_u \cdot \mathbf{n}_v)\, dudv + (\mathbf{n}_v \cdot \mathbf{n}_v)\, dvdv \end{aligned}$$

where the subscripts $u, v$ mean partial derivative with respect to these surface coordinates. Using Weingarten equations we have:

$$\mathbf{n}_u \cdot \mathbf{n}_u = \left(\frac{fF - eG}{a}\mathbf{E}_1 + \frac{eF - fE}{a}\mathbf{E}_2\right) \cdot \left(\frac{fF - eG}{a}\mathbf{E}_1 + \frac{eF - fE}{a}\mathbf{E}_2\right)$$

$$
\begin{aligned}
&= \frac{(fF-eG)^2}{a^2}E + \frac{2\left(efF^2 - f^2EF - e^2FG + efEG\right)}{a^2}F + \frac{(eF-fE)^2}{a^2}G \\
&= \frac{e^2EG^2 + 2efF^3 - f^2EF^2 - e^2F^2G - 2efEFG + f^2E^2G}{a^2} \\
&= \frac{\left(e^2G - 2efF + f^2E\right)\left(EG - F^2\right)}{a^2} \\
&= \frac{e^2G - 2efF + f^2E}{a} \\
&= \frac{e^2G - 2efF + f^2E + egE - egE}{a} \\
&= \frac{\left(egE - 2efF + e^2G\right) + \left(f^2E - egE\right)}{a} \\
&= \frac{\left(gE - 2fF + eG\right)e - \left(eg - f^2\right)E}{a} \\
&= \frac{\left(gE - 2fF + eG\right)}{a}e - \frac{\left(eg - f^2\right)}{a}E \\
&= 2He - KE
\end{aligned}
$$

$$
\begin{aligned}
\mathbf{n}_u \cdot \mathbf{n}_v &= \left(\frac{fF-eG}{a}\mathbf{E}_1 + \frac{eF-fE}{a}\mathbf{E}_2\right)\cdot\left(\frac{gF-fG}{a}\mathbf{E}_1 + \frac{fF-gE}{a}\mathbf{E}_2\right) \\
&= \frac{(fF-eG)(gF-fG)}{a^2}E + \frac{(fF-eG)(fF-gE)}{a^2}F + \\
&\quad \frac{(eF-fE)(gF-fG)}{a^2}F + \frac{(eF-fE)(fF-gE)}{a^2}G \\
&= \frac{-f^2EFG + efEG^2 + f^2F^3 - efF^2G}{a^2} + \\
&\quad \frac{egF^3 - fgEF^2 - egEFG + fgE^2G}{a^2} \\
&= \frac{\left(fgE - f^2F + efG - egF\right)\left(EG - F^2\right)}{a^2} \\
&= \frac{fgE - f^2F + efG - egF}{a} \\
&= \frac{fgE - 2f^2F + efG - egF + f^2F}{a} \\
&= \frac{\left(gE - 2fF + eG\right)f - \left(eg - f^2\right)F}{a} \\
&= 2Hf - KF
\end{aligned}
$$

$$
\begin{aligned}
\mathbf{n}_v \cdot \mathbf{n}_v &= \left(\frac{gF-fG}{a}\mathbf{E}_1 + \frac{fF-gE}{a}\mathbf{E}_2\right)\cdot\left(\frac{gF-fG}{a}\mathbf{E}_1 + \frac{fF-gE}{a}\mathbf{E}_2\right) \\
&= \frac{(gF-fG)^2}{a^2}E + \frac{2\left(gF-fG\right)\left(fF-gE\right)}{a^2}F + \frac{(fF-gE)^2}{a^2}G
\end{aligned}
$$

$$\begin{aligned}
&= \frac{f^2EG^2 + 2fgF^3 - f^2F^2G - g^2EF^2 - 2fgEFG + g^2E^2G}{a^2} \\
&= \frac{(g^2E - 2fgF + f^2G)(EG - F^2)}{a^2} \\
&= \frac{g^2E - 2fgF + f^2G}{a} \\
&= \frac{g^2E - 2fgF + f^2G + egG - egG}{a} \\
&= \frac{(gE - 2fF + eG)g - (eg - f^2)G}{a} \\
&= 2Hg - KG
\end{aligned}$$

Hence:

$$\begin{aligned}
III_S &= (\mathbf{n}_u \cdot \mathbf{n}_u)\,dudu + 2(\mathbf{n}_u \cdot \mathbf{n}_v)\,dudv + (\mathbf{n}_v \cdot \mathbf{n}_v)\,dvdv \\
&= (2He - KE)\,dudu + 2(2Hf - KF)\,dudv + (2Hg - KG)\,dvdv \\
&= 2H(e\,dudu + 2f\,dudv + g\,dvdv) - K(E\,dudu + 2F\,dudv + G\,dvdv) \\
&= 2H\,II_S - K\,I_S
\end{aligned}$$

that is:

$$K I_S - 2H\,II_S + III_S = 0$$

119. Starting from the equation: $Ka_{\alpha\beta} - 2Hb_{\alpha\beta} + c_{\alpha\beta} = 0$, derive, with full explanation, the following relation: $\text{tr}\left(c_\alpha^\beta\right) = 4H^2 - 2K$.
**Answer**:

$$\begin{aligned}
Ka_{\alpha\beta} - 2Hb_{\alpha\beta} + c_{\alpha\beta} &= 0 \\
Ka_{\alpha\beta}a^{\alpha\beta} - 2Hb_{\alpha\beta}a^{\alpha\beta} + c_{\alpha\beta}a^{\alpha\beta} &= 0 \\
Ka_\alpha^\alpha - 2Hb_\alpha^\alpha + c_\alpha^\alpha &= 0 \\
c_\alpha^\alpha &= 2Hb_\alpha^\alpha - Ka_\alpha^\alpha \\
c_\alpha^\alpha &= 2H(2H) - Ka_\alpha^\alpha \\
\text{tr}\left(c_\alpha^\beta\right) &= 4H^2 - 2K
\end{aligned}$$

where in line 2 the two sides are multiplied by $a^{\alpha\beta}$, in lin3 $a^{\alpha\beta}$ is used as an index raising operator, in line 5 the relation $H = \frac{b_\alpha^\alpha}{2}$ is used, and in line 6 the definition of trace and the relation $a_\alpha^\alpha = \delta_\alpha^\alpha = \delta_1^1 + \delta_2^2 = 1 + 1 = 2$ are used.

120. Explain the correspondence between the Frenet-Serret formulae for space curves and the equations of Gauss and Weingarten for space surfaces.
**Answer**: The Frenet-Serret formulae express the derivatives of $\mathbf{T}, \mathbf{N}, \mathbf{B}$ as combinations of these vectors using $\kappa$ and $\tau$ as coefficients. Similarly, the equations of Gauss and Weingarten express the derivatives of $\mathbf{E}_1, \mathbf{E}_2, \mathbf{n}$ as combinations of these vectors with coefficients based on the first and second fundamental forms.

121. Why the partial derivatives of the surface basis vectors, $\mathbf{E}_1$ and $\mathbf{E}_2$, and the unit normal vector to the surface, $\mathbf{n}$, with respect to the surface coordinates, $u^1$ and $u^2$, can be expressed as combinations of these vectors?
**Answer**: Because $\mathbf{E}_1$, $\mathbf{E}_2$ and $\mathbf{n}$ form a set of linearly independent vectors that can serve as a basis for the 3D embedding space, and hence any vector in this pace, (including the partial derivatives of these vectors which are also vectors in the embedding space) can be expressed as a combination of this set of basis vectors.

122. Prove the following equation using the Weingarten equations:

$$\frac{\partial \mathbf{n}}{\partial u^1} \times \frac{\partial \mathbf{n}}{\partial u^2} = \frac{eg - f^2}{EG - F^2} (\mathbf{E}_1 \times \mathbf{E}_2) = K (\mathbf{E}_1 \times \mathbf{E}_2)$$

**Answer**: We have:

$$\begin{aligned}
\frac{\partial \mathbf{n}}{\partial u^1} \times \frac{\partial \mathbf{n}}{\partial u^2} &= \left(\frac{fF - eG}{a}\mathbf{E}_1 + \frac{eF - fE}{a}\mathbf{E}_2\right) \times \left(\frac{gF - fG}{a}\mathbf{E}_1 + \frac{fF - gE}{a}\mathbf{E}_2\right) \\
&= \frac{(fF - eG)(gF - fG)}{a^2}\mathbf{E}_1 \times \mathbf{E}_1 + \frac{(fF - eG)(fF - gE)}{a^2}\mathbf{E}_1 \times \mathbf{E}_2 + \\
&\quad \frac{(eF - fE)(gF - fG)}{a^2}\mathbf{E}_2 \times \mathbf{E}_1 + \frac{(eF - fE)(fF - gE)}{a^2}\mathbf{E}_2 \times \mathbf{E}_2 \\
&= \frac{(fF - eG)(fF - gE)}{a^2}\mathbf{E}_1 \times \mathbf{E}_2 + \frac{(eF - fE)(gF - fG)}{a^2}\mathbf{E}_2 \times \mathbf{E}_1 \\
&= \frac{(fF - eG)(fF - gE)}{a^2}\mathbf{E}_1 \times \mathbf{E}_2 - \frac{(eF - fE)(gF - fG)}{a^2}\mathbf{E}_1 \times \mathbf{E}_2 \\
&= \frac{(fF - eG)(fF - gE) - (eF - fE)(gF - fG)}{a^2}\mathbf{E}_1 \times \mathbf{E}_2 \\
&= \frac{f^2F^2 + egEG - egF^2 - f^2EG}{a^2}\mathbf{E}_1 \times \mathbf{E}_2 \\
&= \frac{(eg - f^2)(EG - F^2)}{a^2}\mathbf{E}_1 \times \mathbf{E}_2 \\
&= \frac{(eg - f^2)(EG - F^2)}{(EG - F^2)^2}\mathbf{E}_1 \times \mathbf{E}_2 \\
&= \frac{eg - f^2}{EG - F^2}\mathbf{E}_1 \times \mathbf{E}_2 \\
&= K (\mathbf{E}_1 \times \mathbf{E}_2)
\end{aligned}$$

where equality 1 is from the Weingarten equations, equality 2 is distributivity, equality 3 is because $\mathbf{E}_1 \times \mathbf{E}_1 = \mathbf{E}_2 \times \mathbf{E}_2 = \mathbf{0}$, equality 4 because $\mathbf{E}_2 \times \mathbf{E}_1 = -\mathbf{E}_1 \times \mathbf{E}_2$ (anti-commutativity), equalities 5-7 are algebraic manipulations, equality 8 is because $a = EG - F^2$, and equality 10 is based on the definition of Gaussian curvature $K$.

123. Write the derivatives of the surface basis vectors (i.e. $\partial_\beta \mathbf{E}_\alpha$) in terms of the surface vectors (i.e. $\mathbf{E}_\gamma$ and $\mathbf{n}$) in their vector and tensor forms.

**Answer**:

$$\begin{aligned} \partial_\beta \mathbf{E}_\alpha &= \Gamma^\gamma_{\alpha\beta} \mathbf{E}_\gamma + b_{\alpha\beta} \mathbf{n} \\ x^i{}_{\alpha,\beta} &= \Gamma^\gamma_{\alpha\beta} x^i_\gamma + b_{\alpha\beta} n^i \end{aligned}$$

124. What is the essence of the equations of Weingarten? Provide qualitative and quantitative descriptions of these equations. Also write these equations in a matrix form involving the covariant curvature tensor and the contravariant metric tensor of the surface.
**Answer**: The essence of the Weingarten equations is that the partial derivatives of the normal unit vector **n** with respect to the surface coordinates are represented by combinations of the surface basis vectors, $\mathbf{E}_1$ and $\mathbf{E}_2$, where the coefficients are obtained from the coefficients of the first and second fundamental forms.[21] This can be easily seen from their mathematical form which is given by:

$$\begin{aligned} \frac{\partial \mathbf{n}}{\partial u^1} &= \frac{fF - eG}{EG - F^2} \mathbf{E}_1 + \frac{eF - fE}{EG - F^2} \mathbf{E}_2 \\ \frac{\partial \mathbf{n}}{\partial u^2} &= \frac{gF - fG}{EG - F^2} \mathbf{E}_1 + \frac{fF - gE}{EG - F^2} \mathbf{E}_2 \end{aligned}$$

where $E, F, G, e, f, g$ are the coefficients of the first and second fundamental forms. These equations may be expressed in matrix form as:

$$\begin{bmatrix} \frac{\partial \mathbf{n}}{\partial u^1} \\ \frac{\partial \mathbf{n}}{\partial u^2} \end{bmatrix} = -\mathbf{II}_S \mathbf{I}_S^{-1} \begin{bmatrix} \mathbf{E}_1 \\ \mathbf{E}_2 \end{bmatrix} \tag{27}$$

where $\mathbf{II}_S$ is the surface covariant curvature tensor and $\mathbf{I}_S^{-1}$ is the surface contravariant metric tensor.

125. Derive Weingarten equations for a Monge patch of the form $\mathbf{r}(u, v) = (u, v, f(u, v))$.
**Answer**: For a Monge patch of this from, we have (refer to Exercises 66 and 88):

$$\begin{aligned} [a_{\alpha\beta}] &\equiv \begin{bmatrix} E & F \\ F & G \end{bmatrix} = \begin{bmatrix} 1 + f_u^2 & f_u f_v \\ f_u f_v & 1 + f_v^2 \end{bmatrix} \\ [b_{\alpha\beta}] &\equiv \begin{bmatrix} e & f \\ f & g \end{bmatrix} = \frac{1}{\sqrt{1 + f_u^2 + f_v^2}} \begin{bmatrix} f_{uu} & f_{uv} \\ f_{vu} & f_{vv} \end{bmatrix} \\ a &\equiv EG - F^2 = 1 + f_u^2 + f_v^2 \end{aligned} \tag{28}$$

Hence, by substitution from the above equations into the Weingarten equations in their general form we get:

$$\frac{\partial \mathbf{n}}{\partial u} = \frac{fF - eG}{a} \mathbf{E}_1 + \frac{eF - fE}{a} \mathbf{E}_2$$

[21] In brief, these partial derivatives belong to the tangent space.

$$
\begin{aligned}
&= \frac{f_{uv}f_uf_v - f_{uu}(1+f_v^2)}{\sqrt{1+f_u^2+f_v^2}\,(1+f_u^2+f_v^2)}\mathbf{E}_1 + \frac{f_{uu}f_uf_v - f_{uv}(1+f_u^2)}{\sqrt{1+f_u^2+f_v^2}\,(1+f_u^2+f_v^2)}\mathbf{E}_2 \\
&= \frac{f_{uv}f_uf_v - f_{uu} - f_{uu}f_v^2}{(1+f_u^2+f_v^2)^{3/2}}\mathbf{E}_1 + \frac{f_{uu}f_uf_v - f_{uv} - f_{uv}f_u^2}{(1+f_u^2+f_v^2)^{3/2}}\mathbf{E}_2 \\
\frac{\partial \mathbf{n}}{\partial v} &= \frac{gF - fG}{a}\mathbf{E}_1 + \frac{fF - gE}{a}\mathbf{E}_2 \\
&= \frac{f_{vv}f_uf_v - f_{uv}(1+f_v^2)}{\sqrt{1+f_u^2+f_v^2}\,(1+f_u^2+f_v^2)}\mathbf{E}_1 + \frac{f_{uv}f_uf_v - f_{vv}(1+f_u^2)}{\sqrt{1+f_u^2+f_v^2}\,(1+f_u^2+f_v^2)}\mathbf{E}_2 \\
&= \frac{f_{vv}f_uf_v - f_{uv} - f_{uv}f_v^2}{(1+f_u^2+f_v^2)^{3/2}}\mathbf{E}_1 + \frac{f_{uv}f_uf_v - f_{vv} - f_{vv}f_u^2}{(1+f_u^2+f_v^2)^{3/2}}\mathbf{E}_2
\end{aligned}
$$

126. Express the partial derivatives of **n** with respect to the surface coordinates, $u$ and $v$, in terms of the Gaussian curvature $K$ and the mean curvature $H$ and the coefficients of the first and second fundamental forms $E, F, G, e, f, g$.
**Answer**: We have (see Exercise 118):

$$
\begin{aligned}
\frac{\partial \mathbf{n}}{\partial u}\cdot\frac{\partial \mathbf{n}}{\partial u} &= 2eH - EK \\
\frac{\partial \mathbf{n}}{\partial u}\cdot\frac{\partial \mathbf{n}}{\partial v} &= 2fH - FK \\
\frac{\partial \mathbf{n}}{\partial v}\cdot\frac{\partial \mathbf{n}}{\partial v} &= 2gH - GK
\end{aligned}
$$

127. Give the equations of Gauss for a Monge patch of the form $\mathbf{r}(u,v) = (u, v, f(u,v))$.
**Answer**:

$$
\begin{aligned}
\frac{\partial \mathbf{E}_1}{\partial u} &= \frac{1}{1+f_u^2+f_v^2}\left(f_uf_{uu}\mathbf{E}_1 + f_vf_{uu}\mathbf{E}_2 + f_{uu}\sqrt{1+f_u^2+f_v^2}\,\mathbf{n}\right) \\
\frac{\partial \mathbf{E}_1}{\partial v} &= \frac{1}{1+f_u^2+f_v^2}\left(f_uf_{uv}\mathbf{E}_1 + f_vf_{uv}\mathbf{E}_2 + f_{uv}\sqrt{1+f_u^2+f_v^2}\,\mathbf{n}\right) = \frac{\partial \mathbf{E}_2}{\partial u} \\
\frac{\partial \mathbf{E}_2}{\partial v} &= \frac{1}{1+f_u^2+f_v^2}\left(f_uf_{vv}\mathbf{E}_1 + f_vf_{vv}\mathbf{E}_2 + f_{vv}\sqrt{1+f_u^2+f_v^2}\,\mathbf{n}\right)
\end{aligned}
$$

128. State, using full tensor notation, the equations of Codazzi-Mainardi explaining all the symbols involved.
**Answer**:

$$
\frac{\partial b_{\alpha\beta}}{\partial u^\gamma} - \frac{\partial b_{\alpha\gamma}}{\partial u^\beta} = b_{\delta\beta}\Gamma^\delta_{\alpha\gamma} - b_{\delta\gamma}\Gamma^\delta_{\alpha\beta}
$$

where the indexed $b$ represent surface covariant curvature tensor, the indexed $\Gamma$ symbolize Christoffel symbols of the second kind, and $u^\gamma$ and $u^\beta$ are surface coordinates. All the indices represent surface coordinates and hence they range over 1,2.

129. Derive, using the Codazzi-Mainardi equations, the following relation: $b_{\alpha\beta;\gamma} = b_{\alpha\gamma;\beta}$.
**Answer**:

$$\begin{aligned}
\frac{\partial b_{\alpha\beta}}{\partial u^{\gamma}} - \frac{\partial b_{\alpha\gamma}}{\partial u^{\beta}} &= b_{\delta\beta}\Gamma^{\delta}_{\alpha\gamma} - b_{\delta\gamma}\Gamma^{\delta}_{\alpha\beta} \\
\frac{\partial b_{\alpha\beta}}{\partial u^{\gamma}} - b_{\delta\beta}\Gamma^{\delta}_{\alpha\gamma} &= \frac{\partial b_{\alpha\gamma}}{\partial u^{\beta}} - b_{\delta\gamma}\Gamma^{\delta}_{\alpha\beta} \\
\frac{\partial b_{\alpha\beta}}{\partial u^{\gamma}} - b_{\delta\beta}\Gamma^{\delta}_{\alpha\gamma} - b_{\alpha\delta}\Gamma^{\delta}_{\gamma\beta} &= \frac{\partial b_{\alpha\gamma}}{\partial u^{\beta}} - b_{\delta\gamma}\Gamma^{\delta}_{\alpha\beta} - b_{\alpha\delta}\Gamma^{\delta}_{\gamma\beta} \\
b_{\alpha\beta;\gamma} &= b_{\alpha\gamma;\beta}
\end{aligned}$$

where line 1 is the tensor form of the Codazzi-Mainardi equations, in line 2 we transform terms between the two sides, in line 3 we subtract $b_{\alpha\delta}\Gamma^{\delta}_{\gamma\beta}$ from both sides, and in line 4 we use the definition of covariant derivative.

130. Explain how the relation in the previous question indicates that there are only two independent components for the Codazzi-Mainardi equations. What are these two independent components?
**Answer**: Because the indices $\alpha, \beta, \gamma$ range over 1,2 then we should have 8 possibilities. However, because the covariant curvature tensor is symmetric in its two indices the 8 possibilities will reduce to 4. Moreover, because the covariant derivative according to the relation $b_{\alpha\beta;\gamma} = b_{\alpha\gamma;\beta}$ is symmetric in its last two indices the remaining 4 possibilities will reduce to just 2 possibilities, i.e. there are only two independent components for the Codazzi-Mainardi equations which the relation $b_{\alpha\beta;\gamma} = b_{\alpha\gamma;\beta}$ is based upon, as shown in the previous exercise. These two independent components are given by $b_{\alpha\alpha;\beta} = b_{\alpha\beta;\alpha}$ where $\alpha \neq \beta$ and there is no summation over $\alpha$, i.e. $b_{11;2} = b_{12;1}$ and $b_{22;1} = b_{21;2}$.

131. Describe sphere mapping in qualitative and technical terms. Also, explain the meaning of the following equation:

$$\lim_{\mathfrak{R} \to P} \frac{\sigma(\bar{\mathfrak{R}})}{\sigma(\mathfrak{R})} = |K_P|$$

**Answer**: Sphere mapping is a correlation between the points of a surface and the unit sphere where each point on the surface is projected onto its unit normal as a point on the unit sphere which is centered at the origin of coordinates. The above equation means that the limit of the ratio of the area of a region $\bar{\mathfrak{R}}$ on the spherical image to the area of the corresponding region $\mathfrak{R}$ on the surface $S$ in the neighborhood of a given point $P$ on $S$ equals the absolute value of the Gaussian curvature at $P$ as $\mathfrak{R}$ shrinks to the point $P$.

132. Prove the theorem represented by the equation in the previous question.
**Answer**: In Exercise 122 we proved the following relation:

$$\partial_u \mathbf{n} \times \partial_v \mathbf{n} = K\left(\mathbf{E}_1 \times \mathbf{E}_2\right)$$

On taking the modulus of the two sides we obtain:

$$|\partial_u \mathbf{n} \times \partial_v \mathbf{n}| = |K|\,|\mathbf{E}_1 \times \mathbf{E}_2|$$

$$\frac{|\partial_u \mathbf{n} \times \partial_v \mathbf{n}|}{|\mathbf{E}_1 \times \mathbf{E}_2|} = |K|$$

Now, $\partial_u \mathbf{n}$ and $\partial_v \mathbf{n}$ are perpendicular to $\mathbf{n}$ and hence they are surface basis vectors for the spherical image.[22] Accordingly, $|\partial_u \mathbf{n} \times \partial_v \mathbf{n}|\, dudv$ represents an infinitesimal surface area $d\sigma_I$ on the spherical image. Similarly, $|\mathbf{E}_1 \times \mathbf{E}_2|\, dudv$ represents an infinitesimal surface area $d\sigma_S$ on the surface that is spherically mapped on the unit sphere. Hence, in the limit we will have:

$$\frac{d\sigma_I}{d\sigma_S} = |K|$$

$$\lim_{\mathfrak{R} \to P} \frac{\sigma(\bar{\mathfrak{R}})}{\sigma(\mathfrak{R})} = |K_P|$$

which is the required result.

133. Prove that at a given point $P$ on a surface with $K \neq 0$, there exists a neighborhood $\mathcal{N}$ of $P$ where an injective mapping can be established between $\mathcal{N}$ and its spherical image $\bar{\mathcal{N}}$.
**Answer**: This can be easily concluded from the proof of the previous question.

134. State one of the global theorems of space surface and explain why it is global.
**Answer**: "Planes are the only connected surfaces of class $C^2$ whose all points are flat". This theorem is global because it is about the character of plane surface as a whole and not about individual points or particular patches or parts of the surface with a local nature.

---

[22] As indicated before, the derivative of unit vector (or indeed any vector of constant magnitude) is perpendicular to the vector. This can be demonstrated as follows:

$$\begin{aligned} \mathbf{n} \cdot \mathbf{n} &= 1 \\ \partial (\mathbf{n} \cdot \mathbf{n}) &= 0 \\ 2 (\mathbf{n} \cdot \partial \mathbf{n}) &= 0 \\ \mathbf{n} \cdot \partial \mathbf{n} &= 0 \end{aligned}$$

where $\partial$ symbolizes partial derivative operator with respect to a surface coordinate. The above fact can also be deduced from the Weingarten equations where the the partial derivatives of $\mathbf{n}$ are expressed as linear combinations of the surface basis vectors, $\mathbf{E}_1$ and $\mathbf{E}_2$, and hence they belong to the tangent space (see Exercise 124).

# Chapter 4
# Curvature

1. Discuss the similarities and differences between the curvature of curves and the curvature of surfaces.
   **Answer**: The curvature of both curves and surfaces is an attribute that is originally based on their local shape although it may be extended to have a global significance. However, while surface curvature can be intrinsic as well as extrinsic, curve curvature can only be extrinsic since curves have no sufficient dimensionality to show their curvature intrinsically.

2. Define, descriptively and mathematically, the curvature vector $\mathbf{K}$ of surface curves and its relation to the principal normal vector $\mathbf{N}$ of the curve.
   **Answer**: The curvature vector $\mathbf{K}$ of a surface curve $C$ at a given point $P$ on the curve is defined as the derivative of the tangent unit vector $\mathbf{T}$ of $C$ at $P$ with respect to a natural parameter $s$ of the curve, i.e. $\mathbf{K} = \mathbf{T}'$. The curvature vector $\mathbf{K}$ is related to the principal normal vector $\mathbf{N}$ of $C$ by the relation $\mathbf{K} = \kappa\mathbf{N}$ where $\kappa$ is the curvature of $C$ at $P$.

3. Compare the vectors $\mathbf{n}$ and $\mathbf{N}$ at a point on a surface curve outlining their similarities and differences.
   **Answer**: We note the following:
   • While $\mathbf{n}$ is a vector normal to the surface, $\mathbf{N}$ is a vector normal to the curve.
   • Both $\mathbf{n}$ and $\mathbf{N}$ are unit vectors.
   • Both $\mathbf{n}$ and $\mathbf{N}$ represent extrinsic attribute to the concerned entity.
   • Both $\mathbf{n}$ and $\mathbf{N}$ belong to a point on the concerned entity.
   • Both $\mathbf{n}$ and $\mathbf{N}$ can be expressed as cross product of two other vectors that belong to the concerned entity, i.e. $\mathbf{E}_1$ and $\mathbf{E}_2$ for $\mathbf{n}$ and $\mathbf{B}$ and $\mathbf{T}$ for $\mathbf{N}$.

4. Discuss the dependency of the curvature vector of a surface curve at a given point of the curve on the following parameters: curve orientation, curve parameterization, surface orientation as indicated by the direction of $\mathbf{n}$, surface parameterization, tangential direction and position of the point on the surface.
   **Answer**: The curvature vector of a surface curve is independent of the curve orientation, curve parameterization, surface orientation and surface parameterization. However, the curvature vector depends on the tangential direction to the surface[23] and the position of the point on the surface.

[23] This tangential direction to the surface should not be confused with the curve orientation which the curvature vector is independent of. However, considering the tangential direction to the surface makes the curve undetermined apart from being passing through a given point on the surface and this difference with the other factors (i.e. curve orientation, etc.) should be noticed.

5. What "inflection point" on a surface curve means?
**Answer**: It is a point on the curve at which the curvature vector $\mathbf{K}$ vanishes.

6. What is the radius of curvature at a point of inflection?
**Answer**: It is infinite.

7. Resolve the curvature vector of a surface curve into its tangential and normal components and name these components. Express these components in terms of the unit vectors $\mathbf{n}$ and $\mathbf{u}$ explaining all the symbols involved in this expression.
**Answer**:

$$\mathbf{K} = \mathbf{K}_g + \mathbf{K}_n$$

where $\mathbf{K}$ is the curvature vector, $\mathbf{K}_g$ and $\mathbf{K}_n$ are its tangential (or geodesic) and normal components which can be expressed in terms of $\mathbf{n}$ and $\mathbf{u}$ as:

$$\mathbf{K}_g = \kappa_g \mathbf{u} \qquad \text{and} \qquad \mathbf{K}_n = \kappa_n \mathbf{n}$$

where $\kappa_g$ and $\kappa_n$ are the geodesic and normal curvatures of the curve, $\mathbf{n}$ is the normal unit vector to the surface and $\mathbf{u}$ is the geodesic normal vector to the curve.

8. Find the curvature vector, $\mathbf{K}$, of a space curve represented by: $\mathbf{r}(t) = (3t^2, t, 2\sin t)$.
**Answer**:

$$\begin{aligned}
\dot{\mathbf{r}} &= \frac{d\mathbf{r}}{dt} = (6t, 1, 2\cos t) \\
\ddot{\mathbf{r}} &= \frac{d\dot{\mathbf{r}}}{dt} = (6, 0, -2\sin t) \\
\ddot{\mathbf{r}} \times \dot{\mathbf{r}} &= (2\sin t, -12\cos t - 12t\sin t, 6) \\
\dot{\mathbf{r}} \times (\ddot{\mathbf{r}} \times \dot{\mathbf{r}}) &= (6 + 24\cos^2 t + 24t\cos t\sin t, \\
&\quad 4\cos t\sin t - 36t,\ -72t\cos t - 72t^2\sin t - 2\sin t) \\
\mathbf{N} &= \frac{\dot{\mathbf{r}} \times (\ddot{\mathbf{r}} \times \dot{\mathbf{r}})}{|\dot{\mathbf{r}}|\,|\ddot{\mathbf{r}} \times \dot{\mathbf{r}}|} \\
\kappa &= \frac{|\dot{\mathbf{r}} \times \ddot{\mathbf{r}}|}{|\dot{\mathbf{r}}|^3} \\
\mathbf{K} &= \kappa\mathbf{N} \\
&= \frac{|\dot{\mathbf{r}} \times \ddot{\mathbf{r}}|}{|\dot{\mathbf{r}}|^3} \frac{\dot{\mathbf{r}} \times (\ddot{\mathbf{r}} \times \dot{\mathbf{r}})}{|\dot{\mathbf{r}}|\,|\ddot{\mathbf{r}} \times \dot{\mathbf{r}}|} \\
&= \frac{\dot{\mathbf{r}} \times (\ddot{\mathbf{r}} \times \dot{\mathbf{r}})}{|\dot{\mathbf{r}}|^4} \\
&= \frac{1}{(36t^2 + 1 + 4\cos^2 t)^2}(6 + 24\cos^2 t + 24t\cos t\sin t, \\
&\quad 4\cos t\sin t - 36t,\ -72t\cos t - 72t^2\sin t - 2\sin t)
\end{aligned}$$

9. Define, descriptively and quantitatively, the normal and geodesic curvatures $\kappa_n$ and $\kappa_g$.
**Answer**: The normal curvature $\kappa_n$ is the projection of the curvature vector $\mathbf{K}$ on the normal unit vector to the surface $\mathbf{n}$ while the geodesic curvatures $\kappa_g$ is the projection of the curvature vector $\mathbf{K}$ on the geodesic normal vector to the curve $\mathbf{u}$, that is:

$$\kappa_n = \mathbf{n} \cdot \mathbf{K} \qquad \text{and} \qquad \kappa_g = \mathbf{u} \cdot \mathbf{K} \tag{29}$$

10. Which of $\kappa_n$ and $\kappa_g$ is an intrinsic property and which is an extrinsic property? Explain why.
**Answer**: $\kappa_g$ is an intrinsic property because it depends on the first fundamental form only (as can be seen for example from Eq. 31 in Exercise 14), while $\kappa_n$ is an extrinsic property because it depends on the second fundamental form as well as the first fundamental form (as can be seen for example from the relation $\kappa_n = II_S/I_S$ which is derived in Exercise 20).

11. Compare the following four moving frames: $(\mathbf{T}, \mathbf{N}, \mathbf{B})$, $(\mathbf{E}_1, \mathbf{E}_2, \mathbf{n})$, $(\mathbf{n}, \mathbf{T}, \mathbf{u})$ and $(\mathbf{d}_1, \mathbf{d}_2, \mathbf{n})$ outlining their similarities and dissimilarities.
**Answer**: We note the following:
• All these moving frames consist of 3 linearly independent vectors and hence they can be used to reference the 3D embedding space.
• They all vary in space (since they are moving frames) and hence they are spatially dependent.
• $(\mathbf{T}, \mathbf{N}, \mathbf{B})$ is associated with space curve while $(\mathbf{E}_1, \mathbf{E}_2, \mathbf{n})$, $(\mathbf{n}, \mathbf{T}, \mathbf{u})$ and $(\mathbf{d}_1, \mathbf{d}_2, \mathbf{n})$ are associated with space surface.[24]
• All these bases are orthonormal (i.e. made of mutually-orthogonal unit vectors) except $(\mathbf{E}_1, \mathbf{E}_2, \mathbf{n})$ which is not because $\mathbf{E}_1$ and $\mathbf{E}_2$ are not necessarily orthogonal or unit vectors although $\mathbf{n}$ is orthogonal to both $\mathbf{E}_1$ and $\mathbf{E}_2$ and it is a unit vector.

12. Which of the frames in the previous question employ both surface and curve vectors? Which of these frames are orthonormal by definition and which are not?
**Answer**: The frame $(\mathbf{n}, \mathbf{T}, \mathbf{u})$ employs both surface and curve vectors since $\mathbf{n}$ is normal to surface and $\mathbf{T}$ is tangent to curve while $\mathbf{u}$ $(= \mathbf{n} \times \mathbf{T})$ belongs to both.
The frames $(\mathbf{T}, \mathbf{N}, \mathbf{B})$, $(\mathbf{n}, \mathbf{T}, \mathbf{u})$ and $(\mathbf{d}_1, \mathbf{d}_2, \mathbf{n})$ are orthonormal by definition. However, this is not the case with $(\mathbf{E}_1, \mathbf{E}_2, \mathbf{n})$ which may be or may be not.

13. How is the curvature $\kappa$ of a surface curve $C$ at a given point $P$ related to its normal and geodesic curvatures $\kappa_n$ and $\kappa_g$ at that point? Can you make sense of this considering the normal and tangential components of the curvature vector $\mathbf{K}$?
**Answer**: We have:

$$\kappa_n = \kappa \cos \phi \qquad \text{and} \qquad \kappa_g = \kappa \sin \phi \tag{30}$$

[24] In fact, $(\mathbf{n}, \mathbf{T}, \mathbf{u})$ is associated with surface curve.

where $\kappa$ is the curvature of $C$ at $P$ and $\phi$ is the angle between the vector $\mathbf{N}$ of $C$ at $P$ and the vector $\mathbf{n}$ of the surface at $P$. Hence:

$$\kappa_n^2 + \kappa_g^2 = \kappa^2 \cos^2 \phi + \kappa^2 \sin^2 \phi = \kappa^2 \left(\cos^2 \phi + \sin^2 \phi\right) = \kappa^2$$

Because $\kappa_n$ and $\kappa_g$ are obtained from the projection of $\mathbf{K}$ onto $\mathbf{n}$ and $\mathbf{u}$ (i.e. $\kappa_n = \mathbf{n} \cdot \mathbf{K}$ and $\kappa_g = \mathbf{u} \cdot \mathbf{K}$) and because $\mathbf{n}$ and $\mathbf{u}$ are orthogonal then the obtained relation (i.e. $\kappa^2 = \kappa_n^2 + \kappa_g^2$) makes full sense. In other words, the relation $\kappa^2 = \kappa_n^2 + \kappa_g^2$ is no more than the Pythagoras theorem.

14. Prove that the geodesic curvature of a naturally parameterized curve is given by the relation:

$$\begin{aligned} \kappa_g =& \sqrt{a}\left[\Gamma_{11}^2 \left(\frac{du^1}{ds}\right)^3 + \left(2\Gamma_{12}^2 - \Gamma_{11}^1\right)\left(\frac{du^1}{ds}\right)^2 \frac{du^2}{ds} + \right. \\ & \left. \left(\Gamma_{22}^2 - 2\Gamma_{12}^1\right) \frac{du^1}{ds}\left(\frac{du^2}{ds}\right)^2 - \Gamma_{22}^1 \left(\frac{du^2}{ds}\right)^3 + \frac{du^1}{ds}\frac{d^2u^2}{ds^2} - \frac{d^2u^1}{ds^2}\frac{du^2}{ds}\right] \end{aligned} \qquad (31)$$

**Answer**: If the curve is spatially represented by $\mathbf{r}(s)$ where $s$ is a natural parameter, then we have:

$$\kappa_g = \mathbf{u} \cdot \mathbf{K} = (\mathbf{n} \times \mathbf{T}) \cdot \mathbf{K} = (\mathbf{n} \times \mathbf{r}') \cdot \mathbf{T}' = (\mathbf{n} \times \mathbf{r}') \cdot \mathbf{r}''$$

where the prime symbolizes derivative with respect to $s$. Now, if we use subscripts to symbolize partial derivative and employ the chain and product rules of differentiation (noting that the surface coordinates are mutually independent and the partial differential operators are commutative) then we have:

$$\begin{aligned}
\mathbf{r}' &= \mathbf{r}_u u' + \mathbf{r}_v v' \\
\mathbf{r}'' &= \left(\mathbf{r}_u u' + \mathbf{r}_v v'\right)' \\
&= \left(\mathbf{r}_u u' + \mathbf{r}_v v'\right)_u u' + \left(\mathbf{r}_u u' + \mathbf{r}_v v'\right)_v v' \\
&= \mathbf{r}_{uu} \left(u'\right)^2 + \mathbf{r}_u u'' + \mathbf{r}_{vu} v' u' + 0 + \mathbf{r}_{uv} u' v' + 0 + \mathbf{r}_{vv} \left(v'\right)^2 + \mathbf{r}_v v'' \\
&= \mathbf{r}_{uu} \left(u'\right)^2 + \mathbf{r}_u u'' + 2\mathbf{r}_{uv} u' v' + \mathbf{r}_v v'' + \mathbf{r}_{vv} \left(v'\right)^2 \\
\mathbf{n} \times \mathbf{r}' &= \mathbf{n} \times \left(\mathbf{r}_u u' + \mathbf{r}_v v'\right) \\
&= \left(\mathbf{n} \times \mathbf{r}_u\right) u' + \left(\mathbf{n} \times \mathbf{r}_v\right) v' \\
\kappa_g &= \left(\mathbf{n} \times \mathbf{r}'\right) \cdot \mathbf{r}'' \\
&= \left(\mathbf{n} \times \mathbf{r}_u u' + \mathbf{n} \times \mathbf{r}_v v'\right) \cdot \mathbf{r}'' \\
&= \left(\mathbf{n} \times \mathbf{r}_u\right) \cdot \mathbf{r}'' u' + \left(\mathbf{n} \times \mathbf{r}_v\right) \cdot \mathbf{r}'' v' \\
&= \left(\mathbf{n} \times \mathbf{r}_u\right) \cdot \left[\mathbf{r}_{uu} \left(u'\right)^2 + \mathbf{r}_u u'' + 2\mathbf{r}_{uv} u' v' + \mathbf{r}_v v'' + \mathbf{r}_{vv} \left(v'\right)^2\right] u' + \\
&\quad \left(\mathbf{n} \times \mathbf{r}_v\right) \cdot \left[\mathbf{r}_{uu} \left(u'\right)^2 + \mathbf{r}_u u'' + 2\mathbf{r}_{uv} u' v' + \mathbf{r}_v v'' + \mathbf{r}_{vv} \left(v'\right)^2\right] v' \\
&= \left(\mathbf{n} \times \mathbf{r}_u\right) \cdot \left[\mathbf{r}_{uu} \left(u'\right)^3 + \mathbf{r}_u u' u'' + 2\mathbf{r}_{uv} \left(u'\right)^2 v' + \mathbf{r}_v u' v'' + \mathbf{r}_{vv} u' \left(v'\right)^2\right] +
\end{aligned}$$

$$
\begin{aligned}
& (\mathbf{n}\times\mathbf{r}_v)\cdot\left[\mathbf{r}_{uu}(u')^2 v' + \mathbf{r}_u u'' v' + 2\mathbf{r}_{uv} u' (v')^2 + \mathbf{r}_v v' v'' + \mathbf{r}_{vv}(v')^3\right] \\
= & (\mathbf{n}\times\mathbf{r}_u)\cdot\mathbf{r}_{uu}(u')^3 + (\mathbf{n}\times\mathbf{r}_u)\cdot\mathbf{r}_u u' u'' + 2(\mathbf{n}\times\mathbf{r}_u)\cdot\mathbf{r}_{uv}(u')^2 v' + \\
& (\mathbf{n}\times\mathbf{r}_u)\cdot\mathbf{r}_v u' v'' + (\mathbf{n}\times\mathbf{r}_u)\cdot\mathbf{r}_{vv} u' (v')^2 + \\
& (\mathbf{n}\times\mathbf{r}_v)\cdot\mathbf{r}_{uu}(u')^2 v' + (\mathbf{n}\times\mathbf{r}_v)\cdot\mathbf{r}_u u'' v' + 2(\mathbf{n}\times\mathbf{r}_v)\cdot\mathbf{r}_{uv} u' (v')^2 + \\
& (\mathbf{n}\times\mathbf{r}_v)\cdot\mathbf{r}_v v' v'' + (\mathbf{n}\times\mathbf{r}_v)\cdot\mathbf{r}_{vv}(v')^3 \\
= & (\mathbf{n}\times\mathbf{r}_u)\cdot\mathbf{r}_{uu}(u')^3 + 0 + 2(\mathbf{n}\times\mathbf{r}_u)\cdot\mathbf{r}_{uv}(u')^2 v' + \\
& (\mathbf{n}\times\mathbf{r}_u)\cdot\mathbf{r}_v u' v'' + (\mathbf{n}\times\mathbf{r}_u)\cdot\mathbf{r}_{vv} u' (v')^2 + (\mathbf{n}\times\mathbf{r}_v)\cdot\mathbf{r}_{uu}(u')^2 v' + \\
& (\mathbf{n}\times\mathbf{r}_v)\cdot\mathbf{r}_u u'' v' + 2(\mathbf{n}\times\mathbf{r}_v)\cdot\mathbf{r}_{uv} u' (v')^2 + 0 + (\mathbf{n}\times\mathbf{r}_v)\cdot\mathbf{r}_{vv}(v')^3
\end{aligned}
$$

that is:

$$
\begin{aligned}
\kappa_g = & (\mathbf{n}\times\mathbf{r}_u)\cdot\mathbf{r}_{uu}(u')^3 + 2(\mathbf{n}\times\mathbf{r}_u)\cdot\mathbf{r}_{uv}(u')^2 v' + (\mathbf{n}\times\mathbf{r}_u)\cdot\mathbf{r}_v u' v'' + \\
& (\mathbf{n}\times\mathbf{r}_u)\cdot\mathbf{r}_{vv} u' (v')^2 + (\mathbf{n}\times\mathbf{r}_v)\cdot\mathbf{r}_{uu}(u')^2 v' + (\mathbf{n}\times\mathbf{r}_v)\cdot\mathbf{r}_u u'' v' + \\
& 2(\mathbf{n}\times\mathbf{r}_v)\cdot\mathbf{r}_{uv} u' (v')^2 + (\mathbf{n}\times\mathbf{r}_v)\cdot\mathbf{r}_{vv}(v')^3
\end{aligned} \tag{32}
$$

Now, if we note that $\mathbf{r}_u \equiv \mathbf{E}_1$, $\mathbf{r}_v \equiv \mathbf{E}_2$, $\mathbf{r}_{uu} \equiv \partial_u \mathbf{E}_1$, $\mathbf{r}_{vv} \equiv \partial_v \mathbf{E}_2$ and $\mathbf{r}_{uv} \equiv \partial_v \mathbf{E}_1 = \partial_u \mathbf{E}_2 \equiv \mathbf{r}_{vu}$ and we use the Gauss equations,[25] then we have:

$$
\begin{aligned}
\mathbf{r}_{uu} &= \Gamma^1_{11}\mathbf{r}_u + \Gamma^2_{11}\mathbf{r}_v + e\mathbf{n} \\
\mathbf{r}_{uv} &= \Gamma^1_{12}\mathbf{r}_u + \Gamma^2_{12}\mathbf{r}_v + f\mathbf{n} \\
\mathbf{r}_{vv} &= \Gamma^1_{22}\mathbf{r}_u + \Gamma^2_{22}\mathbf{r}_v + g\mathbf{n} \\
(\mathbf{n}\times\mathbf{r}_u)\cdot\mathbf{r}_{uu} &= (\mathbf{n}\times\mathbf{r}_u)\cdot\left(\Gamma^1_{11}\mathbf{r}_u + \Gamma^2_{11}\mathbf{r}_v + e\mathbf{n}\right) \\
&= \Gamma^2_{11}(\mathbf{n}\times\mathbf{r}_u)\cdot\mathbf{r}_v \\
(\mathbf{n}\times\mathbf{r}_u)\cdot\mathbf{r}_{uv} &= (\mathbf{n}\times\mathbf{r}_u)\cdot\left(\Gamma^1_{12}\mathbf{r}_u + \Gamma^2_{12}\mathbf{r}_v + f\mathbf{n}\right) \\
&= \Gamma^2_{12}(\mathbf{n}\times\mathbf{r}_u)\cdot\mathbf{r}_v \\
(\mathbf{n}\times\mathbf{r}_u)\cdot\mathbf{r}_{vv} &= (\mathbf{n}\times\mathbf{r}_u)\cdot\left(\Gamma^1_{22}\mathbf{r}_u + \Gamma^2_{22}\mathbf{r}_v + g\mathbf{n}\right) \\
&= \Gamma^2_{22}(\mathbf{n}\times\mathbf{r}_u)\cdot\mathbf{r}_v \\
(\mathbf{n}\times\mathbf{r}_v)\cdot\mathbf{r}_{uu} &= (\mathbf{n}\times\mathbf{r}_v)\cdot\left(\Gamma^1_{11}\mathbf{r}_u + \Gamma^2_{11}\mathbf{r}_v + e\mathbf{n}\right) \\
&= \Gamma^1_{11}(\mathbf{n}\times\mathbf{r}_v)\cdot\mathbf{r}_u \\
&= -\Gamma^1_{11}(\mathbf{n}\times\mathbf{r}_u)\cdot\mathbf{r}_v \\
(\mathbf{n}\times\mathbf{r}_v)\cdot\mathbf{r}_{uv} &= (\mathbf{n}\times\mathbf{r}_v)\cdot\left(\Gamma^1_{12}\mathbf{r}_u + \Gamma^2_{12}\mathbf{r}_v + f\mathbf{n}\right) \\
&= \Gamma^1_{12}(\mathbf{n}\times\mathbf{r}_v)\cdot\mathbf{r}_u
\end{aligned}
$$

[25] Gauss equations are:

$$
\begin{aligned}
\partial_u \mathbf{E}_1 &= \Gamma^1_{11}\mathbf{E}_1 + \Gamma^2_{11}\mathbf{E}_2 + e\mathbf{n} \\
\partial_v \mathbf{E}_1 &= \Gamma^1_{12}\mathbf{E}_1 + \Gamma^2_{12}\mathbf{E}_2 + f\mathbf{n} = \partial_u \mathbf{E}_2 \\
\partial_v \mathbf{E}_2 &= \Gamma^1_{22}\mathbf{E}_1 + \Gamma^2_{22}\mathbf{E}_2 + g\mathbf{n}
\end{aligned}
$$

$$\begin{aligned}
&= -\Gamma_{12}^{1}\left(\mathbf{n}\times\mathbf{r}_u\right)\cdot\mathbf{r}_v\\
\left(\mathbf{n}\times\mathbf{r}_v\right)\cdot\mathbf{r}_{vv} &= \left(\mathbf{n}\times\mathbf{r}_v\right)\cdot\left(\Gamma_{22}^{1}\mathbf{r}_u+\Gamma_{22}^{2}\mathbf{r}_v+g\mathbf{n}\right)\\
&= \Gamma_{22}^{1}\left(\mathbf{n}\times\mathbf{r}_v\right)\cdot\mathbf{r}_u\\
&= -\Gamma_{22}^{1}\left(\mathbf{n}\times\mathbf{r}_u\right)\cdot\mathbf{r}_v
\end{aligned}$$

Moreover, we have:

$$\begin{aligned}
\left(\mathbf{n}\times\mathbf{r}_u\right)\cdot\mathbf{r}_v &\equiv \left(\mathbf{n}\times\mathbf{E}_1\right)\cdot\mathbf{E}_2\\
&= \mathbf{n}\cdot\left(\mathbf{E}_1\times\mathbf{E}_2\right)\\
&= \sqrt{a}\\
&= -\left(\mathbf{n}\times\mathbf{r}_v\right)\cdot\mathbf{r}_u
\end{aligned}$$

On substituting from the last equations into Eq. 32 we obtain:

$$\begin{aligned}
\kappa_g &= \sqrt{a}\Big[\Gamma_{11}^{2}\left(u'\right)^3+2\Gamma_{12}^{2}\left(u'\right)^2v'+u'v''+\Gamma_{22}^{2}u'\left(v'\right)^2\\
&\quad -\Gamma_{11}^{1}\left(u'\right)^2v'-u''v'-2\Gamma_{12}^{1}u'\left(v'\right)^2-\Gamma_{22}^{1}\left(v'\right)^3\Big]\\
&= \sqrt{a}\Big[\Gamma_{11}^{2}\left(u'\right)^3+\left(2\Gamma_{12}^{2}-\Gamma_{11}^{1}\right)\left(u'\right)^2v'+\\
&\quad \left(\Gamma_{22}^{2}-2\Gamma_{12}^{1}\right)u'\left(v'\right)^2-\Gamma_{22}^{1}\left(v'\right)^3+u'v''-u''v'\Big]
\end{aligned}$$

which is the required result (with minor notational differences).

15. Show that in any two orthogonal directions at a given point $P$ on a sufficiently smooth surface, the sum of the normal curvatures corresponding to these directions at $P$ is constant.
**Answer**: This is obviously true at umbilical points. So, we need to show that this is also valid at non-umbilical points.
According to Euler theorem (see Exercise 46) the normal curvature $\kappa_n$ at a given point $P$ on a surface of class $C^2$ in a given direction can be expressed as a combination of the principal curvatures, $\kappa_1$ and $\kappa_2$, at $P$ as:

$$\kappa_n = \kappa_1\cos^2\theta+\kappa_2\sin^2\theta$$

where $\theta$ is the angle between the principal direction of $\kappa_1$ at $P$ and the given direction. Now, if the angle between the principal direction of $\kappa_1$ at a given point and the given direction is $\theta$, then the angle between the principal direction of $\kappa_1$ at that point and the orthogonal direction to the given direction is $\frac{\pi}{2}-\theta$. Hence, if we label the normal curvature in the given direction as $\kappa_{n1}$ and in the orthogonal direction as $\kappa_{n2}$ then we have:

$$\begin{aligned}
\kappa_{n1}+\kappa_{n2} &= \kappa_1\cos^2\theta+\kappa_2\sin^2\theta+\kappa_1\cos^2\left(\frac{\pi}{2}-\theta\right)+\kappa_2\sin^2\left(\frac{\pi}{2}-\theta\right)\\
&= \kappa_1\cos^2\theta+\kappa_2\sin^2\theta+\kappa_1\sin^2\theta+\kappa_2\cos^2\theta
\end{aligned}$$

$$= \kappa_1 \left(\cos^2 \theta + \sin^2 \theta\right) + \kappa_2 \left(\cos^2 \theta + \sin^2 \theta\right)$$
$$= \kappa_1 + \kappa_2$$

which is a constant, as required.

16. Give a brief statement of the theorem of Meusnier outlining its significance. State this theorem in a second alternative form.
**Answer**: The theorem of Meusnier states that if $P$ is a given point on a sufficiently smooth surface $S$, then all curves on $S$ that pass through $P$ with the same tangential non-asymptotic direction at $P$ have the same normal curvature at $P$. The theorem may also be stated as: the curvature of any surface curve at a given point $P$ on the curve is equal in magnitude to the curvature of the normal section which is tangent to the curve at $P$ divided by the cosine of the angle between the principal normal vector to the curve at $P$ and the normal vector to the surface at $P$. The significance of Meusnier theorem is that the normal curvature at a given point on a smooth surface and in a given tangential direction is a property of the surface and not only a property of a particular curve passing through the given point and in the given direction.

17. Show that the osculating circles of all curves on a surface that pass through a given point and in a specific non-asymptotic direction are on a sphere.
**Answer**: Let have an arbitrary curve $C$ on a surface $S$ passing through a given point $P$ in a specific direction $d$ where $\mathbf{N}$ is the normal unit vector of $C$ at $P$, $\mathbf{n}$ is the normal unit vector of $S$ at $P$, $\kappa$ is the curvature of $C$ at $P$ and $\kappa_n$ is the normal curvature of $S$ at $P$ in the direction $d$. Now, let orient (with no loss of generality) $\mathbf{n}$ such that the angle $\phi$ between $\mathbf{N}$ and $\mathbf{n}$ is between 0 and $\pi/2$ (i.e. $0 \le \phi < \pi/2$). Since the direction is non-asymptotic then $\kappa_n \neq 0$, and hence from the relations:

$$\kappa_n = \mathbf{n} \cdot \mathbf{K} = \mathbf{n} \cdot (\kappa \mathbf{N}) = \kappa (\mathbf{n} \cdot \mathbf{N}) = \kappa \cos\phi \neq 0$$

we conclude that $\mathbf{K} \neq \mathbf{0}$, $\kappa \neq 0$ and $\cos\phi \neq 0$. Also, from the fact that $\kappa$ is non-negative and $0 \le \phi < \pi/2$ we conclude that $\kappa_n > 0$. So, if $R_\kappa = 1/\kappa$ and $R_n = 1/\kappa_n$ then from the relation $\kappa_n = \kappa \cos\phi$ we obtain:

$$R_\kappa = R_n \cos\phi$$

Now, $R_n$ is a constant (since the point $P$ and the direction $d$ are fixed and hence $\kappa_n$ is constant) that represents the radius of curvature of the normal section of $S$ at $P$ in the direction $d$. Hence, $R_n$ is the radius of a great circle of a sphere $S_s$ tangent to the tangent plane of $S$ at $P$. Moreover, $R_\kappa$ is the radius of curvature of $C$ at $P$ and hence it is the radius of the osculating circle of $C$ at $P$. The relation $R_\kappa = R_n \cos\phi$ then means that the osculating circle of $C$ at $P$ is the projection of that great circle onto the osculating plane of $C$ at $P$ and hence the osculating circle of $C$ at $P$ is the intersection of the osculating plane of $C$ at $P$ with the sphere $S_s$. Now, since $C$ is an arbitrary curve (within the given conditions) then this applies to all curves with those conditions, i.e. the osculating circles of all curves on a surface that pass through a given point and in a specific non-asymptotic direction are on a sphere, as required.

18. Define, descriptively and mathematically, the normal component $\mathbf{K}_n$ of the curvature vector $\mathbf{K}$ of a surface curve outlining its relation to the curvature vector and the normal vector to the surface, $\mathbf{n}$.
**Answer**: $\mathbf{K}_n$ is the projection of the curvature vector $\mathbf{K}$ of a surface curve $C$ at a given point $P$ onto the orientation of the normal unit vector $\mathbf{n}$ of the surface at $P$. Accordingly, it is given by:

$$\mathbf{K}_n = (\mathbf{n} \cdot \mathbf{K})\,\mathbf{n} = \kappa_n \mathbf{n} = (\kappa \cos\phi)\,\mathbf{n}$$

i.e. it is a vector whose magnitude is $|\mathbf{n} \cdot \mathbf{K}| = |\kappa_n| = \kappa\,|\cos\phi|$ and whose direction is $\operatorname{sgn}(\cos\phi)\,\mathbf{n}$ where $\kappa$ is the curvature of $C$ at $P$, $\phi$ is the angle between $\mathbf{n}$ and the normal unit vector $\mathbf{N}$ of $C$ at $P$, $\kappa_n$ is the normal curvature of $C$ at $P$, and $\operatorname{sgn}(\cos\phi)$ is the sign function of $\cos\phi$.

19. Define, descriptively and mathematically, the geodesic component $\mathbf{K}_g$ of the curvature vector $\mathbf{K}$ of a surface curve outlining its relation to the curvature vector and the surface basis vectors $\mathbf{E}_1$ and $\mathbf{E}_2$.
**Answer**: $\mathbf{K}_g$ is the projection of the curvature vector $\mathbf{K}$ of a surface curve $C$ at a given point $P$ onto the orientation of the geodesic normal vector $\mathbf{u}$ to the curve at $P$. Accordingly, it is given by:

$$\mathbf{K}_g = (\mathbf{u} \cdot \mathbf{K})\,\mathbf{u} = \kappa_g \mathbf{u} = (\kappa \sin\phi)\,\mathbf{u}$$

i.e. it is a vector whose magnitude is $|\mathbf{u} \cdot \mathbf{K}| = |\kappa_g| = \kappa\,|\sin\phi|$ and whose direction is $\operatorname{sgn}(\sin\phi)\,\mathbf{u}$ where $\kappa$ is the curvature of $C$ at $P$, $\phi$ is the angle between $\mathbf{n}$ and the normal unit vector $\mathbf{N}$ of $C$ at $P$, $\kappa_g$ is the geodesic curvature of $C$ at $P$, and $\operatorname{sgn}(\sin\phi)$ is the sign function of $\sin\phi$. Now, $\mathbf{u}$ is a normalized projection of $\mathbf{K}$ onto the tangent space of the surface at $P$ and hence it can be expressed as a linear combination of the surface basis vectors $\mathbf{E}_1$ and $\mathbf{E}_2$ at $P$. This means that $\mathbf{K}_g$ is also a linear combination of $\mathbf{E}_1$ and $\mathbf{E}_2$ and hence it belongs to the tangent space.

20. Derive the formula for the normal curvature $\kappa_n$ as a ratio of the second fundamental form to the first fundamental form.
**Answer**: If we parameterize the surface curve with a natural parameter $s$ then we have:

$$\begin{aligned}
\kappa_n &= \mathbf{n} \cdot \mathbf{K} \\
&= -\mathbf{T} \cdot \frac{d\mathbf{n}}{ds} \\
&= -\left(\frac{\partial \mathbf{r}}{\partial u^\alpha}\frac{du^\alpha}{ds}\right) \cdot \left(\frac{\partial \mathbf{n}}{\partial u^\beta}\frac{du^\beta}{ds}\right) \\
&= -\left(\frac{\partial \mathbf{r}}{\partial u^\alpha} \cdot \frac{\partial \mathbf{n}}{\partial u^\beta}\right)\frac{du^\alpha}{ds}\frac{du^\beta}{ds} \\
&= \frac{-\left(\frac{\partial \mathbf{r}}{\partial u^\alpha} \cdot \frac{\partial \mathbf{n}}{\partial u^\beta}\right) du^\alpha du^\beta}{ds\,ds}
\end{aligned}$$

$$= \frac{b_{\alpha\beta}du^{\alpha}du^{\beta}}{(ds)^2}$$

$$= \frac{b_{\alpha\beta}du^{\alpha}du^{\beta}}{a_{\gamma\delta}du^{\gamma}du^{\delta}}$$

$$= \frac{II_S}{I_S}$$

where line 1 is a definition, line 2 is based on the fact that:

$$(\mathbf{n}\cdot\mathbf{T})' = (0)' = 0 = \mathbf{n}'\cdot\mathbf{T} + \mathbf{n}\cdot\mathbf{T}' = \mathbf{n}'\cdot\mathbf{T} + \mathbf{n}\cdot\mathbf{K}$$

and hence $\mathbf{n}\cdot\mathbf{K} = -\mathbf{n}'\cdot\mathbf{T} = -\mathbf{T}\cdot\mathbf{n}' \equiv -\mathbf{T}\cdot\frac{d\mathbf{n}}{ds}$, line 3 is based on the definition of $\mathbf{T}$ (i.e. $\mathbf{T} = \mathbf{r}'$) and the chain rule of differentiation, lines 4 and 5 are algebraic manipulation, line 6 is based on the definition of the covariant curvature tensor, line 7 is based on the expression of the infinitesimal line element $ds$, and line 8 is based on the definition of the first and second fundamental forms.

21. Why the sign of the normal curvature $\kappa_n$ is determined only by the sign of the second fundamental form?
**Answer**: Because $\kappa_n = II_S/I_s$ and the first fundamental form $I_S$ is positive definite. Hence, the sign of $\kappa_n$ is solely determined by the sign of the second fundamental form $II_S$.

22. Show that at any point $P$ of a smooth surface $S$ there exists a paraboloid tangent to $S$ at $P$ such that the normal curvature of the paraboloid in any direction is equal to the normal curvature of $S$ at $P$ in that direction.
**Answer**: We can assume (with no loss of generality) that the surface $S$ is coordinated by an orthonormal Cartesian coordinate system where the point $P$ is at the origin $O$ and the normal unit vector $\mathbf{n}$ is in the positive $z$ direction, i.e. $\mathbf{n}(0,0) = (0,0,1)$. Moreover, the surface is spatially represented by the parameterization $\mathbf{r}(u,v) = (u,v,f(u,v))$ such that at $O$ we have $\mathbf{r}_u(0,0) = (1,0,0)$ and $\mathbf{r}_v(0,0) = (0,1,0)$.[26] Now, in the close proximity of $P$ (or $O$) we can use the Taylor approximation and hence $S$ can be represented as:

$$\mathbf{r} \simeq \left(u, v, \frac{Au^2 + 2Buv + Cv^2}{2}\right) \tag{33}$$

where $\mathbf{r}_{uu}(0,0) = (0,0,A)$, $\mathbf{r}_{uv}(0,0) = (0,0,B)$ and $\mathbf{r}_{vv}(0,0) = (0,0,C)$. We note that the surface represented by Eq. 33 as an approximation to $S$ is a paraboloid tangent to the $xy$ plane and hence it is also tangent to $S$ at $P$ (i.e. $O$). The normal curvature $\kappa_n$ of this paraboloid at $P$ in any particular direction (which should also be the normal curvature of $S$ at $P$) is given by:

$$\kappa_n = \frac{II_S}{I_S} = \frac{e(du)^2 + 2f\,dudv + g(dv)^2}{E(du)^2 + 2F\,dudv + G(dv)^2} \tag{34}$$

[26] The direction of coordinates is a matter of choice (within certain restrictions) while the magnitude can be achieved by scaling. Alternatively, 1 can be replaced by a scalar with minor amendments to the subsequent formulation.

Now, at $P$ we have:

$$\begin{aligned}
E &= \mathbf{E}_1 \cdot \mathbf{E}_1 \equiv \mathbf{r}_u \cdot \mathbf{r}_u = (1,0,0) \cdot (1,0,0) = 1 \\
F &= \mathbf{E}_1 \cdot \mathbf{E}_2 \equiv \mathbf{r}_u \cdot \mathbf{r}_v = (1,0,0) \cdot (0,1,0) = 0 \\
G &= \mathbf{E}_2 \cdot \mathbf{E}_2 \equiv \mathbf{r}_v \cdot \mathbf{r}_v = (0,1,0) \cdot (0,1,0) = 1 \\
e &= \mathbf{n} \cdot \partial_u \mathbf{E}_1 \equiv \mathbf{n} \cdot \mathbf{r}_{uu} = (0,0,1) \cdot (0,0,A) = A \\
f &= \mathbf{n} \cdot \partial_v \mathbf{E}_1 \equiv \mathbf{n} \cdot \mathbf{r}_{uv} = (0,0,1) \cdot (0,0,B) = B \\
g &= \mathbf{n} \cdot \partial_v \mathbf{E}_2 \equiv \mathbf{n} \cdot \mathbf{r}_{vv} = (0,0,1) \cdot (0,0,C) = C
\end{aligned}$$

On substituting from the last equations into Eq. 34 we get:

$$\kappa_n = \frac{A(du)^2 + 2B\,dudv + C(dv)^2}{(du)^2 + (dv)^2}$$

So, in the close proximity of $P$ the normal curvature of the paraboloid should approximates the normal curvature of $S$ (since the paraboloid itself approximates $S$ there) and this approximation should become exact (as a limit) at $P$ (i.e. $\kappa_n$ as given by the last equation is the normal curvature of $S$ as well as the paraboloid).
To sum up, we conclude that there exists a paraboloid (i.e. the surface represented by Eq. 33) tangent to $S$ at $P$ such that the normal curvature of the paraboloid in any direction is equal to the normal curvature of $S$ at $P$ in that direction, as required.

23. Discuss, in detail, the following statement: "The normal curvature at a given point on a surface and in a given tangential direction to the surface is a property of the surface". Can you link this to the Meusnier theorem?
**Answer**: Because all the surface curves that pass through a given point on the surface and in a given tangential direction to the surface have the same normal curvature, then the normal curvature is not a property of the individual curves only but it is also a property of the surface itself at that point and in that direction. In fact, this is the essence of the Meusnier theorem which states "if $P$ is a given point on a sufficiently smooth surface $S$, then all curves on $S$ that pass through $P$ with the same tangential non-asymptotic direction at $P$ have the same normal curvature at $P$" (see Exercise 16).

24. For what type of surface curve the following relation is true: $|\kappa_n| = \kappa$? Explain why this is so.
**Answer**: This relation is true for normal sections because these curves have no geodesic curvature and hence their curvature $\kappa$ is totally normal, i.e. $\kappa = |\kappa_n|$. Taking the absolute value of $\kappa_n$ is because by definition $\kappa$ is non-negative while the sign of $\kappa_n$ depends on the orientation of the surface which depends on the choice of the direction of the normal unit vector $\mathbf{n}$.

25. What is the significance of having a paraboloid at the points of a smooth surface whose normal curvature in a given direction is equal to the normal curvature of the surface in that direction?

**Answer**: The significance is that a smooth surface can be locally approximated as a quadratic surface which is inline with the Taylor approximation as explained in Exercise 22.

26. What is the sign of $b$ (i.e. being greater than, less than or equal to zero) at flat, elliptic, parabolic and hyperbolic points on a surface, where $b$ is the determinant of the surface covariant curvature tensor?
**Answer**: $b > 0$ at elliptic points, $b < 0$ at hyperbolic points, and $b = 0$ at flat and parabolic points.

27. Classify the local shape of a surface at a given point $P$ according to the values of $K$ and $H$ at $P$.
**Answer**: If $K = H = 0$ the point is flat. If $K = 0$ and $H \neq 0$ the point is parabolic. If $K > 0$ the point is elliptic. If $K < 0$ the point is hyperbolic.

28. Using one of the mathematical definitions of the geodesic curvature $\kappa_g$, explain why $\kappa_g$ should be classified as an intrinsic or extrinsic property.
**Answer**: If we look to Eq. 31 we see that $\kappa_g$ is defined in terms of the determinant of the metric tensor and the Christoffel symbols. Since both of these are exclusively defined in terms of the coefficients of the metric tensor (which are the same as the coefficients of the first fundamental form) and their derivatives, then $\kappa_g$ should be an intrinsic property.

29. At what type of surface points the following relation is true: $\frac{e}{E} = \frac{f}{F} = \frac{g}{G} = c$ where $c$ is constant for all directions? What $c$ stands for?
**Answer**: At umbilical points. The constant $c$ stands for the normal curvature $\kappa_n$.

30. Express the equalities in the previous question in terms of the coefficients of the covariant metric and covariant curvature tensors, $a_{\alpha\beta}$ and $b_{\alpha\beta}$, of the surface.
**Answer**:

$$\frac{b_{11}}{a_{11}} = \frac{b_{12}}{a_{12}} = \frac{b_{22}}{a_{22}}$$

31. Outline two direct consequences of Meusnier theorem.
**Answer**: Referring to the statement of the Meusnier theorem (see Exercises 16 and 23) and to Exercise 17, we note the following:
• The center of the sphere $S_s$ is the center of curvature of the normal section at $P$ in the given direction.
• The osculating circles of all the tangent curves are the intersection of the sphere $S_s$ with the osculating planes of these curves at $P$.

32. Write a mathematical relation linking the geodesic component $\mathbf{K}_g$ of the curvature vector to the surface basis vectors $\mathbf{E}_1$ and $\mathbf{E}_2$.
**Answer**:

$$\mathbf{K}_g = \left(\frac{d^2u^1}{ds^2} + \Gamma^1_{\alpha\beta}\frac{du^\alpha}{ds}\frac{du^\beta}{ds}\right)\mathbf{E}_1 + \left(\frac{d^2u^2}{ds^2} + \Gamma^2_{\alpha\beta}\frac{du^\alpha}{ds}\frac{du^\beta}{ds}\right)\mathbf{E}_2$$

33. What is the relation between $\mathbf{K}_g$ and the tangent space $T_PS$ of the surface at a given point?
**Answer**: The geodesic component $\mathbf{K}_g$ of the curvature vector $\mathbf{K}$ of a surface curve at a given point $P$ is the projection of $\mathbf{K}$ onto the tangent space $T_PS$ of the surface at $P$. Hence, $\mathbf{K}_g$ is embedded in the tangent space.

34. Give the formulae of the geodesic curvature $\kappa_g$ of the coordinate curves. Simplify these formulae in the case of having orthogonal coordinate curves.
**Answer**:

$$\begin{aligned}\kappa_{gu} &= \frac{\sqrt{a}}{E^{3/2}}\Gamma^2_{11} \\ \kappa_{gv} &= -\frac{\sqrt{a}}{G^{3/2}}\Gamma^1_{22}\end{aligned}$$

where $\kappa_{gu}, \kappa_{gv}$ are the geodesic curvatures of the $u, v$ coordinate curves, $a$ is the determinant of the surface covariant metric tensor, $E, G$ are coefficients of the first fundamental form, and $\Gamma^2_{11}, \Gamma^1_{22}$ are Christoffel symbols of the second kind for the surface.
In the case of orthogonal coordinate curves, the above formulae will simplify to:

$$\begin{aligned}\kappa_{gu} &= -\frac{E_v}{2E\sqrt{G}} \\ \kappa_{gv} &= \frac{G_u}{2G\sqrt{E}}\end{aligned}$$

35. State a mathematical relation between the curvature $\kappa$ and the geodesic curvature $\kappa_g$ of a surface curve at a given point $P$ on the curve explaining all the symbols involved.
**Answer**:

$$\kappa_g = \kappa \sin\phi \tag{35}$$

where $\phi$ is the angle between the principal normal vector $\mathbf{N}$ of the curve at $P$ and the unit normal vector $\mathbf{n}$ of the surface at $P$.

36. Give a formula for the geodesic curvature $\kappa_g$ in which extrinsic entities are involved. Does this mean that $\kappa_g$ is an extrinsic property of the surface?
**Answer**:

$$\kappa_g = \frac{\ddot{\mathbf{r}} \cdot (\mathbf{n} \times \dot{\mathbf{r}})}{\dot{\mathbf{r}} \cdot \dot{\mathbf{r}}}$$

This does not mean that $\kappa_g$ is an extrinsic property because $\kappa_g$ can be defined by purely intrinsic parameters, as we saw for example in Exercise 28.

37. Explain, in detail, all the symbols used in the following formula:

$$\kappa_g = \frac{d\theta}{ds} + \kappa_{gu}\cos\theta + \kappa_{gv}\sin\theta$$

What is the name of this formula?
**Answer**: $\kappa_g$ is the geodesic curvature of a surface curve $C$, $\kappa_{gu}$ and $\kappa_{gv}$ are the geodesic

curvatures of the $u$ and $v$ coordinate curves, $s$ is a natural parameter of $C$, and $\theta$ is a parameter defined by:

$$\mathbf{T} = \frac{\mathbf{E}_1}{|\mathbf{E}_1|}\cos\theta + \frac{\mathbf{E}_2}{|\mathbf{E}_2|}\sin\theta \tag{36}$$

where $\mathbf{T}$ is the tangent unit vector of $C$, and $\mathbf{E}_1$ and $\mathbf{E}_2$ are the surface basis vectors. All the given quantities are evaluated at a given point on the curve and the coordinate curves are assumed orthogonal. This formula is known as Liouville formula.

38. Prove the relation given in the last question.
**Answer**:[27] Let $s$, $s_1$ and $s_2$ symbolize natural parameters along the curve, along the $u$ coordinate curve and along the $v$ coordinate curve. Also, to simplify the notation we use (in this exercise only) $\mathbf{e}_1$ to symbolize the unit vector $\mathbf{E}_1/|\mathbf{E}_1|$ and $\mathbf{e}_2$ to symbolize the unit vector $\mathbf{E}_2/|\mathbf{E}_2|$. We also note that $\mathbf{E}_1 \equiv \mathbf{r}_u$ and $\mathbf{E}_2 \equiv \mathbf{r}_v$ (with $\mathbf{r}$ being the spatial representation of the surface and the subscripts $u$ and $v$ representing partial derivative with respect to these coordinates) and therefore we use these alternative notations as convenient. Hence, we have:

$$\begin{aligned}
\mathbf{e}_1 &\equiv \frac{\mathbf{E}_1}{|\mathbf{E}_1|} = \frac{\mathbf{E}_1}{\sqrt{\mathbf{E}_1\cdot\mathbf{E}_1}} = \frac{\mathbf{E}_1}{\sqrt{E}} \\
\mathbf{e}_2 &\equiv \frac{\mathbf{E}_2}{|\mathbf{E}_2|} = \frac{\mathbf{E}_2}{\sqrt{\mathbf{E}_2\cdot\mathbf{E}_2}} = \frac{\mathbf{E}_2}{\sqrt{G}}
\end{aligned}$$

Since the coordinate curves are orthogonal (see the previous exercise and the book), then from Eq. 36 and the last two equations we get:

$$\begin{aligned}
\cos\theta &= \frac{\mathbf{E}_1}{|\mathbf{E}_1|}\cdot\mathbf{T} = \frac{\mathbf{r}_u}{|\mathbf{r}_u|}\cdot\left(\mathbf{r}_u\frac{du}{ds}+\mathbf{r}_v\frac{dv}{ds}\right) = \frac{\mathbf{r}_u\cdot\mathbf{r}_u}{|\mathbf{r}_u|}\frac{du}{ds} = |\mathbf{r}_u|\frac{du}{ds} \\
\sin\theta &= \frac{\mathbf{E}_2}{|\mathbf{E}_2|}\cdot\mathbf{T} = \frac{\mathbf{r}_v}{|\mathbf{r}_v|}\cdot\left(\mathbf{r}_u\frac{du}{ds}+\mathbf{r}_v\frac{dv}{ds}\right) = \frac{\mathbf{r}_v\cdot\mathbf{r}_v}{|\mathbf{r}_v|}\frac{dv}{ds} = |\mathbf{r}_v|\frac{dv}{ds}
\end{aligned}$$

On differentiating $\mathbf{e}_1$ and $\mathbf{e}_2$ with respect to $s$ we get:[28]

$$\begin{aligned}
\frac{d\mathbf{e}_1}{ds} &= \frac{\partial\mathbf{e}_1}{\partial u}\frac{du}{ds} + \frac{\partial\mathbf{e}_1}{\partial v}\frac{dv}{ds} \\
&= \frac{\partial\mathbf{e}_1}{\partial s_1}\frac{ds_1}{du}\frac{du}{ds} + \frac{\partial\mathbf{e}_1}{\partial s_2}\frac{ds_2}{dv}\frac{dv}{ds} \\
&= \frac{\partial\mathbf{e}_1}{\partial s_1}|\mathbf{r}_u|\frac{du}{ds} + \frac{\partial\mathbf{e}_1}{\partial s_2}|\mathbf{r}_v|\frac{dv}{ds}
\end{aligned}$$

[27] In this answer (like in several other answers in this book) I follow Lipschutz book which is cited in the References of my book.

[28] We note that the relations $\frac{ds_1}{du} = |\mathbf{r}_u|$ and $\frac{ds_2}{dv} = |\mathbf{r}_v|$ are essentially the same as the relation $\left|\frac{ds}{dt}\right| = \left|\frac{d\mathbf{r}}{dt}\right|$ (which we proved in Exercise 20) with $u$ and $v$ standing for $t$ and noting that along the $u$ coordinate curves $\mathbf{r}$ solely depends on $u$ and along the $v$ coordinate curves $\mathbf{r}$ solely depends on $v$. We also note that by proper parameterization $\frac{ds_1}{du}$ and $\frac{ds_2}{dv}$ can be made positive.

$$= \frac{\partial \mathbf{e}_1}{\partial s_1}\cos\theta + \frac{\partial \mathbf{e}_1}{\partial s_2}\sin\theta$$

Similarly:

$$\frac{d\mathbf{e}_2}{ds} = \frac{\partial \mathbf{e}_2}{\partial s_1}\cos\theta + \frac{\partial \mathbf{e}_2}{\partial s_2}\sin\theta$$

Hence:

$$\begin{aligned}
\mathbf{K} &= \frac{d\mathbf{T}}{ds} \\
&= \frac{d}{ds}\left(\mathbf{e}_1\cos\theta + \mathbf{e}_2\sin\theta\right) \\
&= \frac{d\mathbf{e}_1}{ds}\cos\theta + \mathbf{e}_1\frac{d\cos\theta}{ds} + \frac{d\mathbf{e}_2}{ds}\sin\theta + \mathbf{e}_2\frac{d\sin\theta}{ds} \\
&= \left(\frac{\partial \mathbf{e}_1}{\partial s_1}\cos\theta + \frac{\partial \mathbf{e}_1}{\partial s_2}\sin\theta\right)\cos\theta - \mathbf{e}_1\sin\theta\frac{d\theta}{ds} + \\
&\quad \left(\frac{\partial \mathbf{e}_2}{\partial s_1}\cos\theta + \frac{\partial \mathbf{e}_2}{\partial s_2}\sin\theta\right)\sin\theta + \mathbf{e}_2\cos\theta\frac{d\theta}{ds} \\
&= \frac{\partial \mathbf{e}_1}{\partial s_1}\cos^2\theta + \frac{\partial \mathbf{e}_1}{\partial s_2}\sin\theta\cos\theta - \mathbf{e}_1\sin\theta\frac{d\theta}{ds} + \\
&\quad \frac{\partial \mathbf{e}_2}{\partial s_1}\sin\theta\cos\theta + \frac{\partial \mathbf{e}_2}{\partial s_2}\sin^2\theta + \mathbf{e}_2\cos\theta\frac{d\theta}{ds} \\
&= \frac{\partial \mathbf{e}_1}{\partial s_1}\cos^2\theta + \left(\frac{\partial \mathbf{e}_1}{\partial s_2} + \frac{\partial \mathbf{e}_2}{\partial s_1}\right)\sin\theta\cos\theta + \frac{\partial \mathbf{e}_2}{\partial s_2}\sin^2\theta + \\
&\quad \left(-\mathbf{e}_1\sin\theta + \mathbf{e}_2\cos\theta\right)\frac{d\theta}{ds} \\
&= \frac{\partial \mathbf{e}_1}{\partial s_1}\cos^2\theta + \left(\frac{\partial \mathbf{e}_1}{\partial s_2} + \frac{\partial \mathbf{e}_2}{\partial s_1}\right)\sin\theta\cos\theta + \frac{\partial \mathbf{e}_2}{\partial s_2}\sin^2\theta + \mathbf{u}\frac{d\theta}{ds}
\end{aligned}$$

The last equality is justified by:

$$\begin{aligned}
\mathbf{u} &= \mathbf{n}\times\mathbf{T} \\
&= \left(\mathbf{e}_1\times\mathbf{e}_2\right)\times\mathbf{T} \\
&= -\left[\mathbf{T}\times\left(\mathbf{e}_1\times\mathbf{e}_2\right)\right] \\
&= -\left[\mathbf{e}_1\left(\mathbf{T}\cdot\mathbf{e}_2\right) - \mathbf{e}_2\left(\mathbf{T}\cdot\mathbf{e}_1\right)\right] \\
&= -\mathbf{e}_1\sin\theta + \mathbf{e}_2\cos\theta
\end{aligned}$$

where we used Eq. 36 plus some identities from vector calculus $\left[\text{i.e. } \mathbf{A}\times(\mathbf{B}\times\mathbf{C}) = \mathbf{B}(\mathbf{A}\cdot\mathbf{C}) - \mathbf{C}(\mathbf{A}\cdot\mathbf{B})\right]$. Now, the geodesic curvature is given by:

$$\begin{aligned}
\kappa_g &= \mathbf{u}\cdot\mathbf{K} \\
&= \mathbf{u}\cdot\left[\frac{\partial \mathbf{e}_1}{\partial s_1}\cos^2\theta + \left(\frac{\partial \mathbf{e}_1}{\partial s_2} + \frac{\partial \mathbf{e}_2}{\partial s_1}\right)\sin\theta\cos\theta + \frac{\partial \mathbf{e}_2}{\partial s_2}\sin^2\theta + \mathbf{u}\frac{d\theta}{ds}\right]
\end{aligned}$$

$$
\begin{aligned}
&= \mathbf{u}\cdot\left(\frac{\partial \mathbf{e}_1}{\partial s_1}\right)\cos^2\theta + \mathbf{u}\cdot\left(\frac{\partial \mathbf{e}_1}{\partial s_2}+\frac{\partial \mathbf{e}_2}{\partial s_1}\right)\sin\theta\cos\theta + \mathbf{u}\cdot\left(\frac{\partial \mathbf{e}_2}{\partial s_2}\right)\sin^2\theta + \frac{d\theta}{ds}\\
&= (-\mathbf{e}_1\sin\theta + \mathbf{e}_2\cos\theta)\cdot\left(\frac{\partial \mathbf{e}_1}{\partial s_1}\right)\cos^2\theta +\\
&\quad (-\mathbf{e}_1\sin\theta + \mathbf{e}_2\cos\theta)\cdot\left(\frac{\partial \mathbf{e}_1}{\partial s_2}+\frac{\partial \mathbf{e}_2}{\partial s_1}\right)\sin\theta\cos\theta +\\
&\quad (-\mathbf{e}_1\sin\theta + \mathbf{e}_2\cos\theta)\cdot\left(\frac{\partial \mathbf{e}_2}{\partial s_2}\right)\sin^2\theta + \frac{d\theta}{ds}\\
&= 0 + \mathbf{e}_2\cdot\left(\frac{\partial \mathbf{e}_1}{\partial s_1}\right)\cos^3\theta +\\
&\quad 0 - \mathbf{e}_1\cdot\left(\frac{\partial \mathbf{e}_2}{\partial s_1}\right)\sin^2\theta\cos\theta + \mathbf{e}_2\cdot\left(\frac{\partial \mathbf{e}_1}{\partial s_2}\right)\sin\theta\cos^2\theta + 0\\
&\quad -\mathbf{e}_1\cdot\left(\frac{\partial \mathbf{e}_2}{\partial s_2}\right)\sin^3\theta + 0 + \frac{d\theta}{ds}
\end{aligned}
$$

where the vanishing terms are justified by:[29]

$$
\mathbf{e}_1\cdot\left(\frac{\partial \mathbf{e}_1}{\partial s_1}\right) = \mathbf{e}_1\cdot\left(\frac{\partial \mathbf{e}_1}{\partial s_2}\right) = \mathbf{e}_2\cdot\left(\frac{\partial \mathbf{e}_2}{\partial s_1}\right) = \mathbf{e}_2\cdot\left(\frac{\partial \mathbf{e}_2}{\partial s_2}\right) = 0
$$

So:

$$
\begin{aligned}
\kappa_g &= \mathbf{e}_2\cdot\left(\frac{\partial \mathbf{e}_1}{\partial s_1}\right)\cos^3\theta - \mathbf{e}_1\cdot\left(\frac{\partial \mathbf{e}_2}{\partial s_1}\right)\sin^2\theta\cos\theta + \mathbf{e}_2\cdot\left(\frac{\partial \mathbf{e}_1}{\partial s_2}\right)\sin\theta\cos^2\theta \qquad (37)\\
&\quad -\mathbf{e}_1\cdot\left(\frac{\partial \mathbf{e}_2}{\partial s_2}\right)\sin^3\theta + \frac{d\theta}{ds}
\end{aligned}
$$

Now, if we note that:[30]

$$
\begin{aligned}
\kappa_{gu} &= \mathbf{e}_2\cdot\left(\frac{\partial \mathbf{e}_1}{\partial s_1}\right) = -\mathbf{e}_1\cdot\left(\frac{\partial \mathbf{e}_2}{\partial s_1}\right)\\
\kappa_{gv} &= \mathbf{e}_2\cdot\left(\frac{\partial \mathbf{e}_1}{\partial s_2}\right) = -\mathbf{e}_1\cdot\left(\frac{\partial \mathbf{e}_2}{\partial s_2}\right)
\end{aligned}
$$

then Eq. 37 becomes:

$$
\kappa_g = \kappa_{gu}\cos^3\theta + \kappa_{gu}\sin^2\theta\cos\theta + \kappa_{gv}\sin\theta\cos^2\theta + \kappa_{gv}\sin^3\theta + \frac{d\theta}{ds}
$$

[29] We note that the derivative of unit vector is perpendicular to the vector.

[30] We note that these equations are from Eq. 37 with $\theta = 0$ and $\theta = \pi/2$ respectively. We also note that the relations $\mathbf{e}_2\cdot\left(\frac{\partial \mathbf{e}_1}{\partial s_1}\right) = -\mathbf{e}_1\cdot\left(\frac{\partial \mathbf{e}_2}{\partial s_1}\right)$ and $\mathbf{e}_2\cdot\left(\frac{\partial \mathbf{e}_1}{\partial s_2}\right) = -\mathbf{e}_1\cdot\left(\frac{\partial \mathbf{e}_2}{\partial s_2}\right)$ are justified by the fact that $\mathbf{e}_1\cdot\mathbf{e}_2 = 0$ (since the coordinates are orthogonal) and hence by taking the derivative of both sides of this equation (with respect to $s_1$ or $s_2$) these results are obtained.

$$= \kappa_{gu}\cos\theta\left(\cos^2\theta+\sin^2\theta\right)+\kappa_{gv}\sin\theta\left(\cos^2\theta++\sin^2\theta\right)+\frac{d\theta}{ds}$$
$$= \kappa_{gu}\cos\theta+\kappa_{gv}\sin\theta+\frac{d\theta}{ds}$$

as required.

39. Give a mathematical formula in which $\kappa_n$ is expressed in terms of the coefficients of the first and second fundamental forms $E, F, G, e, f, g$.
**Answer**:
$$\kappa_n = \frac{e(du)^2+2f\,dudv+g(dv)^2}{E(du)^2+2F\,dudv+G(dv)^2}$$

40. What the two "principal curvatures" of a surface at a given point mean?
**Answer**: The two principal curvatures, $\kappa_1$ and $\kappa_2$, of a surface $S$ at a given point $P$ represent respectively the maximum and minimum values of the normal curvature $\kappa_n$ of $S$ at $P$.

41. Find analytical expressions for the principal curvatures on a surface represented by the equation: $\xi_2\cos\xi_3-\xi_1\sin\xi_3=0$ where $\xi_1,\xi_2,\xi_3$ are mutually independent real variables.
**Answer**: This surface is represented spatially by $(\xi_1,\xi_2,\xi_3)$ such that $\xi_2\cos\xi_3-\xi_1\sin\xi_3=0$, and hence $\xi_1,\xi_2,\xi_3$ are space coordinates. From the equation $\xi_2\cos\xi_3-\xi_1\sin\xi_3=0$, we get:

$$\begin{aligned}\xi_2\cos\xi_3 &= \xi_1\sin\xi_3\\ \tan\xi_3 &= \xi_2/\xi_1\\ \xi_3 &= \arctan(\xi_2/\xi_1)\end{aligned}$$

Hence, the surface can be represented spatially as $\mathbf{r}(\xi_1,\xi_2)=(\xi_1,\xi_2,\arctan(\xi_2/\xi_1))$. Accordingly, we have:

$$\begin{aligned}
\mathbf{E}_1 &= \partial_1\mathbf{r}=\left(1,0,\frac{-\xi_2}{\xi_1^2\left[1+(\xi_2/\xi_1)^2\right]}\right)=\left(1,0,\frac{-\xi_2}{\xi_1^2+\xi_2^2}\right)\\
\mathbf{E}_2 &= \partial_2\mathbf{r}=\left(0,1,\frac{1}{\xi_1\left[1+(\xi_2/\xi_1)^2\right]}\right)=\left(0,1,\frac{\xi_1}{\xi_1^2+\xi_2^2}\right)\\
\mathbf{n} &= \frac{\mathbf{E}_1\times\mathbf{E}_2}{|\mathbf{E}_1\times\mathbf{E}_2|}=\frac{\left(\frac{\xi_2}{\xi_1^2+\xi_2^2},\frac{-\xi_1}{\xi_1^2+\xi_2^2},1\right)}{\left(\frac{\xi_2^2+\xi_1^2+\xi_1^4+2\xi_1^2\xi_2^2+\xi_2^4}{\left(\xi_1^2+\xi_2^2\right)^2}\right)^{1/2}}\\
&= \frac{\left(\xi_2,-\xi_1,\xi_1^2+\xi_2^2\right)}{\left(\xi_2^2+\xi_1^2+\xi_1^4+2\xi_1^2\xi_2^2+\xi_2^4\right)^{1/2}}\\
\partial_1\mathbf{E}_1 &= \left(0,0,\frac{2\xi_1\xi_2}{\left(\xi_1^2+\xi_2^2\right)^2}\right)
\end{aligned}$$

$$\begin{aligned}
\partial_2 \mathbf{E}_1 &= \left(0, 0, \frac{-\xi_1^2 + \xi_2^2}{\left(\xi_1^2 + \xi_2^2\right)^2}\right) \\
\partial_2 \mathbf{E}_2 &= \left(0, 0, \frac{-2\xi_1 \xi_2}{\left(\xi_1^2 + \xi_2^2\right)^2}\right) \\
E &= \mathbf{E}_1 \cdot \mathbf{E}_1 = 1 + \frac{\xi_2^2}{\left(\xi_1^2 + \xi_2^2\right)^2} \\
F &= \mathbf{E}_1 \cdot \mathbf{E}_2 = \frac{-\xi_1 \xi_2}{\left(\xi_1^2 + \xi_2^2\right)^2} \\
G &= \mathbf{E}_2 \cdot \mathbf{E}_2 = 1 + \frac{\xi_1^2}{\left(\xi_1^2 + \xi_2^2\right)^2} \\
e &= \mathbf{n} \cdot \partial_1 \mathbf{E}_1 = \frac{\left(\xi_2, -\xi_1, \xi_1^2 + \xi_2^2\right)}{\left(\xi_2^2 + \xi_1^2 + \xi_1^4 + 2\xi_1^2 \xi_2^2 + \xi_2^4\right)^{1/2}} \cdot \left(0, 0, \frac{2\xi_1 \xi_2}{\left(\xi_1^2 + \xi_2^2\right)^2}\right) \\
&= \frac{2\xi_1 \xi_2}{\left(\xi_2^2 + \xi_1^2 + \xi_1^4 + 2\xi_1^2 \xi_2^2 + \xi_2^4\right)^{1/2} \left(\xi_1^2 + \xi_2^2\right)} \\
f &= \mathbf{n} \cdot \partial_2 \mathbf{E}_1 = \frac{\left(\xi_2, -\xi_1, \xi_1^2 + \xi_2^2\right)}{\left(\xi_2^2 + \xi_1^2 + \xi_1^4 + 2\xi_1^2 \xi_2^2 + \xi_2^4\right)^{1/2}} \cdot \left(0, 0, \frac{-\xi_1^2 + \xi_2^2}{\left(\xi_1^2 + \xi_2^2\right)^2}\right) \\
&= \frac{-\xi_1^2 + \xi_2^2}{\left(\xi_2^2 + \xi_1^2 + \xi_1^4 + 2\xi_1^2 \xi_2^2 + \xi_2^4\right)^{1/2} \left(\xi_1^2 + \xi_2^2\right)} \\
g &= \mathbf{n} \cdot \partial_2 \mathbf{E}_2 = \frac{\left(\xi_2, -\xi_1, \xi_1^2 + \xi_2^2\right)}{\left(\xi_2^2 + \xi_1^2 + \xi_1^4 + 2\xi_1^2 \xi_2^2 + \xi_2^4\right)^{1/2}} \cdot \left(0, 0, \frac{-2\xi_1 \xi_2}{\left(\xi_1^2 + \xi_2^2\right)^2}\right) \\
&= \frac{-2\xi_1 \xi_2}{\left(\xi_2^2 + \xi_1^2 + \xi_1^4 + 2\xi_1^2 \xi_2^2 + \xi_2^4\right)^{1/2} \left(\xi_1^2 + \xi_2^2\right)} \\
EG - F^2 &= \left[1 + \frac{\xi_2^2}{\left(\xi_1^2 + \xi_2^2\right)^2}\right] \left[1 + \frac{\xi_1^2}{\left(\xi_1^2 + \xi_2^2\right)^2}\right] - \left[\frac{-\xi_1 \xi_2}{\left(\xi_1^2 + \xi_2^2\right)^2}\right]^2 \\
&= 1 + \frac{1}{\left(\xi_1^2 + \xi_2^2\right)} + \frac{\xi_1^2 \xi_2^2}{\left(\xi_1^2 + \xi_2^2\right)^4} - \frac{\xi_1^2 \xi_2^2}{\left(\xi_1^2 + \xi_2^2\right)^4} \\
&= 1 + \frac{1}{\xi_1^2 + \xi_2^2} \\
&= \frac{\xi_1^2 + \xi_2^2 + 1}{\xi_1^2 + \xi_2^2} \\
eg - f^2 &= \frac{\left(2\xi_1 \xi_2\right)\left(-2\xi_1 \xi_2\right) - \left(-\xi_1^2 + \xi_2^2\right)^2}{\left(\xi_2^2 + \xi_1^2 + \xi_1^4 + 2\xi_1^2 \xi_2^2 + \xi_2^4\right)\left(\xi_1^2 + \xi_2^2\right)^2} \\
&= \frac{-4\xi_1^2 \xi_2^2 - \xi_1^4 + 2\xi_1^2 \xi_2^2 - \xi_2^4}{\left(\xi_2^2 + \xi_1^2 + \xi_1^4 + 2\xi_1^2 \xi_2^2 + \xi_2^4\right)\left(\xi_1^2 + \xi_2^2\right)^2} \\
&= \frac{-\left(\xi_1^4 + 2\xi_1^2 \xi_2^2 + \xi_2^4\right)}{\left(\xi_2^2 + \xi_1^2 + \xi_1^4 + 2\xi_1^2 \xi_2^2 + \xi_2^4\right)\left(\xi_1^2 + \xi_2^2\right)^2}
\end{aligned}$$

$$
\begin{aligned}
&= \frac{-\left(\xi_1^2+\xi_2^2\right)^2}{\left(\xi_2^2+\xi_1^2+\xi_1^4+2\xi_1^2\xi_2^2+\xi_2^4\right)\left(\xi_1^2+\xi_2^2\right)^2} \\
&= \frac{-1}{\xi_2^2+\xi_1^2+\xi_1^4+2\xi_1^2\xi_2^2+\xi_2^4} \\
K &= \frac{eg-f^2}{EG-F^2} = \left[\frac{-1}{\xi_2^2+\xi_1^2+\xi_1^4+2\xi_1^2\xi_2^2+\xi_2^4}\right]\left[\frac{\xi_1^2+\xi_2^2}{\xi_1^2+\xi_2^2+1}\right] \\
&= \frac{-\left(\xi_1^2+\xi_2^2\right)}{\left(\xi_2^2+\xi_1^2+\xi_1^4+2\xi_1^2\xi_2^2+\xi_2^4\right)\left(\xi_1^2+\xi_2^2+1\right)} \\
eG-2fF+gE &= \left[\frac{2\xi_1\xi_2}{\left(\xi_2^2+\xi_1^2+\xi_1^4+2\xi_1^2\xi_2^2+\xi_2^4\right)^{1/2}\left(\xi_1^2+\xi_2^2\right)}\right]\left[1+\frac{\xi_1^2}{\left(\xi_1^2+\xi_2^2\right)^2}\right] - \\
&\quad 2\left[\frac{-\xi_1^2+\xi_2^2}{\left(\xi_2^2+\xi_1^2+\xi_1^4+2\xi_1^2\xi_2^2+\xi_2^4\right)^{1/2}\left(\xi_1^2+\xi_2^2\right)}\right]\left[\frac{-\xi_1\xi_2}{\left(\xi_1^2+\xi_2^2\right)^2}\right] + \\
&\quad \left[\frac{-2\xi_1\xi_2}{\left(\xi_2^2+\xi_1^2+\xi_1^4+2\xi_1^2\xi_2^2+\xi_2^4\right)^{1/2}\left(\xi_1^2+\xi_2^2\right)}\right]\left[1+\frac{\xi_2^2}{\left(\xi_1^2+\xi_2^2\right)^2}\right] \\
&= 0 \\
H &= \frac{eG-2fF+gE}{2\left(EG-F^2\right)} = 0
\end{aligned}
$$

Hence, the principal curvatures are:

$$
\kappa_{1,2} = H \pm \sqrt{H^2-K} = \pm\sqrt{-K} = \pm\sqrt{\frac{\xi_1^2+\xi_2^2}{\left(\xi_2^2+\xi_1^2+\xi_1^4+2\xi_1^2\xi_2^2+\xi_2^4\right)\left(\xi_1^2+\xi_2^2+1\right)}}
$$

42. The principal curvatures of a surface at a given point correspond to the two directions represented by $\lambda_1$ and $\lambda_2$ which are the roots of the following quadratic equation:

$$
(gF-fG)\lambda^2 + (gE-eG)\lambda + (fE-eF) = 0
$$

From the rules of polynomial equations, find the sum and product of these roots.
**Answer**: A quadratic equation is given in terms of its roots, $x_1$ and $x_2$, by the following standard form:

$$
x^2 - (x_1+x_2)\,x + x_1x_2 = 0
$$

So, if we put the equation that is given in the question into this standard form by dividing by $gF-fG$ (which is not zero since the equation is assumed quadratic) then we get:

$$
\lambda^2 + \frac{(gE-eG)}{(gF-fG)}\lambda + \frac{(fE-eF)}{(gF-fG)} = 0
$$

On comparing the last equation with the above standard form of quadratic we obtain:

$$
\lambda_1+\lambda_2 = -\frac{gE-eG}{gF-fG} \qquad \text{and} \qquad \lambda_1\lambda_2 = \frac{fE-eF}{gF-fG}
$$

43. Define the "principal directions" descriptively and mathematically.
**Answer**: The principal directions of a surface $S$ at a given point $P$ are the tangential directions to $S$ that correspond to the principal curvatures of $S$ at $P$. Mathematically, the principal directions are identified spatially by the following vectors:

$$\begin{aligned}\mathbf{D}_1 &= \mathbf{E}_1 + \lambda_1 \mathbf{E}_2 \\ \mathbf{D}_2 &= \mathbf{E}_1 + \lambda_2 \mathbf{E}_2\end{aligned}$$

where $\mathbf{E}_1$ and $\mathbf{E}_2$ are the surface basis vectors at $P$ and $\lambda_1$ and $\lambda_2$ represent surface "directions", i.e. $\lambda = du^2/du^1$ (see previous exercise).

44. Show that $\kappa$ is a principal curvature with a principal direction $\frac{dv}{du}$ *iff* the following conditions are satisfied:

$$(e - \kappa E)du + (f - \kappa F)dv = 0 \qquad (38)$$

$$(f - \kappa F)du + (g - \kappa G)dv = 0 \qquad (39)$$

**Answer**: We have two parts to this proof:
(a) If $\kappa$ is a principal curvature associated with a principal direction $dv/du$ then Eqs. 38 and 39 are satisfied: let $\kappa_p$ be a principal curvature with a principal direction $dv_p/du_p$. Now, since principal curvatures are extremum of normal curvatures (where normal curvature is given by $\kappa_n = II_S/I_S$ as derived in Exercise 20) then in the direction $dv_p/du_p$ we should have:

$$\begin{aligned}\frac{\partial \kappa_n}{\partial (du)} &= \frac{I_S (II_S)_{du} - II_S (I_S)_{du}}{I_S^2} = 0 \\ \frac{\partial \kappa_n}{\partial (dv)} &= \frac{I_S (II_S)_{dv} - II_S (I_S)_{dv}}{I_S^2} = 0\end{aligned}$$

where the subscripts $du$ and $dv$ mean partial derivative with respect to $du$ and $dv$ and where we used the quotient rule of differentiation. On multiplying with $I_S$ we get:

$$(II_S)_{du} - \frac{II_S}{I_S} (I_S)_{du} = 0 \qquad (40)$$

$$(II_S)_{dv} - \frac{II_S}{I_S} (I_S)_{dv} = 0 \qquad (41)$$

Now:

$$\begin{aligned}\frac{II_S}{I_S} &= \kappa_p \\ (I_S)_{du} &= \frac{\partial}{\partial (du)} \left[E(du)^2 + 2F\, dudv + G(dv)^2\right] = 2E\, du + 2F\, dv \\ (I_S)_{dv} &= \frac{\partial}{\partial (dv)} \left[E(du)^2 + 2F\, dudv + G(dv)^2\right] = 2F\, du + 2G\, dv\end{aligned}$$

$$\begin{aligned}(II_S)_{du} &= \frac{\partial}{\partial (du)}\left[e(du)^2 + 2f\,dudv + g(dv)^2\right] = 2e\,du + 2f\,dv \\ (II_S)_{dv} &= \frac{\partial}{\partial (dv)}\left[e(du)^2 + 2f\,dudv + g(dv)^2\right] = 2f\,du + 2g\,dv\end{aligned}$$

On substituting from the last equations into Eqs. 40 and 41, we get:

$$\begin{aligned}2e\,du + 2f\,dv - \kappa_p\,(2E\,du + 2F\,dv) &= 0 \\ 2f\,du + 2g\,dv - \kappa_p\,(2F\,du + 2G\,dv) &= 0\end{aligned}$$

that is:

$$\begin{aligned}(e - \kappa_p E)\,du + (f - \kappa_p F)\,dv &= 0 \\ (f - \kappa_p F)\,du + (g - \kappa_p G)\,dv &= 0\end{aligned}$$

as required.
(b) If Eqs. 38 and 39 are satisfied then $\kappa$ is a principal curvature associated with a principal direction $dv/du$: in this case Eqs. 38 and 39 can be put in the following matrix form:

$$\begin{bmatrix} e - \kappa E & f - \kappa F \\ f - \kappa F & g - \kappa G \end{bmatrix} \begin{bmatrix} du \\ dv \end{bmatrix} = \begin{bmatrix} 0 \\ 0 \end{bmatrix}$$

This system of homogeneous linear equations has a non-trivial solution $(du, dv)$ *iff* the determinant of the coefficient matrix is zero, that is:

$$\begin{aligned}\begin{vmatrix} e - \kappa E & f - \kappa F \\ f - \kappa F & g - \kappa G \end{vmatrix} &= 0 \\ (e - \kappa E)(g - \kappa G) - (f - \kappa F)^2 &= 0 \\ \left(EG - F^2\right)\kappa^2 - (eG - 2fF + gE)\,\kappa + \left(eg - f^2\right) &= 0\end{aligned}$$

As explained in the book, the two roots of the last equation are the principal curvatures of the surface at the given point, and hence $\kappa_p$ must be one of these principal curvatures associated with the principal direction $dv_p/du_p$. If the point is umbilical then the last equation will have two equal roots (i.e. repeated root) which should be $\kappa_p$ with a principal direction $dv_p/du_p$ (although this equally applies to any other direction), as required.

45. Find the principal curvatures and the principal directions on a surface represented parametrically by: $\mathbf{r}(u, v) = (u, v, 2u^2 + 5v^2)$ at the point with $(u, v) = (2.3, 1.6)$.
**Answer**:

$$\begin{aligned}\mathbf{E}_1 &= \partial_u \mathbf{r} = (1, 0, 4u) \\ \mathbf{E}_2 &= \partial_v \mathbf{r} = (0, 1, 10v) \\ \mathbf{n} &= \frac{\mathbf{E}_1 \times \mathbf{E}_2}{|\mathbf{E}_1 \times \mathbf{E}_2|} = \frac{(-4u, -10v, 1)}{\sqrt{16u^2 + 100v^2 + 1}}\end{aligned}$$

$$
\begin{aligned}
\partial_u \mathbf{E}_1 &= (0,0,4) \\
\partial_v \mathbf{E}_1 &= (0,0,0) \\
\partial_v \mathbf{E}_2 &= (0,0,10) \\
E &= \mathbf{E}_1 \cdot \mathbf{E}_1 = 1 + 16u^2 \\
F &= \mathbf{E}_1 \cdot \mathbf{E}_2 = 40uv \\
G &= \mathbf{E}_2 \cdot \mathbf{E}_2 = 1 + 100v^2 \\
e &= \mathbf{n} \cdot \partial_u \mathbf{E}_1 = \frac{4}{\sqrt{16u^2 + 100v^2 + 1}} \\
f &= \mathbf{n} \cdot \partial_v \mathbf{E}_1 = 0 \\
g &= \mathbf{n} \cdot \partial_v \mathbf{E}_2 = \frac{10}{\sqrt{16u^2 + 100v^2 + 1}} \\
EG - F^2 &= 16u^2 + 100v^2 + 1 \\
eg - f^2 &= \frac{40}{16u^2 + 100v^2 + 1} \\
K &= \frac{eg - f^2}{EG - F^2} = \frac{40}{(16u^2 + 100v^2 + 1)^2} \\
eG - 2fF + gE &= \frac{14 + 400v^2 + 160u^2}{\sqrt{16u^2 + 100v^2 + 1}} \\
H &= \frac{eG - 2fF + gE}{2(EG - F^2)} = \frac{7 + 200v^2 + 80u^2}{(16u^2 + 100v^2 + 1)^{3/2}}
\end{aligned}
$$

At the point with $(u,v) = (2.3, 1.6)$ we have:

$$
\begin{aligned}
K &= \frac{40}{\left(16\,(2.3)^2 + 100\,(1.6)^2 + 1\right)^2} \simeq 0.0003427 \\
H &= \frac{7 + 200\,(1.6)^2 + 80\,(2.3)^2}{\left(16\,(2.3)^2 + 100\,(1.6)^2 + 1\right)^{3/2}} \simeq 0.1492
\end{aligned}
$$

Hence, the principal curvatures are:

$$
\kappa_{1,2} = H \pm \sqrt{H^2 - K} \simeq 0.1492 \pm \sqrt{0.1492^2 - 0.0003427}
$$

that is $\kappa_1 \simeq 0.2973$ and $\kappa_2 \simeq 0.001153$.
Regrading the principal directions, they can be determined from the equation:

$$
(gF - fG)\lambda^2 + (gE - eG)\lambda + (fE - eF) = 0 \tag{42}
$$

Now, at point $(u,v) = (2.3, 1.6)$ we have:

$$
gF - fG = \frac{400uv}{\sqrt{16u^2 + 100v^2 + 1}} = \frac{400\,(2.3)\,(1.6)}{\sqrt{16\,(2.3)^2 + 100\,(1.6)^2 + 1}} \simeq 79.6386
$$

$$gE - eG = \frac{6 + 160u^2 - 400v^2}{\sqrt{16u^2 + 100v^2 + 1}} = \frac{6 + 160\,(2.3)^2 - 400\,(1.6)^2}{\sqrt{16\,(2.3)^2 + 100\,(1.6)^2 + 1}} \simeq -9.2840$$

$$fE - eF = \frac{-160uv}{\sqrt{16u^2 + 100v^2 + 1}} = \frac{-160\,(2.3)\,(1.6)}{\sqrt{16\,(2.3)^2 + 100\,(1.6)^2 + 1}} \simeq -31.8554$$

and hence Eq. 42 becomes:

$$79.6386\lambda^2 - 9.2840\lambda - 31.8554 = 0$$

On solving this quadratic equation, we obtain:

$$\lambda_1 \simeq -0.5768 \qquad \lambda_2 \simeq 0.6934$$

that is $dv_1 \simeq -0.5768du_1$ and $dv_2 \simeq 0.6934du_2$.

46. Prove Euler theorem which is given by the following equation:

$$\kappa_n = \kappa_1 \cos^2\theta + \kappa_2 \sin^2\theta$$

where $\theta$ is the angle between the principal direction of $\kappa_1$ and the given direction.
**Answer**: Let choose (with no loss of generality) a surface coordinate system where the $u$ and $v$ coordinate curves are aligned along the principal directions. It was shown in the book that in such a system $f = F = 0$ and hence the normal curvature $\kappa_n$ in any direction will be given by:

$$\kappa_n = \frac{e(du)^2 + g(dv)^2}{E(du)^2 + G(dv)^2} \tag{43}$$

Moreover (also see the book):

$$\kappa_1 = \frac{e}{E} \qquad \text{and} \qquad \kappa_2 = \frac{g}{G}$$

Hence from Eq. 43 we get:

$$\begin{aligned}\kappa_n &= \frac{e(du)^2}{E(du)^2 + G(dv)^2} + \frac{g(dv)^2}{E(du)^2 + G(dv)^2} \\ &= \frac{(e/E)\,E(du)^2}{E(du)^2 + G(dv)^2} + \frac{(g/G)\,G(dv)^2}{E(du)^2 + G(dv)^2} \\ &= \kappa_1 \frac{E(du)^2}{E(du)^2 + G(dv)^2} + \kappa_2 \frac{G(dv)^2}{E(du)^2 + G(dv)^2}\end{aligned} \tag{44}$$

Now, we have:[31]

$$\cos\theta = \frac{\sqrt{E}\,du}{\sqrt{E(du)^2 + G(dv)^2}} \qquad \text{and} \qquad \cos\phi = \frac{\sqrt{G}\,dv}{\sqrt{E(du)^2 + G(dv)^2}}$$

[31] This is justified by the fact that if $d\mathbf{r} = \mathbf{r}_u du + \mathbf{r}_v dv$ and $\delta\mathbf{r} = \mathbf{r}_u \delta u + \mathbf{r}_v \delta v$ are two surface vectors then

where $\theta$ is the angle between the principal direction of $\kappa_1$ and the given direction while $\phi$ is the angle between the principal direction of $\kappa_2$ and the given direction. Hence, Eq. 44 becomes:

$$\kappa_n = \kappa_1 \cos^2 \theta + \kappa_2 \cos^2 \phi$$

Also, because the coordinate curves are orthogonal (since the principal directions are orthogonal) then $\phi = \frac{\pi}{2} - \theta$, and hence $\cos \phi = \cos\left(\frac{\pi}{2} - \theta\right) = \sin \theta$. Therefore, the last equation becomes:

$$\kappa_n = \kappa_1 \cos^2 \theta + \kappa_2 \sin^2 \theta$$

as required.

47. What is Darboux frame? Are the vectors of this frame orthonormal? Is this frame defined at umbilical points on the surface? Fully justify your answer related to the last two parts of the question.
**Answer**: Darboux frame, which is a frame for 3D space associated with a surface $S$, consists of the vector triad $(\mathbf{d}_1, \mathbf{d}_2, \mathbf{n})$ where $\mathbf{d}_1$ and $\mathbf{d}_2$ are the unit vectors corresponding to the principal directions at a given point $P$, and $\mathbf{n} = \mathbf{d}_1 \times \mathbf{d}_2$ is the unit normal vector to $S$ at $P$.
The vectors of this frame are orthonormal because all these vectors are unit vectors. Moreover, $\mathbf{n}$ is orthogonal to $\mathbf{d}_1$ and $\mathbf{d}_2$ since $\mathbf{n}$ is the cross product of $\mathbf{d}_1$ and $\mathbf{d}_2$. Also, $\mathbf{d}_1$ and $\mathbf{d}_2$ are mutually orthogonal because they are along the principal directions which are orthogonal. Hence, all these vectors are mutually orthogonal and of unit length and hence they are orthonormal.
This frame is not defined at umbilical points because the principal directions are not well defined there since any direction can be regarded as a principal direction (or there is no principal direction) and hence $\mathbf{d}_1$ and $\mathbf{d}_2$ are not well defined.

48. Write the formulae for the positions of the centers of curvature of the normal sections corresponding to the two principal curvatures at a given point $P$ on a surface $S$.
**Answer**:

$$x_1^i = x_P^i + \frac{N_1^i}{|\kappa_1|} \qquad \text{and} \qquad x_2^i = x_P^i + \frac{N_2^i}{|\kappa_2|}$$

---

the angle $\alpha$ between them is given by:

$$\begin{aligned}
\cos \alpha &= \frac{d\mathbf{r} \cdot \delta\mathbf{r}}{|d\mathbf{r}|\,|\delta\mathbf{r}|} \\
&= \frac{(\mathbf{r}_u du + \mathbf{r}_v dv) \cdot (\mathbf{r}_u \delta u + \mathbf{r}_v \delta v)}{|\mathbf{r}_u du + \mathbf{r}_v dv|\,|\mathbf{r}_u \delta u + \mathbf{r}_v \delta v|} \\
&= \frac{E\, du\, \delta u + F\,(du \delta v + \delta u dv) + G\, dv\, \delta v}{\sqrt{E\,(du)^2 + 2F\, du\, dv + G\,(dv)^2}\sqrt{E\,(\delta u)^2 + 2F\, \delta u\, \delta v + G\,(\delta v)^2}}
\end{aligned}$$

In our case $F = 0$. Moreover, when $\delta\mathbf{r}$ represents the direction of the $u$ coordinate curve we have $\delta v = 0$, and when $\delta\mathbf{r}$ represents the direction of the $v$ coordinate curve we have $\delta u = 0$, and hence the formulae for $\cos\theta$ and $\cos\phi$ will follow.

where $x_1^i$ and $x_2^i$ are the spatial coordinates of the first and second center of curvature corresponding to the two principal curvatures, $x_P^i$ are the spatial coordinates of $P$, $N_1^i$ and $N_2^i$ are the principal normal vectors of the two normal sections corresponding to the two principal curvatures, $\kappa_1$ and $\kappa_2$ are the principal curvatures of $S$ at $P$, and $i = 1, 2, 3$.

49. Correlate, mathematically with full explanation of all the symbols involved, the normal curvature $\kappa_n$ at a given point $P$ and in a given direction on a smooth surface to the two principal curvatures at that point.
**Answer**: This is Euler theorem (see Exercise 46), that is:

$$\kappa_n = \kappa_1 \cos^2\theta + \kappa_2 \sin^2\theta$$

where $\kappa_1$ and $\kappa_2$ are the principal curvatures at $P$, and $\theta$ is the angle between the principal direction of $\kappa_1$ at $P$ and the given direction.

50. Define, mathematically in terms of the principal curvatures, the following terms: principal radii, mean curvature and Gaussian curvature.
**Answer**:
The principal radii of curvature, $R_1$ and $R_2$, are the magnitude of the reciprocals of the principal curvatures, i.e. $R_1 = \left|\frac{1}{\kappa_1}\right|$ and $R_2 = \left|\frac{1}{\kappa_2}\right|$.
The mean curvature $H$ is the average of the principal curvatures, i.e. $H = \frac{\kappa_1+\kappa_2}{2}$.
The Gaussian curvature $K$ is the product of the principal curvatures, i.e. $K = \kappa_1\kappa_2$.

51. Distinguish between the "total curvature" of a curve and the "total curvature" of a surface. For surface, what are the two meanings of this term?
**Answer**: The total curvature (which is also known as the third curvature) of a curve is the expression $\sqrt{(ds_{\mathbf{T}})^2 + (ds_{\mathbf{B}})^2}$, where $ds_{\mathbf{T}}$ and $ds_{\mathbf{B}}$ are respectively the lengths of the line element components in the tangent and binormal directions.
The total curvature of a surface is commonly used as synonymous to the Gaussian curvature $K$. However, it may also be used for the area integral $\int K d\sigma$. To avoid any confusion, we use total curvature of a surface exclusively for the area integral $\int K d\sigma$.

52. Find the Gaussian curvature $K$ and the mean curvature $H$ of a surface given by: $\mathbf{r}(u, v) = (3u - v, u + 2v, 1.5uv)$ at the point with $(u, v) = (3, 1)$.
**Answer**: At the point with $(u, v) = (3, 1)$ we have:

$$\begin{aligned}
\mathbf{E}_1 &= \partial_u \mathbf{r} = (3, 1, 1.5v) = (3, 1, 1.5) \\
\mathbf{E}_2 &= \partial_v \mathbf{r} = (-1, 2, 1.5u) = (-1, 2, 4.5) \\
\mathbf{n} &= \frac{\mathbf{E}_1 \times \mathbf{E}_2}{|\mathbf{E}_1 \times \mathbf{E}_2|} = \frac{(1.5u - 3v,\ -1.5v - 4.5u,\ 7)}{\sqrt{22.5u^2 + 11.25v^2 + 4.5uv + 49}} = \frac{(1.5,\ -15,\ 7)}{\sqrt{276.25}} \\
\partial_u \mathbf{E}_1 &= (0, 0, 0) \\
\partial_v \mathbf{E}_1 &= (0, 0, 1.5)
\end{aligned}$$

$$
\begin{aligned}
\partial_v \mathbf{E}_2 &= (0,0,0) \\
E &= \mathbf{E}_1 \cdot \mathbf{E}_1 = 10 + 2.25v^2 = 12.25 \\
F &= \mathbf{E}_1 \cdot \mathbf{E}_2 = -1 + 2.25uv = 5.75 \\
G &= \mathbf{E}_2 \cdot \mathbf{E}_2 = 5 + 2.25u^2 = 25.25 \\
e &= \mathbf{n} \cdot \partial_u \mathbf{E}_1 = 0 \\
f &= \mathbf{n} \cdot \partial_v \mathbf{E}_1 = \frac{10.5}{\sqrt{22.5u^2 + 11.25v^2 + 4.5uv + 49}} = \frac{10.5}{\sqrt{276.25}} \\
g &= \mathbf{n} \cdot \partial_v \mathbf{E}_2 = 0 \\
EG - F^2 &= 49 + 22.5u^2 + 11.25v^2 + 4.5uv = 276.25 \\
eg - f^2 &= \frac{-110.25}{22.5u^2 + 11.25v^2 + 4.5uv + 49} = -\frac{110.25}{276.25} \\
eG - 2fF + gE &= 0 - 2 \times \frac{10.5}{\sqrt{276.25}} \times 5.75 + 0 = -\frac{120.75}{\sqrt{276.25}} \\
K &= \frac{eg - f^2}{EG - F^2} = -\frac{110.25}{(276.25)^2} \simeq -0.001445 \\
H &= \frac{eG - 2fF + gE}{2(EG - F^2)} = -\frac{120.75}{2(276.25)^{3/2}} \simeq -0.01315
\end{aligned}
$$

53. State the limiting conditions on the principal curvatures $\kappa_1$ and $\kappa_2$, and hence deduce the conditions on the mean curvature $H$ and the Gaussian curvature $K$, on the surface of sphere and on the surface of hyperboloid of one sheet.
**Answer**: For sphere, we have $\kappa_1 = \kappa_2 < 0$ and hence $H < 0$ and $K > 0$ (assuming $\mathbf{n}$ is in the outside direction). For hyperboloid of one sheet we have $\kappa_1 > 0$ and $\kappa_2 < 0$ and hence $K < 0$ while $H$ depends on the size of $\kappa_1$ and $\kappa_2$.

54. Prove that there is no compact surface of class $C^2$ with non-positive Gaussian curvature $K$ over the whole surface.
**Answer**: Let $S$ be a compact surface of class $C^2$, and we consider the function $d_P^2 = \mathbf{r} \cdot \mathbf{r}$ where $d_P$ is the distance between an arbitrary point $P$ on $S$ and the origin of coordinates and $\mathbf{r}$ is the surface spatial representation that corresponds to $P$. From the given conditions, $d_P^2$ must be a continuous function and hence it must have a maximum $d_o^2$ at a particular point $P_o$ on the surface. It is obvious that $d_o^2 > 0$ since it is maximum because otherwise the surface will be a single point (i.e. the origin). Now, if $\mathbf{r}(u,v)$ is a patch on $S$ that includes $P_o$ such that the $u$ and $v$ coordinate curves are oriented along the principal directions at $P_o$, then at $P_o$ we must have:

$$
\begin{aligned}
\partial_u d_P^2 &= 2\,\mathbf{r} \cdot \mathbf{r}_u = 0 \\
\partial_v d_P^2 &= 2\,\mathbf{r} \cdot \mathbf{r}_v = 0 \\
\partial_{uu} d_P^2 &= 2\,\mathbf{r}_u \cdot \mathbf{r}_u + 2\,\mathbf{r} \cdot \mathbf{r}_{uu} \le 0 \\
\partial_{vv} d_P^2 &= 2\,\mathbf{r}_v \cdot \mathbf{r}_v + 2\,\mathbf{r} \cdot \mathbf{r}_{vv} \le 0
\end{aligned}
$$

where these equations are justified by the fact that at a maximum point the first derivative is zero while the second derivative is non-positive. So, from the first two equations we conclude that at $P_o$ $\mathbf{r}$ is orthogonal to $\mathbf{r}_u$ and $\mathbf{r}_v$ and hence $\mathbf{n} = \pm\mathbf{r}/|\mathbf{r}| = \pm\mathbf{r}/d_o$ (since $\mathbf{r}_u$ and $\mathbf{r}_v$ are tangent to the surface). Because the sign of $\mathbf{n}$ is rather arbitrary we can assume that $\mathbf{n} = \mathbf{r}/d_o$ at $P_o$. On substituting this into the last two equations we obtain:

$$\mathbf{r}_u \cdot \mathbf{r}_u + d_o\mathbf{n} \cdot \mathbf{r}_{uu} \le 0 \qquad \text{and} \qquad \mathbf{r}_v \cdot \mathbf{r}_v + d_o\mathbf{n} \cdot \mathbf{r}_{vv} \le 0$$

that is:

$$E + d_o e \le 0 \qquad \text{and} \qquad G + d_o g \le 0$$

Hence, at $P_o$ we have:

$$\frac{e}{E} \le -\frac{1}{d_o} < 0 \qquad \text{and} \qquad \frac{g}{G} \le -\frac{1}{d_o} < 0$$

Now, since the coordinate curves at $P_o$ are oriented along the principal directions then the principal curvatures $\kappa_1$ and $\kappa_2$ are given by (refer to the book):

$$\kappa_1 = \frac{e}{E} \qquad \text{and} \qquad \kappa_2 = \frac{g}{G}$$

Hence, we conclude that at $P_o$ we have:

$$\kappa_1 \le -\frac{1}{d_o} < 0 \qquad \text{and} \qquad \kappa_2 \le -\frac{1}{d_o} < 0$$

and therefore the Gaussian curvature at $P_o$ is $K \equiv \kappa_1\kappa_2 > 0$. Hence, a compact surface of class $C^2$ must contain at least one point with positive Gaussian curvature, i.e. there is no compact surface of class $C^2$ with non-positive Gaussian curvature over the whole surface, as required.

55. Analyze the following equation outlining its significance:

$$S(x, y) \simeq S(0, 0) + \frac{\kappa_1 x^2}{2} + \frac{\kappa_2 y^2}{2}$$

**Answer**: This equation represents a quadratic approximation of a surface in the neighborhood of a given point on the surface. The background of this equation is that if $P$ is a given point on a sufficiently smooth surface $S$ embedded in a 3D space coordinated by a rectangular Cartesian system $(x, y, z)$ with $P$ being above the origin, the tangent plane of $S(x, y)$ at $P$ being parallel to the $xy$ plane, and the principal directions being along the $x$ and $y$ coordinate lines then in the neighborhood of $P$ the surface can be approximated by the above quadratic form. The significance of this is that any sufficiently smooth surface can be locally represented by a quadratic surface that is solely determined by the principal curvatures of the surface at $P$.

56. State the necessary and sufficient condition for a real number to be a principal curvature of a surface at a given point.

**Answer**: The necessary and sufficient condition for a real number $\kappa$ to be a principal curvature of a smooth surface $S$ at a given point $P$ and in a given direction $\frac{dv}{du}$, where $(du)^2 + (dv)^2 \neq 0$, is that the following equations are satisfied:

$$\begin{aligned}(e - \kappa E)du + (f - \kappa F)dv &= 0 \\ (f - \kappa F)du + (g - \kappa G)dv &= 0\end{aligned}$$

where $E, F, G, e, f, g$ are the coefficients of the first and second fundamental forms at $P$ (see Exercise 44).

57. Investigate the number of roots of the following quadratic equation and the impact of this on the number of principal curvatures of the surface at the point where this equation applies:

$$\left(EG - F^2\right)\kappa^2 - (gE - 2fF + eG)\kappa + \left(eg - f^2\right) = 0 \tag{45}$$

**Answer**: This quadratic equation in $\kappa$ has a non-negative discriminant and hence it possesses either two distinct real roots or a repeated real root. In the former case there are two distinct principal curvatures corresponding to two orthogonal principal directions, while in the latter case the point is umbilical.

58. From the equation in the previous question, obtain an analytical expression for the principal curvatures of the surface at the point where this equation applies.
**Answer**: The principal curvatures, $\kappa_1$ and $\kappa_2$, are the roots of this quadratic equation and hence they are obtained by the quadratic formula,[32] that is:

$$\kappa_{1,2} = \frac{(gE - 2fF + eG) \pm \sqrt{(gE - 2fF + eG)^2 - 4\left(EG - F^2\right)\left(eg - f^2\right)}}{2\left(EG - F^2\right)}$$

59. From the equation in the last two questions, obtain the equation: $\kappa^2 - 2H\kappa + K = 0$ and hence verify that the principal curvatures are given by: $\kappa_{1,2} = H \pm \sqrt{H^2 - K}$.
**Answer**: The mean curvature $H$ and the Gaussian curvature $K$ are defined by:

$$\begin{aligned}H &= \frac{gE - 2fF + eG}{2\left(EG - F^2\right)} \\ K &= \frac{eg - f^2}{EG - F^2}\end{aligned}$$

Hence, on dividing Eq. 45 by $\left(EG - F^2\right)$ we get:

$$\kappa^2 - \frac{(gE - 2fF + eG)}{(EG - F^2)}\kappa + \frac{(eg - f^2)}{(EG - F^2)} = 0$$

[32] For a quadratic equation of the form $ax^2 + bx + c = 0$, the quadratic formula for the roots $x_{1,2}$ is:

$$x_{1,2} = \frac{-b \pm \sqrt{b^2 - 4ac}}{2a}$$

$$\begin{aligned}\kappa^2 - 2\frac{(gE - 2fF + eG)}{2(EG - F^2)}\kappa + \frac{(eg - f^2)}{(EG - F^2)} &= 0 \\ \kappa^2 - 2H\kappa + K &= 0\end{aligned}$$

On applying the quadratic formula on the last equation, we obtain:

$$\begin{aligned}\kappa_{1,2} &= \frac{2H \pm \sqrt{(-2H)^2 - 4K}}{2} \\ &= \frac{2H}{2} \pm \sqrt{\frac{4H^2 - 4K}{4}} \\ &= H \pm \sqrt{H^2 - K}\end{aligned}$$

60. Write down the equations of the principal curvatures when the $u^1$ and $u^2$ coordinate curves are aligned along the principal directions.
**Answer**:

$$\kappa_1 = \frac{b_{11}}{a_{11}} = \frac{e}{E} \qquad \text{and} \qquad \kappa_2 = \frac{b_{22}}{a_{22}} = \frac{g}{G}$$

where $\kappa_1$ and $\kappa_2$ are the principal curvatures, the indexed $a$ and $b$ are the coefficients of the surface covariant metric and covariant curvature tensors, and $E, G, e, g$ are the coefficients of the first and second fundamental forms.

61. Obtain the equation for the Gaussian curvature of a surface with orthogonal coordinate curves by using the following equation:

$$K = \frac{1}{2\sqrt{a}}\left[\partial_u\left(\frac{FE_v}{E\sqrt{a}} - \frac{G_u}{\sqrt{a}}\right) + \partial_v\left(\frac{2F_u}{\sqrt{a}} - \frac{E_v}{\sqrt{a}} - \frac{FE_u}{E\sqrt{a}}\right)\right] \tag{46}$$

**Answer**: If the coordinate curves are orthogonal then $F = 0$. Hence, all the terms containing $F$ and its derivatives in the above formula will vanish and $a$ $(= EG - F^2)$ will be equal to $EG$. Accordingly, the above formula will become:

$$\begin{aligned}K &= \frac{1}{2\sqrt{EG}}\left[\partial_u\left(0 - \frac{G_u}{\sqrt{EG}}\right) + \partial_v\left(0 - \frac{E_v}{\sqrt{EG}} - 0\right)\right] \\ &= -\frac{1}{2\sqrt{EG}}\left[\partial_u\left(\frac{G_u}{\sqrt{EG}}\right) + \partial_v\left(\frac{E_v}{\sqrt{EG}}\right)\right] \qquad (47)\end{aligned}$$

62. Show that the spheres are the only connected, compact and sufficiently smooth surfaces with constant Gaussian curvature.
**Answer**: Since the Gaussian curvature $K$ at any point is constant (say $K = c^2$),[33] then at any point on the surface we should have:

$$K = c^2 = c^2 \times 1 = c^2 \times \frac{EG - F^2}{EG - F^2} = \frac{(cE)(cG) - (cF)^2}{EG - F^2} = \frac{eg - f^2}{EG - F^2}$$

[33] We note that $c$ is a non-zero real number and hence $K$ is positive because the surface is assumed to be connected and compact (see Exercise 54).

i.e. $\frac{e}{E} = \frac{f}{F} = \frac{g}{G} = c$ which means that the point is umbilical (see Exercises 29 and 128). Now, since sphere is the only connected, compact and sufficiently smooth surface whose all points are spherical umbilical (see Exercise 65) then the surface must be a sphere, as required.

63. State the curvature formula of Rodrigues defining all the symbols involved and discussing its significance.
**Answer**: At a given non-umbilical[34] point $P$ on a sufficiently smooth surface $S$, a direction $dv/du$ is a principal direction *iff* for a real number $\kappa$ the following relation holds true:

$$d\mathbf{n} = -\kappa\, d\mathbf{r}$$

where $\mathbf{n}(u, v)$ is the normal unit vector to $S$ at $P$, $\mathbf{r}(u, v)$ is the spatial representation of $S$ at $P$, and $\kappa$ is the principal curvature of $S$ at $P$ corresponding to the principal direction $dv/du$. The significance of the Rodrigues formula is that in any principal direction the two vectors $d\mathbf{n}$ and $d\mathbf{r}$ have the same orientation where the principal curvature $\kappa$ in that direction is the scale factor between the two vectors.

64. Test the validity of the Rodrigues formula for the principal directions at the point with $(u, v) = (2.3, 1.6)$ on a surface parameterized by: $\mathbf{r}(u, v) = (u, v, 2u^2 + 5v^2)$.[35]
**Answer**: In Exercise 45 we found $\kappa_1 \simeq 0.2973$, $\kappa_2 \simeq 0.001153$, $dv_1 \simeq -0.5768du_1$ and $dv_2 \simeq 0.6934du_2$. Now, at the point with $(u, v) = (2.3, 1.6)$ we have:

$$\begin{aligned}
\frac{\partial \mathbf{n}}{\partial u} &= \frac{(-400v^2 - 4, 160uv, -16u)}{(16u^2 + 100v^2 + 1)^{3/2}} \simeq \frac{(-1028, 588.8, -36.8)}{6314.705} \\
\frac{\partial \mathbf{n}}{\partial v} &= \frac{(400uv, -160u^2 - 10, -100v)}{(16u^2 + 100v^2 + 1)^{3/2}} \simeq \frac{(1472, -856.4, -160)}{6314.705} \\
\mathbf{E}_1 &= (1, 0, 4u) = (1, 0, 9.2) \\
\mathbf{E}_2 &= (0, 1, 10v) = (0, 1, 16)
\end{aligned}$$

So, for the principal direction $dv_1 \simeq -0.5768du_1$ with principal curvature $\kappa_1 \simeq 0.2973$ we have:

$$\begin{aligned}
d\mathbf{n}_1 &= \frac{\partial \mathbf{n}}{\partial u} du_1 + \frac{\partial \mathbf{n}}{\partial v} dv_1 \\
&\simeq \frac{(-1028, 588.8, -36.8)}{6314.705}(1) + \frac{(1472, -856.4, -160)}{6314.705}(-0.5768)
\end{aligned}$$

[34] The condition "non-umbilical" is related to the convention about the principal directions and principal curvatures at umbilical points (see for example Exercise 64 of § 5). However, some proofs (in this book and in the literature of differential geometry) are based on applying the Rodrigues curvature theorem at umbilical points and hence this condition will be ignored since there is no inherent contradiction in applying this theorem at umbilical points.

[35] The question in the book is about a surface parameterized by $\mathbf{r}(u, v) = (u, v, u^2 + 3v^2)$ at the point with $(u, v) = (1.4, 3.9)$. However, we changed here to minimize the space and effort since the method is the same.

$$\begin{aligned}
& \simeq (-0.2973, 0.1715, 0.008787) \\
d\mathbf{r}_1 &= \mathbf{E}_1 du_1 + \mathbf{E}_2 dv_1 \\
&= (1, 0, 9.2)(1) + (0, 1, 16)(-0.5768) \\
&\simeq (1, -0.5768, -0.0288) \\
-\kappa_1 \, d\mathbf{r}_1 &\simeq -0.2973 (1, -0.5768, -0.0288) \\
&\simeq (-0.2973, 0.1715, 0.008562)
\end{aligned}$$

Hence, $d\mathbf{n}_1 = -\kappa_1 \, d\mathbf{r}_1$ (noting the numerical approximations).
Similarly, for the principal direction $dv_2 \simeq 0.6934 du_2$ with principal curvature $\kappa_2 \simeq 0.001153$, we have:

$$\begin{aligned}
d\mathbf{n}_2 &= \frac{\partial \mathbf{n}}{\partial u} du_2 + \frac{\partial \mathbf{n}}{\partial v} dv_2 \\
&\simeq \frac{(-1028, 588.8, -36.8)}{6314.705}(1) + \frac{(1472, -856.4, -160)}{6314.705}(0.6934) \\
&\simeq (-0.001158, -0.0007962, -0.02340) \\
d\mathbf{r}_2 &= \mathbf{E}_1 du_2 + \mathbf{E}_2 dv_2 \\
&= (1, 0, 9.2)(1) + (0, 1, 16)(0.6934) \\
&\simeq (1, 0.6934, 20.2944) \\
-\kappa_2 \, d\mathbf{r}_2 &\simeq -0.001153 (1, 0.6934, 20.2944) \\
&\simeq (-0.001153, -0.0007995, -0.02340)
\end{aligned}$$

Hence, $d\mathbf{n}_2 = -\kappa_2 \, d\mathbf{r}_2$ (noting the numerical approximations).

65. Use the Rodrigues curvature formula to prove that spheres are the only connected closed surfaces of class $C^3$ whose all points are spherical umbilical.
**Answer**: Referring to Exercise 63, we should first not follow the convention that excludes umbilical points from having principal directions and principal curvatures and hence we should allow for the inclusion of umbilical points in the statement of the Rodrigues curvature theorem. So, let have a connected closed surface $S$ of class $C^3$ whose all points are spherical umbilical and hence at every point on $S$ the normal curvature $\kappa_n$ (which is equal to the curvature $\kappa$ in the Rodrigues formula) is the same in all directions (including the directions of coordinate curves) although it may vary from point to point. Therefore, from the Rodrigues formula along the coordinate curves at any point on $S$ we have (refer to the book):

$$\partial_u \mathbf{n} = -\kappa \mathbf{E}_1 \qquad \text{and} \qquad \partial_v \mathbf{n} = -\kappa \mathbf{E}_2$$

Now, to show that $S$ is a sphere we need to show that $\kappa$ is not only the same in any direction at any point but it is also the same over the entire surface, i.e. it does not vary from point to point. So, let first have an arbitrary connected patch $S_1$ on $S$ where the above formulae apply. On differentiating the above formulae with respect to $v$ and $u$ respectively we obtain:

$$\partial_{uv} \mathbf{n} = -(\partial_v \kappa) \mathbf{E}_1 - \kappa \partial_v \mathbf{E}_1 \qquad \text{and} \qquad \partial_{vu} \mathbf{n} = -(\partial_u \kappa) \mathbf{E}_2 - \kappa \partial_u \mathbf{E}_2$$

Now, since $\partial_{uv}\mathbf{n} = \partial_{vu}\mathbf{n}$ and $\partial_v \mathbf{E}_1 = \partial_u \mathbf{E}_2$ then on subtracting the last formulae we obtain:

$$-(\partial_v \kappa)\,\mathbf{E}_1 + (\partial_u \kappa)\,\mathbf{E}_2 = 0$$

Because $\mathbf{E}_1$ and $\mathbf{E}_2$ are linearly independent then the last formula implies that $\partial_v \kappa = \partial_u \kappa = 0$ which means that $\kappa$ is constant over $S_1$. This should also apply to any other patch on $S$ because $S_1$ is arbitrary. To complete the proof, we need to show next that $\kappa$ is the same across all patches on $S$. So, let $S_1$ and $S_2$ be two arbitrary neighboring patches on $S$. Now, if we connect an arbitrary point $P_1$ on $S_1$ to an arbitrary point $P_2$ on $S_2$ by a regular curve $C$ (which must exist because $S$ is connected) then because $S_1$ and $S_2$ are arbitrary patches on a connected surface and $\kappa$ is supposed to be constant on each of $S_1$ and $S_2$ then $\kappa$ must be the same along $C$ and hence $\kappa$ must be the same at $P_1$ and $P_2$. This means that $\kappa$ is the same at any point on $S$ (because $P_1$ and $P_2$ are arbitrary points on arbitrary patches) and hence $\kappa$ is constant over the entire surface $S$. We need next to show that the surface $S$ has a spherical shape. To do this we note that since every point on $S$ is spherical umbilical then an arbitrary regular curve $C(t)$ on an arbitrary patch $S_1$ must be a line of curvature along which the Rodrigues formula applies, that is:

$$\frac{d\mathbf{n}}{dt} = -\kappa \frac{d\mathbf{r}}{dt}$$

On integrating this equation with respect to $t$ (noting that $\kappa$ is constant), we obtain:

$$\mathbf{n} = -\kappa \mathbf{r} + \mathbf{v}$$

where $\mathbf{v}$ is a constant vector. On taking the modulus of the two sides of the last equation (noting that $\mathbf{n}$ is a unit vector) we get:

$$1 = |-\kappa \mathbf{r} + \mathbf{v}| = \kappa \left| \frac{\mathbf{v}}{\kappa} - \mathbf{r} \right|$$

Accordingly, the radius of curvature at any point on $C$ is:

$$R_\kappa \equiv \frac{1}{\kappa} = \left| \frac{\mathbf{v}}{\kappa} - \mathbf{r} \right| = \left| \mathbf{r} - \frac{\mathbf{v}}{\kappa} \right|$$

This means that the distance between any point $\mathbf{r}$ on $C$ and a fixed point $\mathbf{v}/\kappa$ in the space is constant ($= 1/\kappa$) and hence $C$ is a spherical curve with center $\mathbf{v}/\kappa$. Now, since $C$ is an arbitrary curve on $S_1$ then this applies to the entire $S_1$ which means that $S_1$ is a spherical surface (i.e. $S_1$ is part of a sphere $S_s$ with center $\mathbf{v}/\kappa$ and radius $1/\kappa$). Also, since $S_1$ is arbitrary then this applies to the entire surface $S$ (i.e. $S$ is part of $S_s$). Finally, to rule out the possibility that $S$ may be just part of $S_s$ and not the entire $S_s$ (i.e. it is not a sphere but part of a sphere) we need to show that $S$ includes the entire $S_s$ (i.e. $S$ is equal to $S_s$).[36] For this, we note that because $S_s$ is connected and $S$ is

[36] In more technical terms, we proved so far that $S$ is a subset of $S_s$ so we need to prove next that $S$ is not a proper subset of $S_s$ but it is the entire $S_s$.

closed then $S_s$ and $S$ are equal as point sets,[37] i.e. $S$ includes the entire $S_s$ and hence $S \equiv S_s$. In other words, $S$ must be a sphere, as required.

66. Give a mathematical expression for the Gaussian curvature $K$ in terms of the coefficients of the surface metric tensor $a_{11}, a_{12}, a_{21}, a_{22}$ and the curvature tensor $b_{11}, b_{12}, b_{21}, b_{22}$.
**Answer**:
$$K = \frac{b_{11}b_{22} - b_{12}b_{21}}{a_{11}a_{22} - a_{12}a_{21}}$$

67. What is the significance of having an intrinsic surface curvature, represented usually by the Gaussian curvature, as a way for a 2D inhabitant to have some perception of the nature of the surface and its shape as seen from the ambient space by a 3D inhabitant?
**Answer**: Having an intrinsic curvature implies having extrinsic curvature (but the opposite is not true). So, when the surface possesses an intrinsic curvature, a 2D inhabitant can conclude that the 2D space is curved externally as well and hence the 2D inhabitant will have some perception of the nature of the surface and its shape as seen from the ambient space by a 3D inhabitant.

68. Discuss the following statement: "The Gaussian curvature along any parallel line of a surface of revolution is constant".
**Answer**: This statement is intuitive because due to the axial symmetry of the surface the principal curvature of the surface along any given parallel is constant. Moreover, the principal curvature of the surface along the direction of any meridian that cuts the given parallel is the same at any point on the given parallel. Hence, the Gaussian curvature (which is the product of the two principal curvatures) must also be constant along the parallel. We remind the reader that parallels and meridians of surface of revolution are lines of curvature (see Exercise 65 of § 5).

69. Starting from the following relation: $K = \frac{b}{a}$, derive the relation: $K = \det(b^{\alpha}_{\beta})$.
**Answer**:
$$\begin{aligned} K &= \frac{b}{a} \\ &= \frac{\det(b_{\gamma\beta})}{\det(a_{\alpha\gamma})} \\ &= \det(a^{\alpha\gamma})\det(b_{\gamma\beta}) \\ &= \det(a^{\alpha\gamma}b_{\gamma\beta}) \\ &= \det\left(b^{\alpha}_{\beta}\right) \end{aligned}$$

where line 1 is the definition of $K$, line 2 is the definition of $a$ and $b$, line 3 is because $a^{\alpha\gamma}$ and $a_{\alpha\gamma}$ are inverses and hence their determinants are reciprocals, line 4 is because the determinant of product matrices is the product of their determinants, and line 5 is because $a^{\alpha\gamma}$ is an index raising operator.

[37] We refer here to the theorem that we stated in the book as: "If $S_1$ and $S_2$ are two simple surfaces where $S_1$ is connected and $S_2$ is closed and contained in $S_1$, then the two surfaces are equal as point sets". We note that $S_1$ and $S_2$ in this statement correspond to $S_s$ and $S$ in our problem.

70. Give a mathematical relation correlating the Gaussian curvature to the following coefficients of the 2D Riemann-Christoffel curvature tensor: $R_{1212}$, $R_{1221}$, $R_{2121}$ and $R_{2112}$.
**Answer**:

$$K = \frac{b}{a} = \frac{R_{1212}}{a} = -\frac{R_{1221}}{a} = \frac{R_{2121}}{a} = -\frac{R_{2112}}{a}$$

These relations are justified by the fact that $b = R_{1212}$ plus the fact that the covariant Riemann-Christoffel curvature tensor is anti-symmetric in its first two indices and is anti-symmetric in its last two indices and hence $R_{1212} = -R_{2112} = -R_{1221} = R_{2121}$.

71. What is the Gaussian curvature $K$ of a Monge patch of the form $\mathbf{r}(u, v) = (u, v, f(u, v))$?
**Answer**:

$$K = \frac{f_{uu}f_{vv} - f_{uv}^2}{\left(1 + f_u^2 + f_v^2\right)^2}$$

where the subscripts $u$ and $v$ stand for partial derivatives of $f$ with respect to these surface coordinates.

72. Why the Gaussian curvature is independent of the orientation of the surface (where orientation is based on the choice of the direction of the unit normal vector to the surface)?
**Answer**: Because a change in the direction of the unit normal vector $\mathbf{n}$ will change the sign of the principal curvatures but not their absolute value and hence the magnitude is preserved. Furthermore, this change of sign will not affect the sign of the Gaussian curvature since both signs will be changed by the reversal of $\mathbf{n}$ direction and hence their product will not be affected, i.e. it is like multiplying by $(-1)^2 = 1$. Therefore, the sign and magnitude of the Gaussian curvature are both preserved under this reversal. We note that the Gaussian curvature is defined as the product of the principal curvatures, i.e. $K \equiv \kappa_1\kappa_2$.

73. Which of the following geometric shapes have identical Gaussian curvatures at their corresponding points and why: plane, sphere, cylinder, catenoid, ellipsoid, hyperbolic paraboloid, helicoid, and cone? Compare, in your answer, each pair of these shapes.
**Answer**: Plane and cylinder, plane and cone, cylinder and cone, and catenoid and helicoid have identical Gaussian curvatures because they are locally isometric.[38] All other pairs do not have identical Gaussian curvatures.

74. Write down an expression for the Gaussian curvature of a surface of revolution generated by revolving a sufficiently differentiable plane curve of the form $x = f(y)$ around the $y$-axis.
**Answer**:

$$K = -\frac{f_{yy}}{f\left(1 + f_y^2\right)^2}$$

where the subscript $y$ represents derivative of $f$ with respect to this variable.

[38] See Exercise 32 of § 6 about catenoid and helicoid.

75. Explain all the symbols of the following equation with discussion of its significance in relation to the intrinsic and extrinsic geometries of the surface:

$$\partial_u \mathbf{n} \times \partial_v \mathbf{n} = K (\mathbf{E}_1 \times \mathbf{E}_2)$$

**Answer**: $\partial_u$ and $\partial_v$ symbolize partial derivative with respect to $u$ and $v$, the symbol $\times$ represents cross product operation of two vectors, $\mathbf{n}$ is the normal unit vector to the surface, $K$ is the Gaussian curvature, $\mathbf{E}_1$ and $\mathbf{E}_2$ are the surface basis vectors, and all these symbols belong to a given point on a sufficiently smooth surface. Now, since $\mathbf{n}\cdot(\partial_u \mathbf{n} \times \partial_v \mathbf{n}) = K\,\mathbf{n}\cdot(\mathbf{E}_1 \times \mathbf{E}_2) = K\sqrt{a}$ we can conclude that the Gaussian curvature, which is an intrinsic property, can be defined by an extrinsic entity which is $\mathbf{n}$. However, the definition is ultimately based on intrinsic parameters since $\mathbf{n}$ is defined in terms of $\mathbf{E}_1$ and $\mathbf{E}_2$ (we should also note that $\partial_u \mathbf{n}$ and $\partial_v \mathbf{n}$ are surface vectors as they belong to the tangent space).

76. State the mathematical expression that correlates the Gaussian curvature $K$ to the Ricci curvature scalar $\mathcal{R}$ of a surface.
**Answer**:

$$|K| = \frac{|\mathcal{R}|}{2}$$

77. Using Eq. 47 and the parametric equations of Beltrami pseudo-sphere (Eqs. 1-3), show that the pseudo-sphere has a negative constant Gaussian curvature and find this curvature.
**Answer**: Noting that $(u, v)$ correspond to $(\theta, \phi)$, we have:

$$\begin{aligned}
\mathbf{E}_1 &= \partial_\theta \mathbf{r} = a\left(\cos\theta\cos\phi,\ \cos\theta\sin\phi,\ -\sin\theta + \frac{1}{2\cos(\theta/2)\sin(\theta/2)}\right) \\
&= a\left(\cos\theta\cos\phi,\ \cos\theta\sin\phi,\ -\sin\theta + \frac{1}{\sin\theta}\right) \\
\mathbf{E}_2 &= \partial_\phi \mathbf{r} = a(-\sin\theta\sin\phi,\ \sin\theta\cos\phi,\ 0) \\
E &= \mathbf{E}_1 \cdot \mathbf{E}_1 = a^2\left(-1 + \csc^2\theta\right) = a^2\cot^2\theta \\
F &= \mathbf{E}_1 \cdot \mathbf{E}_2 = 0 \\
G &= \mathbf{E}_2 \cdot \mathbf{E}_2 = a^2\sin^2\theta \\
\sqrt{EG} &= \sqrt{a^4\cot^2\theta\sin^2\theta} = \sqrt{a^4\cos^2\theta} = a^2\cos\theta \\
G_\theta &= \partial_\theta G = 2a^2\sin\theta\cos\theta \\
E_\phi &= \partial_\phi E = 0 \\
\frac{G_\theta}{\sqrt{EG}} &= \frac{2a^2\sin\theta\cos\theta}{a^2\cos\theta} = 2\sin\theta \\
\partial_\theta\left(\frac{G_\theta}{\sqrt{EG}}\right) &= 2\cos\theta
\end{aligned}$$

Hence:

$$K = -\frac{1}{2\sqrt{EG}}\left[\partial_\theta\left(\frac{G_\theta}{\sqrt{EG}}\right) + \partial_\phi\left(\frac{E_\phi}{\sqrt{EG}}\right)\right]$$

$$= -\frac{1}{2a^2 \cos\theta} [2\cos\theta + 0]$$
$$= -\frac{1}{a^2}$$

which is a negative constant Gaussian curvature. In fact, this is the same as the formula given in the book (i.e. $K = -\frac{1}{\rho^2}$ where $\rho$ is the pseudo-radius of the pseudo-sphere) with $a = \rho$.[39]

78. Classify surfaces with regard to their Gaussian curvature as having constant or variable curvature giving two examples at least for each.
**Answer**: We have:
• Surfaces with constant Gaussian curvature, where the constant curvature can be zero (planes or cylinders) or positive (spheres) or negative (pseudo-spheres).
• Surfaces with variable Gaussian curvature such as ellipsoids, tori and hyperboloids of one sheet.

79. What is the impact of scaling a surface up or down by a constant positive factor $c$ on its Gaussian curvature?
**Answer**: The Gaussian curvature will scale by the reciprocal square of that factor, i.e. by $1/c^2$.

80. Discuss the effect of an isometric mapping of a surface on its Gaussian curvature.
**Answer**: The Gaussian curvature is invariant under isometric transformations and hence two isometric surfaces have identical Gaussian curvature at their corresponding points.

81. State the Hilbert lemma giving examples for its applications from common types of surface.
**Answer**: The lemma states that if $P$ is a point on a sufficiently smooth surface $S$ with $\kappa_1$ and $\kappa_2$ being the principal curvatures of $S$ at $P$ such that: $\kappa_1 > \kappa_2$, $\kappa_1$ is a local maximum, and $\kappa_2$ is a local minimum, then the Gaussian curvature of $S$ at $P$ is non-positive, that is $K \leq 0$. An obvious example is the minimal circle of catenoid where $K < 0$. Another example is the interior circle of torus where $K$ is also negative.

82. Give the conditions for the validity of the following equation:

$$K = -\frac{1}{2\sqrt{EG}} \left[ \partial_u \left( \frac{G_u}{\sqrt{EG}} \right) + \partial_v \left( \frac{E_v}{\sqrt{EG}} \right) \right]$$

Also, give its simplified form in the case of representing the surface by geodesic coordinates stating the other conditions required for this simplification.
**Answer**: This equation applies to the Gaussian curvature of a surface of class $C^3$ with

[39] The reader should not confuse this $a$ with the symbol $a$ that represents the determinant of the surface covariant metric tensor. We also restrict the range of $\theta$ in this derivation to be between 0 and $\pi/2$ where the result of the other half of the pseudo-sphere can be obtained by symmetry.

orthogonal coordinate curves.
The equation will take the following simplified form:

$$K = -\frac{\partial_{uu}\sqrt{G}}{\sqrt{G}}$$

when the surface is represented by geodesic coordinates with the $u$ coordinate curves being geodesics and $u$ is a natural parameter.

83. Express the mean curvature as a function of the Gaussian curvature taking care of the signs.
**Answer**:

$$|H| = \sqrt{K + C^2}$$

where

$$C = \frac{\sqrt{(e^2G^2 + E^2g^2) - 4fF\,(eG + Eg) + 4\,(f^2EG + F^2eg) - 2egEG}}{2\,(EG - F^2)}$$

This can be verified as follows:

$$\begin{aligned}
& K + C^2 \\
= & \frac{eg - f^2}{EG - F^2} + \frac{(e^2G^2 + E^2g^2) - 4fF\,(eG + Eg) + 4\,(f^2EG + F^2eg) - 2egEG}{4\,(EG - F^2)^2} \\
= & \frac{e^2G^2 - 4efFG + 2egEG + 4f^2F^2 - 4fgEF + g^2E^2}{4\,(EG - F^2)^2} \\
= & \frac{(eG - 2fF + gE)^2}{4\,(EG - F^2)^2} \\
= & H^2
\end{aligned}$$

We note that since $K + C^2$ is equal to a square of a real number (i.e. $H^2$) then the root $\sqrt{K + C^2}$ is always real. The sign of $H$ (i.e. being plus or minus) should be determined according to the appropriate factors in the given problem such as the orientation of the surface which determines the signs of the principal curvatures.

84. Show that spheres are the only connected compact surfaces with constant mean curvature $H$ and positive Gaussian curvature $K$.
**Answer**: Since $K \equiv \kappa_1\kappa_2 > 0$ then $\kappa_1 \neq 0$, $\kappa_2 \neq 0$ and $\kappa_1$ and $\kappa_2$ have the same sign. Accordingly, $H \equiv \frac{\kappa_1+\kappa_2}{2} = c \neq 0$ where $c$ is a constant. Now, from the given conditions $\kappa_1$ and $\kappa_2$ must be continuous functions. Therefore, $\kappa_1$ must have a maximum $\kappa_{1M}$ at a given point $P$ and since $\kappa_1 + \kappa_2$ is constant then $\kappa_2$ must have a minimum $\kappa_{2m}$ at the same point $P$. So, from one of Hilbert theorems[40] we conclude that the point $P$

[40] According to this theorem, if $\kappa_1$ and $\kappa_2$ are the principal curvatures at a given point $P$ on a surface $S$ such that $\kappa_1$ is a local maximum, $\kappa_2$ is a local minimum and the Gaussian curvature at $P$ is positive

is spherical umbilical, i.e. $\kappa_{1M} = \kappa_{2m} \neq 0$. Now, for any other point on the surface we have:

$$\kappa_1 \le \kappa_{1M} = \kappa_{2m} \le \kappa_2$$

and hence $\kappa_1 = \kappa_2$,[41] i.e. the other point is also spherical umbilical. In other words, all the points of the surface are spherical umbilical.[42] So, from the result of Exercise 65 we conclude that the surface is a sphere.

85. Explain how the position of the surface in a deleted neighborhood of a given point $P$ relative to the tangent plane of the surface at $P$ is used to classify the nature of the Gaussian curvature at $P$. From this perspective, discuss the sign of the Gaussian curvature on the points of the following surfaces: hyperbolic paraboloid, sphere, torus and cylinder.
**Answer**: We have three main cases:
(a) The Gaussian curvature of a surface $S$ at a given point $P$ on the surface is positive if all the points of the surface in a deleted neighborhood of $P$ on $S$ are on the same side of the tangent plane to $S$ at $P$.
(b) The Gaussian curvature is negative if for all deleted neighborhoods of $P$ on $S$ some points are on one side of the tangent plane and some are on the other side.
(c) The Gaussian curvature is zero if, in a deleted neighborhood, either all the points lie in the tangent plane or all the points are on one side except some which lie on a curve in the tangent plane.
From this perspective, we conclude that hyperbolic paraboloid has negative Gaussian curvature because it belongs to case b, sphere has positive Gaussian curvature because it belongs to case a, and cylinder has zero Gaussian curvature because it belongs to case c (second instance). Regarding torus, it has points with positive Gaussian curvature (outer half) which represent case a, points with negative Gaussian curvature (inner half) which represent case b, and points with zero Gaussian curvature (top and bottom circles) which represent case c (second instance).

86. The Gaussian curvature of a developable surface is identically zero. Why?
**Answer**: Because developable surface is a warped plane without local distortion by compression or stretching and hence its intrinsic properties are preserved. Now, since the Gaussian curvature is an intrinsic property then the Gaussian curvature of developable surface should be the same as the Gaussian curvature of plane which is identically zero.

---

then $P$ is a spherical umbilical point. In fact, this theorem is closely related to the Hilbert lemma which is stated in Exercise 81 of this chapter and proved in Exercise 67 of § 5. In brief, from the lemma we can see that if $\kappa_1 > \kappa_2$ then the Gaussian curvature is non-positive, so we may intuitively conclude that if the Gaussian curvature is positive then the condition $\kappa_1 > \kappa_2$ must be invalid (i.e. $\kappa_1 = \kappa_2$) and hence the point must be spherical umbilical.

[41] This conclusion is based on the fact that by definition $\kappa_1 \ge \kappa_2$ and from the above equation we have $\kappa_1 \le \kappa_2$. Hence, we should have $\kappa_1 = \kappa_2$.

[42] The extension from the locality of one point to the locality of another point will ensure global extension.

87. What is the Gaussian curvature $K$ of a surface parameterized by: $x = (5 + \cos\phi)\cos\theta$, $y = (5 + \cos\phi)\sin\theta$ and $z = \sin\phi$?
**Answer**: Noting that $(u, v)$ correspond to $(\theta, \phi)$, we have:

$$\begin{aligned}
\mathbf{E}_1 &= \partial_\theta \mathbf{r} = (-5\sin\theta - \sin\theta\cos\phi,\ 5\cos\theta + \cos\theta\cos\phi,\ 0) \\
\mathbf{E}_2 &= \partial_\phi \mathbf{r} = (-\cos\theta\sin\phi,\ -\sin\theta\sin\phi,\ \cos\phi) \\
\mathbf{n} &= \frac{\mathbf{E}_1 \times \mathbf{E}_2}{|\mathbf{E}_1 \times \mathbf{E}_2|} \\
&= \frac{(5\cos\theta\cos\phi + \cos\theta\cos^2\phi,\ 5\sin\theta\cos\phi + \sin\theta\cos^2\phi,\ 5\sin\phi + \cos\phi\sin\phi)}{(5+\cos\phi)^2} \\
\partial_\theta \mathbf{E}_1 &= (-5\cos\theta - \cos\theta\cos\phi,\ -5\sin\theta - \sin\theta\cos\phi,\ 0) \\
\partial_\phi \mathbf{E}_1 &= (\sin\theta\sin\phi,\ -\cos\theta\sin\phi,\ 0) \\
\partial_\phi \mathbf{E}_2 &= (-\cos\theta\cos\phi,\ -\sin\theta\cos\phi,\ -\sin\phi) \\
E &= \mathbf{E}_1 \cdot \mathbf{E}_1 = (5+\cos\phi)^2 \\
F &= \mathbf{E}_1 \cdot \mathbf{E}_2 = 0 \\
G &= \mathbf{E}_2 \cdot \mathbf{E}_2 = 1 \\
e &= \mathbf{n} \cdot \partial_\theta \mathbf{E}_1 = -\cos\phi \\
f &= \mathbf{n} \cdot \partial_\phi \mathbf{E}_1 = 0 \\
g &= \mathbf{n} \cdot \partial_\phi \mathbf{E}_2 = \frac{-1}{5+\cos\phi} \\
EG - F^2 &= (5+\cos\phi)^2 \\
eg - f^2 &= \frac{\cos\phi}{5+\cos\phi} \\
K &= \frac{eg - f^2}{EG - F^2} = \frac{\cos\phi}{(5+\cos\phi)^3}
\end{aligned}$$

88. Provide a mathematical definition for the total curvature of a surface explaining all the symbols used in the definition.
**Answer**: The total curvature $K_t$ is the area integral of the Gaussian curvature $K$ over a surface or a patch of a surface, $S$, that is:

$$K_t = \iint_S K \, d\sigma$$

where $d\sigma$ is an infinitesimal area element on the surface.

89. Define all the symbols used in the following equation:

$$\underline{\epsilon}^{\alpha\beta}\underline{\epsilon}^{\gamma\delta}R_{\alpha\beta\gamma\delta} = K\underline{\epsilon}^{\alpha\beta}\underline{\epsilon}^{\gamma\delta}\underline{\epsilon}_{\alpha\beta}\underline{\epsilon}_{\gamma\delta}$$

**Answer**: $R_{\alpha\beta\gamma\delta}$ is the covariant Riemann-Christoffel curvature tensor, $K$ is the Gaussian curvature, and the indexed $\underline{\epsilon}$ are the contravariant and covariant absolute permutation tensors. All these symbols belong to a surface (i.e. 2D space) and hence all the indices range over 1,2.

90. Explain in detail how the following equation implies that the Gaussian curvature is a rank-0 tensor: $K = \frac{1}{4}\underline{\epsilon}^{\alpha\beta}\underline{\epsilon}^{\gamma\delta}R_{\alpha\beta\gamma\delta}$.
**Answer**: All the indices on the right hand side are contracted and hence they are bound indices (i.e. we have no free index). Therefore, the Gaussian curvature $K$ which is equal to this rank-0 tensor (i.e. $\frac{1}{4}\underline{\epsilon}^{\alpha\beta}\underline{\epsilon}^{\gamma\delta}R_{\alpha\beta\gamma\delta}$) should also be a rank-0 tensor.

91. Write the Gaussian curvature $K$ in terms of the surface curvature tensor using the most simple form.
**Answer**:

$$K = \frac{1}{2}\underline{\epsilon}^{\alpha\beta}\underline{\epsilon}^{\gamma\delta}b_{\gamma\alpha}b_{\delta\beta}$$

where $\underline{\epsilon}^{\alpha\beta}$ and $\underline{\epsilon}^{\gamma\delta}$ represent the 2D contravariant absolute permutation tensor, while $b_{\gamma\alpha}$ and $b_{\delta\beta}$ represent the covariant surface curvature tensor.

92. Algebraically manipulate the relation $K = \frac{b}{a}$ to obtain the following relation:

$$K = \frac{(\partial_u \mathbf{E}_1 \cdot \mathbf{E}_1 \times \mathbf{E}_2)(\partial_v \mathbf{E}_2 \cdot \mathbf{E}_1 \times \mathbf{E}_2) - (\partial_v \mathbf{E}_1 \cdot \mathbf{E}_1 \times \mathbf{E}_2)^2}{(EG - F^2)^2} \tag{48}$$

**Answer**: We have:

$$\begin{aligned}
E &= \mathbf{E}_1 \cdot \mathbf{E}_1 = |\mathbf{E}_1|^2 \\
F &= \mathbf{E}_1 \cdot \mathbf{E}_2 \\
G &= \mathbf{E}_2 \cdot \mathbf{E}_2 = |\mathbf{E}_2|^2 \\
e &= \partial_u \mathbf{E}_1 \cdot \mathbf{n} \\
f &= \partial_v \mathbf{E}_1 \cdot \mathbf{n} \\
g &= \partial_v \mathbf{E}_2 \cdot \mathbf{n} \\
|\mathbf{E}_1 \times \mathbf{E}_2| &= \sqrt{|\mathbf{E}_1|^2 |\mathbf{E}_2|^2 - (\mathbf{E}_1 \cdot \mathbf{E}_2)^2} = \sqrt{EG - F^2} \\
\mathbf{n} &= \frac{\mathbf{E}_1 \times \mathbf{E}_2}{|\mathbf{E}_1 \times \mathbf{E}_2|} = \frac{\mathbf{E}_1 \times \mathbf{E}_2}{\sqrt{EG - F^2}} \\
eg - f^2 &= (\partial_u \mathbf{E}_1 \cdot \mathbf{n})(\partial_v \mathbf{E}_2 \cdot \mathbf{n}) - (\partial_v \mathbf{E}_1 \cdot \mathbf{n})^2 \\
&= \frac{(\partial_u \mathbf{E}_1 \cdot \mathbf{E}_1 \times \mathbf{E}_2)(\partial_v \mathbf{E}_2 \cdot \mathbf{E}_1 \times \mathbf{E}_2) - (\partial_v \mathbf{E}_1 \cdot \mathbf{E}_1 \times \mathbf{E}_2)^2}{EG - F^2}
\end{aligned}$$

Hence:

$$\begin{aligned}
K &= \frac{b}{a} \\
&= \frac{eg - f^2}{EG - F^2} \\
&= \frac{(\partial_u \mathbf{E}_1 \cdot \mathbf{E}_1 \times \mathbf{E}_2)(\partial_v \mathbf{E}_2 \cdot \mathbf{E}_1 \times \mathbf{E}_2) - (\partial_v \mathbf{E}_1 \cdot \mathbf{E}_1 \times \mathbf{E}_2)^2}{(EG - F^2)^2}
\end{aligned}$$

93. Express the mean curvature $H$ in terms of the coefficients of the first and second fundamental forms $E, F, G, e, f, g$.
**Answer**:
$$H = \frac{eG - 2fF + gE}{2\,(EG - F^2)}$$

94. What is the relation between the mean curvature $H$ and the mixed type surface curvature tensor $b_\alpha^\beta$?
**Answer**: $H$ is half the trace of $b_\alpha^\beta$, i.e. $H = \frac{\mathrm{tr}\left(b_\alpha^\beta\right)}{2}$.

95. Compare the sign of the mean curvature to the sign of the Gaussian curvature with regard to their dependency on the direction of the unit normal vector to the surface.
**Answer**: The sign of the Gaussian curvature is independent of the direction of the unit normal vector while the sign of the mean curvature is dependent on the direction of the unit normal vector.

96. Give two examples of common types of surface over which the mean curvature is constant. Also, give an example of a surface with variable mean curvature.
**Answer**: Plane and sphere are examples of surface over which the mean curvature is constant. Torus is an example of a surface with variable mean curvature.

97. What is the mean curvature $H$ of a Monge patch of the form $\mathbf{r}(u, v) = (u, v, f(u, v))$?
**Answer**:
$$H = \frac{\left(1 + f_v^2\right) f_{uu} - 2 f_u f_v f_{uv} + \left(1 + f_u^2\right) f_{vv}}{2\left(1 + f_u^2 + f_v^2\right)^{3/2}}$$
where the subscripts $u$ and $v$ stand for partial derivatives of $f$ with respect to these surface coordinates.

98. What is the essence of Gauss *Theorema Egregium*? Give an example of an equation or a theorem that demonstrates this *theorem*.
**Answer**: The essence of *Theorema Egregium* is that 2D spaces (i.e. surfaces) can have an intrinsic curvature which is usually represented by the Gaussian curvature. This can be seen for example from the following equation $K = \frac{R_{1212}}{a}$ where the Gaussian curvature $K$ is expressed in terms of the coefficient of the Riemann-Christoffel curvature tensor $R_{1212}$ and the determinant of the covariant metric tensor $a$ where both of these are intrinsic attributes to the surface.

99. Derive the equation of *Theorema Egregium* (as given in the book by Eq. 385) using Eq. 48.
**Answer**: It is obvious that $(\partial_u \mathbf{E}_1 \cdot \mathbf{E}_1 \times \mathbf{E}_2)$, $(\partial_v \mathbf{E}_2 \cdot \mathbf{E}_1 \times \mathbf{E}_2)$ and $(\partial_v \mathbf{E}_1 \cdot \mathbf{E}_1 \times \mathbf{E}_2)$ are scalar triple products and hence they can be expressed as determinants. Now, if for simplicity we use $\mathbf{r}_u \equiv \mathbf{E}_1$, $\mathbf{r}_v \equiv \mathbf{E}_2$, $a \equiv EG - F^2$, $\det \begin{pmatrix} \mathbf{a} \\ \mathbf{b} \\ \mathbf{c} \end{pmatrix}$ and $\begin{bmatrix} \mathbf{a} \\ \mathbf{b} \\ \mathbf{c} \end{bmatrix}$ as the determinant and matrix whose rows are $\mathbf{a}, \mathbf{b}, \mathbf{c}$ vectors, and $\det \begin{pmatrix} \mathbf{a} & \mathbf{b} & \mathbf{c} \end{pmatrix}$ and

$\begin{bmatrix} \mathbf{a} & \mathbf{b} & \mathbf{c} \end{bmatrix}$ as the determinant and matrix whose columns are $\mathbf{a}, \mathbf{b}, \mathbf{c}$ vectors, then Eq. 48 can be written as:

$$\begin{aligned}
Ka^2 &= \det\begin{pmatrix} \mathbf{r}_{uu} & \mathbf{r}_u & \mathbf{r}_v \end{pmatrix}\det\begin{pmatrix} \mathbf{r}_{vv} & \mathbf{r}_u & \mathbf{r}_v \end{pmatrix} - \left[\det\begin{pmatrix} \mathbf{r}_{uv} & \mathbf{r}_u & \mathbf{r}_v \end{pmatrix}\right]^2 \\
&= \det\begin{pmatrix} \mathbf{r}_{uu} \\ \mathbf{r}_u \\ \mathbf{r}_v \end{pmatrix}\det\begin{pmatrix} \mathbf{r}_{vv} & \mathbf{r}_u & \mathbf{r}_v \end{pmatrix} - \det\begin{pmatrix} \mathbf{r}_{uv} \\ \mathbf{r}_u \\ \mathbf{r}_v \end{pmatrix}\det\begin{pmatrix} \mathbf{r}_{uv} & \mathbf{r}_u & \mathbf{r}_v \end{pmatrix} \\
&= \det\left(\begin{bmatrix} \mathbf{r}_{uu} \\ \mathbf{r}_u \\ \mathbf{r}_v \end{bmatrix}\begin{bmatrix} \mathbf{r}_{vv} & \mathbf{r}_u & \mathbf{r}_v \end{bmatrix}\right) - \det\left(\begin{bmatrix} \mathbf{r}_{uv} \\ \mathbf{r}_u \\ \mathbf{r}_v \end{bmatrix}\begin{bmatrix} \mathbf{r}_{uv} & \mathbf{r}_u & \mathbf{r}_v \end{bmatrix}\right) \\
&= \det\left(\begin{bmatrix} \mathbf{r}_{uu}\cdot\mathbf{r}_{vv} & \mathbf{r}_{uu}\cdot\mathbf{r}_u & \mathbf{r}_{uu}\cdot\mathbf{r}_v \\ \mathbf{r}_u\cdot\mathbf{r}_{vv} & \mathbf{r}_u\cdot\mathbf{r}_u & \mathbf{r}_u\cdot\mathbf{r}_v \\ \mathbf{r}_v\cdot\mathbf{r}_{vv} & \mathbf{r}_v\cdot\mathbf{r}_u & \mathbf{r}_v\cdot\mathbf{r}_v \end{bmatrix}\right) - \det\left(\begin{bmatrix} \mathbf{r}_{uv}\cdot\mathbf{r}_{uv} & \mathbf{r}_{uv}\cdot\mathbf{r}_u & \mathbf{r}_{uv}\cdot\mathbf{r}_v \\ \mathbf{r}_u\cdot\mathbf{r}_{uv} & \mathbf{r}_u\cdot\mathbf{r}_u & \mathbf{r}_u\cdot\mathbf{r}_v \\ \mathbf{r}_v\cdot\mathbf{r}_{uv} & \mathbf{r}_v\cdot\mathbf{r}_u & \mathbf{r}_v\cdot\mathbf{r}_v \end{bmatrix}\right) \\
&= \begin{vmatrix} \mathbf{r}_{uu}\cdot\mathbf{r}_{vv} & \mathbf{r}_u\cdot\mathbf{r}_{vv} & \mathbf{r}_v\cdot\mathbf{r}_{vv} \\ \mathbf{r}_{uu}\cdot\mathbf{r}_u & \mathbf{r}_u\cdot\mathbf{r}_u & \mathbf{r}_v\cdot\mathbf{r}_u \\ \mathbf{r}_{uu}\cdot\mathbf{r}_v & \mathbf{r}_u\cdot\mathbf{r}_v & \mathbf{r}_v\cdot\mathbf{r}_v \end{vmatrix} - \begin{vmatrix} \mathbf{r}_{uv}\cdot\mathbf{r}_{uv} & \mathbf{r}_{uv}\cdot\mathbf{r}_u & \mathbf{r}_{uv}\cdot\mathbf{r}_v \\ \mathbf{r}_u\cdot\mathbf{r}_{uv} & \mathbf{r}_u\cdot\mathbf{r}_u & \mathbf{r}_u\cdot\mathbf{r}_v \\ \mathbf{r}_v\cdot\mathbf{r}_{uv} & \mathbf{r}_v\cdot\mathbf{r}_u & \mathbf{r}_v\cdot\mathbf{r}_v \end{vmatrix} \\
&= \begin{vmatrix} \mathbf{r}_{uu}\cdot\mathbf{r}_{vv} & F_v - \frac{1}{2}G_u & \frac{1}{2}G_v \\ \frac{1}{2}E_u & E & F \\ F_u - \frac{1}{2}E_v & F & G \end{vmatrix} - \begin{vmatrix} \mathbf{r}_{uv}\cdot\mathbf{r}_{uv} & \frac{1}{2}E_v & \frac{1}{2}G_u \\ \frac{1}{2}E_v & E & F \\ \frac{1}{2}G_u & F & G \end{vmatrix} \\
&= \begin{vmatrix} (\mathbf{r}_{uu}\cdot\mathbf{r}_{vv} - \mathbf{r}_{uv}\cdot\mathbf{r}_{uv}) & F_v - \frac{1}{2}G_u & \frac{1}{2}G_v \\ \frac{1}{2}E_u & E & F \\ F_u - \frac{1}{2}E_v & F & G \end{vmatrix} - \begin{vmatrix} 0 & \frac{1}{2}E_v & \frac{1}{2}G_u \\ \frac{1}{2}E_v & E & F \\ \frac{1}{2}G_u & F & G \end{vmatrix} \\
&= \begin{vmatrix} \frac{1}{2}(-E_{vv} + 2F_{uv} - G_{uu}) & F_v - \frac{1}{2}G_u & \frac{1}{2}G_v \\ \frac{1}{2}E_u & E & F \\ F_u - \frac{1}{2}E_v & F & G \end{vmatrix} - \begin{vmatrix} 0 & \frac{1}{2}E_v & \frac{1}{2}G_u \\ \frac{1}{2}E_v & E & F \\ \frac{1}{2}G_u & F & G \end{vmatrix}
\end{aligned}$$

where in line 2 we use the fact that the determinant is invariant to transposition, in line 3 we use the fact that the product of determinants in equal to the determinant of the product, in line 4 we use the definition of matrix product as the dot product of rows of first matrix with columns of second matrix, in line 5 we transpose the first determinant and use our notation for determinant, in line 6 we use relations that are justified in the upcoming note, in line 7 we use the fact that the minor of the first element in both determinants is identical and hence by distributivity we obtain the above result, and in line 8 we substitute for $\mathbf{r}_{uu}\cdot\mathbf{r}_{vv} - \mathbf{r}_{uv}\cdot\mathbf{r}_{uv}$ (which we justified in the upcoming note).
Note: The relations $\mathbf{r}_u\cdot\mathbf{r}_u = E$, $\mathbf{r}_u\cdot\mathbf{r}_v = \mathbf{r}_v\cdot\mathbf{r}_u = F$ and $\mathbf{r}_v\cdot\mathbf{r}_v = G$ are obvious since $\mathbf{r}_u \equiv \mathbf{E}_1$ and $\mathbf{r}_v \equiv \mathbf{E}_2$. Regarding the other relations, we have:

$$\begin{aligned}
\mathbf{r}_u\cdot\mathbf{r}_{vv} &= \mathbf{E}_1\cdot\partial_2\mathbf{E}_2 = [22,1] = F_v - \frac{1}{2}G_u \\
\mathbf{r}_v\cdot\mathbf{r}_{vv} &= \mathbf{E}_2\cdot\partial_2\mathbf{E}_2 = [22,2] = \frac{G_v}{2} \\
\mathbf{r}_{uu}\cdot\mathbf{r}_u &= \partial_1\mathbf{E}_1\cdot\mathbf{E}_1 = [11,1] = \frac{1}{2}E_u
\end{aligned}$$

$$\begin{aligned}
\mathbf{r}_{uu}\cdot\mathbf{r}_v &= \partial_1\mathbf{E}_1\cdot\mathbf{E}_2 = [11,2] = F_u - \frac{1}{2}E_v \\
\mathbf{r}_{uv}\cdot\mathbf{r}_u &= \partial_2\mathbf{E}_1\cdot\mathbf{E}_1 = [12,1] = \frac{1}{2}E_v = \mathbf{r}_u\cdot\mathbf{r}_{uv} \\
\mathbf{r}_{uv}\cdot\mathbf{r}_v &= \partial_2\mathbf{E}_1\cdot\mathbf{E}_2 = [12,2] = \frac{1}{2}G_u = \mathbf{r}_v\cdot\mathbf{r}_{uv}
\end{aligned}$$

Also:

$$\begin{aligned}
(\mathbf{r}_{uu}\cdot\mathbf{r}_v)_v &= \mathbf{r}_{uuv}\cdot\mathbf{r}_v + \mathbf{r}_{uu}\cdot\mathbf{r}_{vv} = \left(F_u - \frac{1}{2}E_v\right)_v = F_{uv} - \frac{1}{2}E_{vv} \\
(\mathbf{r}_{uv}\cdot\mathbf{r}_v)_u &= \mathbf{r}_{uvu}\cdot\mathbf{r}_v + \mathbf{r}_{uv}\cdot\mathbf{r}_{vu} = \left(\frac{1}{2}G_u\right)_u = \frac{1}{2}G_{uu} \\
\mathbf{r}_{uu}\cdot\mathbf{r}_{vv} - \mathbf{r}_{uv}\cdot\mathbf{r}_{uv} &= (\mathbf{r}_{uu}\cdot\mathbf{r}_v)_v - (\mathbf{r}_{uv}\cdot\mathbf{r}_v)_u = F_{uv} - \frac{1}{2}E_{vv} - \frac{1}{2}G_{uu}
\end{aligned}$$

100. Write down the mathematical equation representing the local form of the Gauss-Bonnet theorem explaining all the symbols involved.
**Answer**: If $\mathfrak{S}$ is a simply connected region on a surface of class $C^3$ where $\mathfrak{S}$ is bordered by a finite number $m$ of piecewise regular curves $C_j$ that meet in $n$ corners then we have:

$$\sum_{j=1}^{m}\int_{C_j}\kappa_g + \sum_{k=1}^{n}\phi_k + \iint_{\mathfrak{S}} K d\sigma = 2\pi \tag{49}$$

where the first sum is over the curves while the second sum is over the corners, $\kappa_g$ is the geodesic curvature of the curves $C_j$ as a function of their coordinates, $\phi_k$ are the exterior angles of the corners and $K$ is the Gaussian curvature of $\mathfrak{S}$ as a function of the coordinates over $\mathfrak{S}$.

101. Give an example for the application of the local Gauss-Bonnet theorem using a planar geometric shape and another example using a non-planar shape.
**Answer**: The first example is a semi-circular disc in a plane with radius $R$ where Eq. 49 becomes:

$$\left(\frac{1}{R}\pi R + 0\times 2R\right) + 2\left(\frac{\pi}{2}\right) + 0 = \pi + \pi + 0 \equiv 2\pi$$

The second example is a hemisphere of radius $R$ where Eq. 49 becomes:

$$0\,(2\pi R) + 0 + \frac{1}{R^2}2\pi R^2 = 0 + 0 + 2\pi \equiv 2\pi$$

102. Explain why two geodesic curves on a patch of a surface with negative Gaussian curvature $K$ cannot intersect at two points.
**Answer**: Because on introducing a vertex at a regular point on one curve we will have an artificial corner with zero exterior angle and $\pi$ interior angle, and hence we will have a geodesic triangle whose interior angles add up to more than $\pi$ on a surface over which $K < 0$, which is impossible.

103. Apply the Gauss-Bonnet theorem on the spherical triangle of Figure 18 giving detailed explanations for each step.

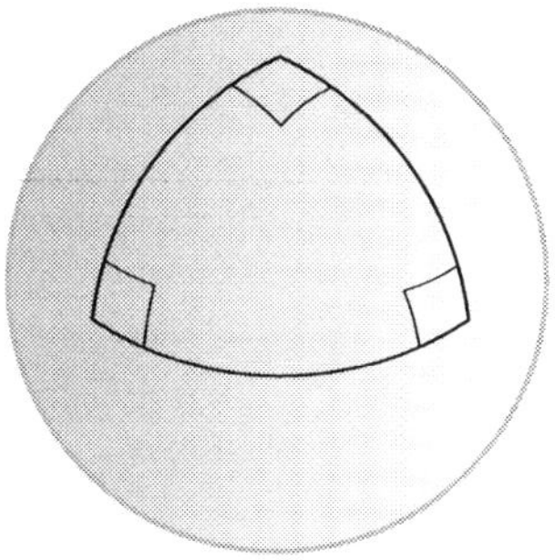

Figure 18: A spherical triangle with three right angles on the surface of a sphere. The three sides of this spherical triangle are arcs of great circles.

**Answer**: Since the three sides of this spherical triangle are arcs of great circles then the geodesic curvature $\kappa_g$ is identically zero over these curves and hence the first term in Eq. 49 is zero. Since the three interior angles are right angles then the second term in Eq. 49 (which represents the sum of the exterior angles) should be $\frac{3\pi}{2}$. Since a sphere of radius $R$ has $K = \frac{1}{R^2}$ and the area of the triangle is 1/8 of the area of the sphere then the third term (which represents the area integral) is $\frac{1}{R^2}\frac{4\pi R^2}{8} = \frac{\pi}{2}$. So, on adding up these three terms we get: $0 + \frac{3\pi}{2} + \frac{\pi}{2} = 2\pi$, as it should be.

104. Use a circular flat disc to demonstrate the application of the local form of the Gauss-Bonnet theorem giving detailed explanations for each step.
**Answer**: Since the normal curvature $\kappa_n$ is zero over any curve on this flat disc then the geodesic curvature $\kappa_g$ over the perimeter of the disc is equal to the curvature $\kappa$ of the perimeter (which is a circle) and hence it is equal to the reciprocal of the disc radius $R$. Moreover, the length of the perimeter is $2\pi R$ and hence the line integral (represented by the first term of Eq. 49) is $\frac{1}{R}2\pi R = 2\pi$. The second term is zero because the exterior angle of any artificial corner is zero and hence the sum in the second term of Eq. 49 should be zero (whatever artificial corners we introduce). The Gaussian curvature $K$ is identically zero over any plane surface and hence $K = 0$ over this flat disc. Therefore, the area integral (represented by the third term of Eq. 49) is zero. So, on adding up these three terms we get: $2\pi + 0 + 0 = 2\pi$, as it should be.

105. What is the global form of the Gauss-Bonnet theorem and what is its significance geometrically and topologically?
**Answer**: On a compact orientable surface $S$ of class $C^3$ the Euler characteristic $\chi$ and the Gaussian curvature $K$ are linked by the following relation:

$$\iint_S K d\sigma = 2\pi\chi \tag{50}$$

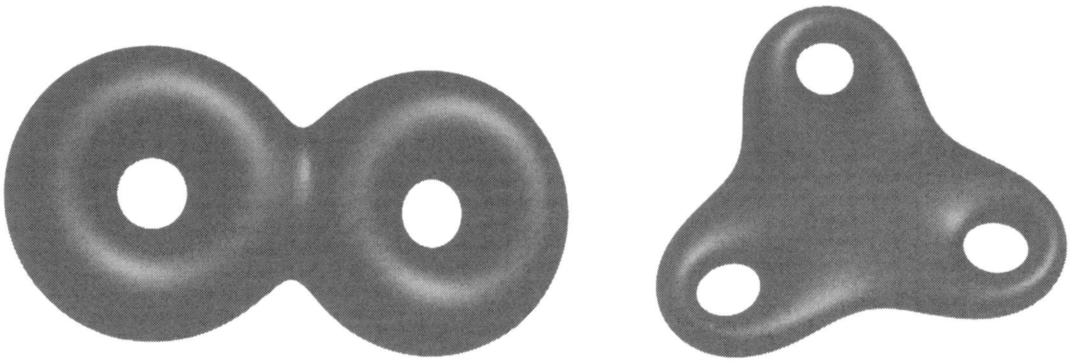

Figure 19: Examples of surfaces of genus 2 (left frame) and genus 3 (right frame).

The significance of this theorem is that it correlates a topological invariant $\chi$ to a geometric invariant $K$ and hence it establishes a relation between the topology and the geometry of the surface.

106. Find the total curvature of the surfaces depicted in Figure 19.
**Answer**: The genus of the surface on the left is $\mathfrak{g} = 2$ while the genus of the surface on the right is $\mathfrak{g} = 3$. Hence, from the relation $\chi = 2(1 - \mathfrak{g})$ we conclude that their Euler characteristic is $\chi = -2$ and $\chi = -4$ respectively. So, from Eq. 50 we conclude that their total curvature is $K_t = -4\pi$ and $K_t = -8\pi$ respectively.

107. Show, mathematically, that the area of a geodesic polygon on a surface with constant non-vanishing Gaussian curvature $K$ is determined by the sum of the internal angles of the polygon.
**Answer**: For geodesic polygon $\kappa_g = 0$ and hence Eq. 49 becomes:

$$\begin{aligned}
\sum_{k=1}^{n} \phi_k + \iint_{\mathfrak{S}} K d\sigma &= 2\pi \\
\iint_{\mathfrak{S}} K d\sigma &= 2\pi - \sum_{k=1}^{n} \phi_k \\
K \iint_{\mathfrak{S}} d\sigma &= 2\pi - \sum_{k=1}^{n} \phi_k \\
\iint_{\mathfrak{S}} d\sigma &= \frac{1}{K}\left(2\pi - \sum_{k=1}^{n} \phi_k\right)
\end{aligned}$$

where lines 3 and 4 are justified by the fact that $K$ is constant and non-vanishing. Now, the left hand side of the last line is the area of the geodesic polygon while everything on the right hand side is constant except $\phi_k$. This means that the area is determined by the sum of the internal angles $\theta_s$ of the polygon since $\theta_s$ is determined by the sum of $\phi_k$, i.e. $\theta_s = n\pi - \sum_{k=1}^{n} \phi_k$.

108. Verify that the total curvatures of ellipsoid and torus are respectively $4\pi$ and 0 by performing detailed surface integral calculations.
**Answer**: The ellipsoid is represented parametrically by:

$$\mathbf{r}(\theta,\phi) = (a\sin\theta\cos\phi, b\sin\theta\sin\phi, c\cos\theta)$$

where $a, b, c$ are constants and $0 \le \theta \le \pi$ and $0 \le \phi \le 2\pi$. Hence, we have:

$$\begin{aligned}
\mathbf{E}_1 &= \partial_\theta \mathbf{r} = (a\cos\theta\cos\phi,\ b\cos\theta\sin\phi,\ -c\sin\theta) \\
\mathbf{E}_2 &= \partial_\phi \mathbf{r} = (-a\sin\theta\sin\phi, b\sin\theta\cos\phi, 0) \\
\partial_\theta \mathbf{E}_1 &= (-a\sin\theta\cos\phi,\ -b\sin\theta\sin\phi,\ -c\cos\theta) \\
\partial_\phi \mathbf{E}_1 &= (-a\cos\theta\sin\phi,\ b\cos\theta\cos\phi,\ 0) \\
\partial_\phi \mathbf{E}_2 &= (-a\sin\theta\cos\phi, -b\sin\theta\sin\phi, 0) \\
E &= \mathbf{E}_1\cdot\mathbf{E}_1 = a^2\cos^2\theta\cos^2\phi + b^2\cos^2\theta\sin^2\phi + c^2\sin^2\theta \\
F &= \mathbf{E}_1\cdot\mathbf{E}_2 = -a^2\cos\theta\sin\theta\cos\phi\sin\phi + b^2\cos\theta\sin\theta\cos\phi\sin\phi \\
G &= \mathbf{E}_2\cdot\mathbf{E}_2 = a^2\sin^2\theta\sin^2\phi + b^2\sin^2\theta\cos^2\phi \\
EG - F^2 &= \sin^2\theta\left(a^2b^2\cos^2\theta + a^2c^2\sin^2\theta\sin^2\phi + b^2c^2\sin^2\theta\cos^2\phi\right) \\
\mathbf{n} &= \frac{\mathbf{E}_1\times\mathbf{E}_2}{\sqrt{EG-F^2}} = \frac{(bc\sin\theta\cos\phi, ac\sin\theta\sin\phi, ab\cos\theta)}{\sqrt{a^2b^2\cos^2\theta + a^2c^2\sin^2\theta\sin^2\phi + b^2c^2\sin^2\theta\cos^2\phi}} \\
e &= \mathbf{n}\cdot\partial_\theta\mathbf{E}_1 = \frac{-abc}{\sqrt{a^2b^2\cos^2\theta + a^2c^2\sin^2\theta\sin^2\phi + b^2c^2\sin^2\theta\cos^2\phi}} \\
f &= \mathbf{n}\cdot\partial_\phi\mathbf{E}_1 = 0 \\
g &= \mathbf{n}\cdot\partial_\phi\mathbf{E}_2 = \frac{-abc\sin^2\theta}{\sqrt{a^2b^2\cos^2\theta + a^2c^2\sin^2\theta\sin^2\phi + b^2c^2\sin^2\theta\cos^2\phi}} \\
eg - f^2 &= \frac{a^2b^2c^2\sin^2\theta}{a^2b^2\cos^2\theta + a^2c^2\sin^2\theta\sin^2\phi + b^2c^2\sin^2\theta\cos^2\phi} \\
K &= \frac{eg-f^2}{EG-F^2} = \frac{a^2b^2c^2}{\left(a^2b^2\cos^2\theta + a^2c^2\sin^2\theta\sin^2\phi + b^2c^2\sin^2\theta\cos^2\phi\right)^2}
\end{aligned}$$

Therefore, the total curvature of ellipsoid is:

$$\begin{aligned}
&\iint_S K\,d\sigma \\
&= \int_{\theta=0}^{\theta=\pi}\int_{\phi=0}^{\phi=2\pi} K\sqrt{EG-F^2}\,d\phi d\theta \\
&= \int_{\theta=0}^{\theta=\pi}\int_{\phi=0}^{\phi=2\pi} \frac{a^2b^2c^2\sin\theta\sqrt{a^2b^2\cos^2\theta + a^2c^2\sin^2\theta\sin^2\phi + b^2c^2\sin^2\theta\cos^2\phi}}{\left(a^2b^2\cos^2\theta + a^2c^2\sin^2\theta\sin^2\phi + b^2c^2\sin^2\theta\cos^2\phi\right)^2} d\phi d\theta \\
&= \int_{\theta=0}^{\theta=\pi}\int_{\phi=0}^{\phi=2\pi} \frac{a^2b^2c^2\sin\theta}{\left(a^2b^2\cos^2\theta + a^2c^2\sin^2\theta\sin^2\phi + b^2c^2\sin^2\theta\cos^2\phi\right)^{3/2}} d\phi d\theta
\end{aligned}$$

However, we could not find an analytical evaluation for this integral. So, we evaluated the integral for special cases of $a, b, c$ values and we obtained correct results. For example, for the case $a = b = c$ (which represents sphere) where the factors $a, b, c$ cancel out we obtained $4\pi$. We also obtained $4\pi$ for many other cases, e.g. $a = 1$, $b = 2$ and $c = 3$.[43]
Regarding the torus, it is represented parametrically by:

$$\mathbf{r}(\theta, \phi) = (R\cos\theta + r\cos\theta\cos\phi,\ R\sin\theta + r\sin\theta\cos\phi,\ r\sin\phi)$$

where $R$ is the torus radius, $r$ is the radius of the generating circle, $\theta \in [0, 2\pi)$ is the angle of variation of $R$ and $\phi \in [0, 2\pi)$ is the angle of variation of $r$. Hence, we have:

$$\begin{aligned}
\mathbf{E}_1 &= \partial_\theta \mathbf{r} = (-R\sin\theta - r\sin\theta\cos\phi,\ R\cos\theta + r\cos\theta\cos\phi,\ 0) \\
\mathbf{E}_2 &= \partial_\phi \mathbf{r} = (-r\cos\theta\sin\phi,\ -r\sin\theta\sin\phi,\ r\cos\phi) \\
\partial_\theta \mathbf{E}_1 &= (-R\cos\theta - r\cos\theta\cos\phi,\ -R\sin\theta - r\sin\theta\cos\phi,\ 0) \\
\partial_\phi \mathbf{E}_1 &= (r\sin\theta\sin\phi,\ -r\cos\theta\sin\phi,\ 0) \\
\partial_\phi \mathbf{E}_2 &= (-r\cos\theta\cos\phi,\ -r\sin\theta\cos\phi,\ -r\sin\phi) \\
E &= \mathbf{E}_1 \cdot \mathbf{E}_1 = (R + r\cos\phi)^2 \\
F &= \mathbf{E}_1 \cdot \mathbf{E}_2 = 0 \\
G &= \mathbf{E}_2 \cdot \mathbf{E}_2 = r^2 \\
EG - F^2 &= r^2 (R + r\cos\phi)^2 \\
\mathbf{n} &= \frac{\mathbf{E}_1 \times \mathbf{E}_2}{\sqrt{EG - F^2}} = (\cos\theta\cos\phi,\ \sin\theta\cos\phi,\ \sin\phi) \\
e &= \mathbf{n} \cdot \partial_\theta \mathbf{E}_1 = -\cos\phi (R + r\cos\phi) \\
f &= \mathbf{n} \cdot \partial_\phi \mathbf{E}_1 = 0 \\
g &= \mathbf{n} \cdot \partial_\phi \mathbf{E}_2 = -r \\
eg - f^2 &= r\cos\phi (R + r\cos\phi) \\
K &= \frac{eg - f^2}{EG - F^2} = \frac{\cos\phi}{r (R + r\cos\phi)}
\end{aligned}$$

Therefore, the total curvature of torus is:

$$\begin{aligned}
\iint_S K d\sigma &= \int_{\phi=0}^{\phi=2\pi} \int_{\theta=0}^{\theta=2\pi} K\sqrt{EG - F^2} d\theta d\phi \\
&= \int_{\phi=0}^{\phi=2\pi} \int_{\theta=0}^{\theta=2\pi} \cos\phi d\theta d\phi
\end{aligned}$$

[43] The reader can check these results using for example the free online Multiple Integral Calculator at: https://www.emathhelp.net/calculators/calculus-3/multiple-double-triple-integral-calculator/.
For example, for the case $a = 1$, $b = 2$ and $c = 3$ use the following inputs:
**Enter a function**: 36 sin (y) /( (4 cos ^2 (y)+9 sin ^2 (y) sin ^2 (x)+36 sin ^2 (y) cos ^2 (x)))^(3/2)
**Enter bounds**: y,0,pi;x,0,2 pi

$$
\begin{aligned}
&= 2\pi \int_{\phi=0}^{\phi=2\pi} \cos\phi d\phi \\
&= 2\pi \left[\sin\phi\right]_0^{2\pi} \\
&= 0
\end{aligned}
$$

109. Outline the usefulness of the global form of the Gauss-Bonnet theorem in obtaining the total curvature of a surface with known topological properties without performing detailed calculations.
**Answer**: From Eq. 50, it can be seen that all we need to obtain the total curvature of a surface is to know its Euler characteristic $\chi$. This is usually much easier than performing lengthy and usually complex area integrals which may not even be possible (see the previous exercise as an example).

110. Using the Gauss-Bonnet theorem, prove that the Gaussian curvature is identically zero on a surface $S$ if at any point $P$ on $S$ there are two families of geodesic curves in the neighborhood of $P$ intersecting at a constant angle.
**Answer**: If $P$ is an arbitrary point on $S$ surrounded by a geodesic quadrilateral then from Eq. 49 we get:

$$\iint_{\mathfrak{S}} K d\sigma = 2\pi - \sum_{k=1}^{4} \phi_k$$

where this is justified by the fact that $\kappa_g$ is identically zero on the quadrilateral. Because the geodesic curves are intersecting at a constant angle then we should have $\sum_{k=1}^{4} \phi_k = 2\pi$ and hence we conclude from the above equation that the total curvature of the part of the surface surrounded by the quadrilateral is zero, i.e. $\iint_{\mathfrak{S}} K d\sigma = 0$. Now, on shrinking the quadrilateral onto the point $P$ the equality $\iint_{\mathfrak{S}} K d\sigma = 0$ can only be true if $K$ at $P$ is zero. Since $P$ is an arbitrary point on the surface then this applies to the whole surface, i.e. the Gaussian curvature is identically zero on the surface, as required.

111. Write down the mathematical relation that links the Euler characteristic $\chi$ of a surface to its topological genus $\mathfrak{g}$.
**Answer**:

$$\chi = 2(1 - \mathfrak{g})$$

112. Use the principal curvatures, $\kappa_1$ and $\kappa_2$, and the mean and Gaussian curvatures, $H$ and $K$, to classify the points with regard to the local shape of the surface as flat, elliptic, parabolic and hyperbolic giving examples of common geometric shapes for each case.
**Answer**: The point is:
• Flat when $\kappa_1 = \kappa_2 = 0$, and hence $H = K = 0$ (since $H = \frac{\kappa_1+\kappa_2}{2}$ and $K = \kappa_1\kappa_2$). Example: plane.
• Elliptic when either $\kappa_1 > 0$ and $\kappa_2 > 0$ or $\kappa_1 < 0$ and $\kappa_2 < 0$, and hence $K > 0$. Example: ellipsoid.
• Parabolic when either $\kappa_1 = 0$ and $\kappa_2 \neq 0$ or $\kappa_2 = 0$ and $\kappa_1 \neq 0$, and hence $K = 0$ and

$H \neq 0$. Example: cylinder.
• Hyperbolic when $\kappa_1 > 0$ and $\kappa_2 < 0$, and hence $K < 0$. Example: catenoid.

113. Repeat the classification of the previous question using this time the coefficients of the second fundamental form of the surface.
**Answer**: The point is:
• Flat when $eg - f^2 = 0$ and $e = f = g = 0$.
• Elliptic when $eg - f^2 > 0$.
• Parabolic when $eg - f^2 = 0$ and $e^2 + f^2 + g^2 \neq 0$.
• Hyperbolic when $eg - f^2 < 0$.

114. Prove that on a circular cylinder all points are parabolic.
**Answer**: Circular cylinder can be parameterized spatially as $\mathbf{r}(\phi, t) = (a\cos\phi, a\sin\phi, t)$ where $a$ is a non-zero constant and $0 \leq \phi < 2\pi$ and $-\infty < t < \infty$. So, we have:

$$\begin{aligned}
\mathbf{E}_1 &= \partial_\phi \mathbf{r} = (-a\sin\phi, a\cos\phi, 0) \\
\mathbf{E}_2 &= \partial_t \mathbf{r} = (0, 0, 1) \\
\mathbf{n} &= \frac{\mathbf{E}_1 \times \mathbf{E}_2}{|\mathbf{E}_1 \times \mathbf{E}_2|} = (\cos\phi, \sin\phi, 0) \\
\partial_\phi \mathbf{E}_1 &= (-a\cos\phi, -a\sin\phi, 0) \\
\partial_t \mathbf{E}_1 &= (0, 0, 0) \\
\partial_t \mathbf{E}_2 &= (0, 0, 0) \\
e &= \mathbf{n} \cdot \partial_\phi \mathbf{E}_1 = -a \\
f &= \mathbf{n} \cdot \partial_t \mathbf{E}_1 = 0 \\
g &= \mathbf{n} \cdot \partial_t \mathbf{E}_2 = 0 \\
eg - f^2 &= 0 \\
e^2 + f^2 + g^2 &= a^2 \neq 0
\end{aligned}$$

Hence, from the answer of the previous question we conclude that all the points on a circular cylinder are parabolic, as required.

115. Show that in the neighborhood of an elliptic point on a surface, the surface lies on one side of its tangent plane at that point.
**Answer**: By definition, a point $P$ is elliptic if either $\kappa_1 > 0$ and $\kappa_2 > 0$ or $\kappa_1 < 0$ and $\kappa_2 < 0$ (see Exercise 112). Now, since the principal curvatures $\kappa_1$ and $\kappa_2$ are curvatures (i.e. $\kappa$ with a given sign) of normal sections then this means that the curvatures (none of which is zero) of all the normal sections have the same sign (i.e. greater than zero or less than zero) and hence all the curvature vectors at $P$ point in the same direction, i.e. all the normal sections curve on the same side of the tangent plane. Since the surface in the neighborhood of $P$ is made of the points of its normal sections in that neighborhood, then this means that the surface lies on one side of its tangent plane at $P$, as required. We remind the reader that $\kappa_1$ and $\kappa_2$ are the maximum and minimum of the normal curvature $\kappa_n$.

116. A surface is represented parametrically by: $\mathbf{r}(u, v) = (u, v, u^2 + v^3)$. Determine the conditions that identify the parabolic, hyperbolic and elliptic points on the surface.
**Answer**:

$$\begin{aligned}
\mathbf{E}_1 &= \partial_u \mathbf{r} = (1, 0, 2u) \\
\mathbf{E}_2 &= \partial_v \mathbf{r} = \left(0, 1, 3v^2\right) \\
\mathbf{n} &= \frac{\mathbf{E}_1 \times \mathbf{E}_2}{|\mathbf{E}_1 \times \mathbf{E}_2|} = \frac{(-2u, -3v^2, 1)}{\sqrt{4u^2 + 9v^4 + 1}} \\
\partial_u \mathbf{E}_1 &= (0, 0, 2) \\
\partial_v \mathbf{E}_1 &= (0, 0, 0) \\
\partial_v \mathbf{E}_2 &= (0, 0, 6v) \\
e &= \mathbf{n} \cdot \partial_u \mathbf{E}_1 = \frac{2}{\sqrt{4u^2 + 9v^4 + 1}} \\
f &= \mathbf{n} \cdot \partial_v \mathbf{E}_1 = 0 \\
g &= \mathbf{n} \cdot \partial_v \mathbf{E}_2 = \frac{6v}{\sqrt{4u^2 + 9v^4 + 1}} \\
eg - f^2 &= \frac{12v}{4u^2 + 9v^4 + 1}
\end{aligned}$$

Now, $(4u^2 + 9v^4 + 1) > 0$ and hence the sign of $eg - f^2$ is determined by the sign of $v$. Referring to Exercise 113, we see that the points are hyperbolic for $v < 0$, parabolic for $v = 0$ and elliptic for $v > 0$.

117. Give an example of a surface having elliptic, parabolic and hyperbolic points at different locations.
**Answer**: Torus has elliptic points on its outside half, parabolic points on its top and bottom parallels, and hyperbolic points on its inside half.

118. Why the point type (i.e. being flat, elliptic, hyperbolic or parabolic) on a surface is an invariant property with respect to changes in the surface representation and parameterization?
**Answer**: Because the point type is a real geometric property of the surface in the neighborhood of the point. This can be seen for example from the characterization of the point type that is based on the position of the surface relative to its tangent plane (i.e. being on one side of the tangent plane, etc.) since the relative position is obviously independent of the representation and parameterization. This can also be inferred for example from the invariance of the Gaussian curvature and the principal curvatures (which determine the point type) with respect to changes in representation and parameterization.[44]

119. Why the point type is invariant with respect to a change of the surface orientation by reversing the direction of the normal vector to the surface?

[44] We note that the principal curvatures may reverse their sign under certain changes, but this does not affect the argument.

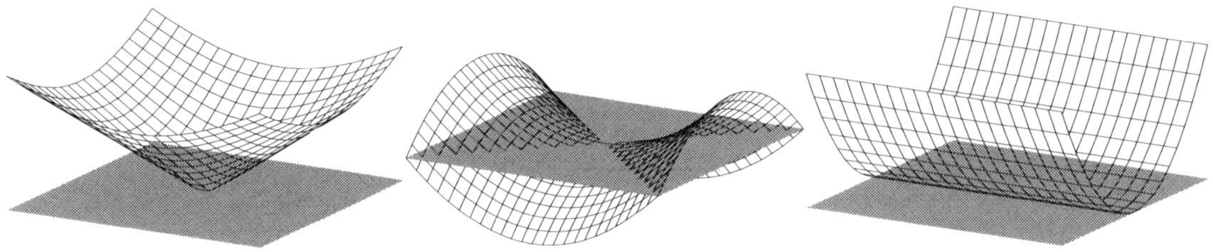

Figure 20: Tangent plane at elliptic (left), hyperbolic (middle) and parabolic (right) points.

**Answer**: As in the answer of the previous question, the point type is a real geometric property and hence it should be independent of conventional factors like surface orientation. This can also be shown more formally using surface parameters that determine the point type such as the Gaussian curvature which is invariant under a change of surface orientation, as explained in Exercise 72.

120. Make a simple sketch outlining the position of a surface relative to the tangent plane at elliptic, hyperbolic and parabolic tangency points.
**Answer**: The sketch should look similar to Figure 20.

121. Demonstrate that the surface represented parametrically by: $\mathbf{r}(u, v) = (u, v, u^2 + v^3)$ lies on both sides of its tangent plane at the point $(u, v) = (0, 0)$.[45]
**Answer**: Referring to Exercise 116, at the point $(u, v) = (0, 0)$ we have $\mathbf{E}_1 = (1, 0, 0)$ and $\mathbf{E}_2 = (0, 1, 0)$. Hence, the tangent plane at this point is the $xy$ plane. Now, the $v$ coordinate curve that passes through this point should have a constant $u$ (i.e. $u = 0$) and hence it is the curve $\mathbf{r}(0, v) = (0, v, 0 + v^3) = (0, v, v^3)$ (i.e. it is the cubic curve $z = y^3$) which is obviously on both sides of the tangent plane since it is negative for $v < 0$ and positive for $v > 0$. Accordingly, this surface lies on both sides of its tangent plane at the point $(u, v) = (0, 0)$, as required.

122. For a surface represented parametrically by: $\mathbf{r} = (u, v, v^4)$, find the equation of a curve on the surface whose points have a common tangent plane.
**Answer**: The necessary and sufficient conditions for two points on a surface curve to have a common tangent plane is that the two points have identical normal unit vector $\mathbf{n}$ and their tangent planes have a common point. So, we need to find the curve (or curves) that satisfy these conditions. Now, we have:

$$\begin{aligned}
\mathbf{E}_1 &= \partial_u \mathbf{r} = (1, 0, 0) \\
\mathbf{E}_2 &= \partial_v \mathbf{r} = \left(0, 1, 4v^3\right) \\
\mathbf{n} &= \frac{\mathbf{E}_1 \times \mathbf{E}_2}{|\mathbf{E}_1 \times \mathbf{E}_2|} = \frac{(0, -4v^3, 1)}{\sqrt{16v^6 + 1}}
\end{aligned}$$

[45] This question is related to footnote 24 in the book, i.e. there are exceptional cases in which the surface in the neighborhood of a parabolic point lies on both sides of the tangent plane.

As we see, **n** is a function of $v$ only (i.e. it is independent of $u$) and hence all the points on any $u$ coordinate curve (i.e. curves along which $v$ is constant, say $v_0$) have identical **n**. Moreover, since the $u$ coordinate curves are straight lines, as can be seen from the fact that $\mathbf{E}_1 = (1, 0, 0)$ which is constant, then all the tangent planes of any $u$ coordinate curve should have a common point, say the point with $(u, v) = (0, v_0)$. Accordingly, all the points of any $u$ coordinate curve have a common tangent plane.

123. Describe how Dupin indicatrix can be used to classify the points of a surface with regard to the local shape (i.e. flat, elliptic, parabolic and hyperbolic).
**Answer**: The Dupin indicatrix is:
• Not defined if the point is flat.
• Ellipse or circle if the point is elliptic.
• Two parallel lines if the point is parabolic.
• Two conjugate hyperbolas if the point is hyperbolic.

124. What are the prototypical geometric shapes that provide the best approximation for the local shape of a sufficiently smooth surface at its: flat, elliptic, hyperbolic and parabolic points?
**Answer**: The best approximation is:
• Plane at flat point.
• Elliptic paraboloid at elliptic point.
• Hyperbolic paraboloid at hyperbolic point.
• Parabolic cylinder at parabolic point.

125. What "umbilical point" means? What are the other terms used to label such a point?
**Answer**: Umbilical point is a surface point at which all the normal sections of the surface have the same normal curvature $\kappa_n$. Umbilical point may also be called "umbilic" or "navel" point.

126. What are the characteristic features of umbilical points?
**Answer**: Umbilical point may be characterized by the following:
• All the normal sections of the surface at the point have the same curvature $\kappa$.
• All the curves of the surface at the point have the same normal curvature $\kappa_n$.
• The two principal curvatures of the surface at the point are equal, i.e. $\kappa_1 = \kappa_2$.
• Umbilical point cannot be a hyperbolic point.
• The Gaussian curvature at umbilical point is non-negative, i.e. $K \ge 0$.
• At umbilical point the mean curvature $H$ and the Gaussian curvature $K$ are related by $H^2 = K$.

127. Give five examples of umbilical points on common geometric surfaces such as spheres and paraboloids.
**Answer**: Examples of umbilical points are:
• Every point of plane.
• Every point of sphere.
• The vertex of elliptic paraboloid of revolution.

- The two vertices of ellipsoid of revolution.
- The two vertices of hyperboloid of two sheets of revolution.

128. State the mathematical relation between the coefficients of the metric and curvature tensors at umbilical points.
**Answer**:

$$b_{\alpha\beta} = \kappa_n \, a_{\alpha\beta}$$

where $b_{\alpha\beta}$ are the coefficients of the covariant curvature tensor, $a_{\alpha\beta}$ are the coefficients of the covariant metric tensor, and $\kappa_n$ is the normal curvature of the surface at the point.

129. Demonstrate that at an umbilical point of a surface we have: $K = H^2$ where $K$ and $H$ are the Gaussian and mean curvatures at the point.
**Answer**: The Gaussian curvature $K$ and the mean curvature $H$ are given by $K = \kappa_1\kappa_2$ and $H = \frac{\kappa_1+\kappa_2}{2}$ where $\kappa_1$ and $\kappa_2$ are the principal curvatures. Now, at umbilical points we have $\kappa_1 = \kappa_2$ and hence:

$$K = \kappa_1\kappa_2 = \kappa_1\kappa_1 = (\kappa_1)^2 = \left(\frac{2\kappa_1}{2}\right)^2 = \left(\frac{\kappa_1+\kappa_2}{2}\right)^2 = H^2$$

where all the steps are justified by the equality $\kappa_1 = \kappa_2$ plus simple algebraic manipulations.

130. Show that the relation: $K = H^2$ can also be written as:

$$\left(a^{\alpha\beta} b_{\alpha\beta}\right)^2 = \frac{4}{a}\left(b_{11}b_{22} - b_{12}^2\right)$$

**Answer**:

$$\begin{aligned}
K &= H^2 \\
\frac{b}{a} &= \left(\frac{a^{\alpha\beta} b_{\alpha\beta}}{2}\right)^2 \\
\frac{b_{11}b_{22} - b_{12}^2}{a} &= \frac{\left(a^{\alpha\beta} b_{\alpha\beta}\right)^2}{4} \\
\frac{4}{a}\left(b_{11}b_{22} - b_{12}^2\right) &= \left(a^{\alpha\beta} b_{\alpha\beta}\right)^2
\end{aligned}$$

where line 2 is justified by the equations $K = \frac{b}{a}$ and $H = \frac{\operatorname{tr}\left(b_{\beta}^{\alpha}\right)}{2}$, and line 3 is justified by the definition of $b$ as the determinant of the covariant curvature tensor.

131. Explain why at umbilical points we have $b = c^2 a$ where $a$ and $b$ are the determinants of the covariant metric and covariant curvature tensors and $c$ is a proportionality factor.
**Answer**: At umbilical point we have $b_{\alpha\beta} = c \, a_{\alpha\beta}$ (with $c = \kappa_n$ as seen in Exercise 128).

On taking the determinant of both sides of this equation we get $b = c^2 a$ where $c$ is being squared because it is a common factor for the elements of a $2 \times 2$ matrix.[46]

132. Give two examples of surfaces whose all points are umbilical, and two other examples of surfaces with no umbilical point at all. Also, give an example of a surface with only one umbilical point, and another example of a surface with only two umbilical points.
**Answer**: Plane and sphere are surfaces whose all points are umbilical, while cylinder and catenoid are surfaces with no umbilical point at all. Elliptic paraboloid of revolution has only one umbilical point (its vertex on the axis of revolution), while ellipsoid of revolution has only two umbilical points (its two vertices on the axis of revolution).

[46] As it is known from linear algebra, if $\mathbf{M}$ and $\mathbf{N}$ are two $n \times n$ matrices such that $\mathbf{M} = c\mathbf{N}$ with $c$ being a scalar, then $\det(\mathbf{M}) = c^n \det(\mathbf{N})$.

# Chapter 5
# Special Curves

1. State two criteria for a space curve to be straight.
   **Answer**: One criterion is that the curvature $\kappa$ vanishes identically over the entire curve. Another criterion is that all the tangent lines to the curve are parallel.

2. Prove that a curve represented by $\mathbf{r}(t)$ is a straight line if $\dot{\mathbf{r}}$ and $\ddot{\mathbf{r}}$ are linearly dependent over the whole curve.
   **Answer**: If $\dot{\mathbf{r}}$ and $\ddot{\mathbf{r}}$ are linearly dependent then $\dot{\mathbf{r}} \times \ddot{\mathbf{r}} = \mathbf{0}$, and hence:
   $$\kappa = \frac{|\dot{\mathbf{r}} \times \ddot{\mathbf{r}}|}{|\dot{\mathbf{r}}|^3} = 0$$
   over the entire curve, i.e. the curve is straight according to the answer of the previous question.

3. Show that a space curve whose all tangent lines are parallel is a straight line.
   **Answer**: The tangent line of a space curve at an arbitrary point $P$ on the curve is given by $\mathbf{r} = \mathbf{r}_P + k\mathbf{T}_P$ where $\mathbf{r}_P$ is the spatial representation of $P$, $\mathbf{T}_P$ is the tangent unit vector to the curve at $P$ and $k$ is a real variable ($-\infty < k < \infty$). Now, since all the tangent lines are parallel then $\mathbf{T}_P$ is the same over the entire curve, i.e. the tangent vector to the curve is constant (say $\mathbf{T}_P = \mathbf{T}$). Accordingly, its derivative is zero (i.e. $\dot{\mathbf{T}} = \mathbf{0}$) and hence
   $$\kappa = \frac{\left|\dot{\mathbf{T}}\right|}{|\dot{\mathbf{r}}|} = 0$$
   over the entire curve, which is the criterion of a straight line (see Exercise 1).

4. Correct, if necessary, the following statement: "All straight lines on a surface are geodesic curves and vice versa".
   **Answer**: All straight lines on a surface are geodesic curves but not vice versa, i.e. not all geodesic curves on a surface are necessarily straight lines.

5. What is the characteristic feature of plane curves? From this, explain why the torsion of plane curves is identically zero.
   **Answer**: The characteristic feature of plane curve is that the whole curve can be contained in a plane which is its osculating plane. The torsion of plane curve is identically zero because the torsion represents the rate of change of the osculating plane, and since the plane curve has a single osculating plane then its osculating plane is constant, and hence the rate of change of the osculating plane is identically zero (also see Exercise 24 of § 2).

6. Prove that a curve is a plane curve if its osculating planes have a common intersection point.
**Answer**: Let have a space curve $C$ that is represented spatially by $\mathbf{r}(s)$ where $s$ is a natural parameter. The osculating plane of any point $P$ on $C$ is the set of all space points $\mathbf{X}$ that satisfy the condition $(\mathbf{X}-\mathbf{r})\cdot\mathbf{B}=0$ where $\mathbf{r}$ is the spatial representation of $P$ while $\mathbf{B}$ is the binormal vector of $C$ at $P$. So, if $\mathbf{X}_c$ is the common intersection point of all the osculating planes of $C$ then we should have $(\mathbf{X}_c-\mathbf{r})\cdot\mathbf{B}=0$ at any point on $C$. On differentiating this equation with respect to $s$ we get:
$$\begin{aligned}[(\mathbf{X}_c-\mathbf{r})\cdot\mathbf{B}]' &= 0\\ (\mathbf{X}_c-\mathbf{r})'\cdot\mathbf{B}+(\mathbf{X}_c-\mathbf{r})\cdot\mathbf{B}' &= 0\\ -\mathbf{r}'\cdot\mathbf{B}+(\mathbf{X}_c-\mathbf{r})\cdot\mathbf{B}' &= 0\\ -\mathbf{T}\cdot\mathbf{B}-\tau(\mathbf{X}_c-\mathbf{r})\cdot\mathbf{N} &= 0\\ \tau(\mathbf{X}_c-\mathbf{r})\cdot\mathbf{N} &= 0\end{aligned}$$
where line 2 is based on the product rule of differentiation, line 3 is based on the fact that $\mathbf{X}_c$ is independent of $s$ since it represents a constant point, line 4 is based on Eqs. 5 and 8, and line 5 is based on the fact that $\mathbf{T}$ and $\mathbf{B}$ are orthogonal. Now, let assume that $C$ is not a plane curve and hence we should have $\tau\neq 0$ at a point $P_0$ on the curve corresponding to $s=s_0$. Due to continuity, the condition $\tau\neq 0$ should also apply in a neighborhood of $P_0$. Hence, in this neighborhood we should have $\tau\neq 0$ and $\tau(\mathbf{X}_c-\mathbf{r})\cdot\mathbf{N}=0$ and hence $(\mathbf{X}_c-\mathbf{r})\cdot\mathbf{N}=0$ which means that $(\mathbf{X}_c-\mathbf{r})$ and $\mathbf{N}$ are orthogonal. But the above condition $(\mathbf{X}_c-\mathbf{r})\cdot\mathbf{B}=0$ also means that $(\mathbf{X}_c-\mathbf{r})$ and $\mathbf{B}$ are orthogonal. So, $(\mathbf{X}_c-\mathbf{r})$ is orthogonal to both $\mathbf{N}$ and $\mathbf{B}$ and hence it should be parallel to $\mathbf{T}$, i.e. $\mathbf{X}_c-\mathbf{r}=k\mathbf{T}$ (with $k$ being a real variable) and hence $\mathbf{X}_c=\mathbf{r}+k\mathbf{T}$. This means that in the neighborhood of $P_0$ the tangent lines to $C$ have a common point (i.e. point $\mathbf{X}_c$) and hence in this neighborhood $C$ is a straight line[47] which contradicts the assumption that $\tau\neq 0$ in this neighborhood (because straight lines have identically vanishing torsion). Hence, the assumption that $\tau\neq 0$ (which is equivalent to having non-plane curve) is invalid and therefore the curve should be plane, as required (see Exercise 24 of § 2).

7. Show that a space curve represented by $\mathbf{r}(t)$ is a plane curve *iff* $\dot{\mathbf{r}}\cdot(\ddot{\mathbf{r}}\times\dddot{\mathbf{r}})$ vanishes identically.
**Answer**: Plane curve is characterized by having an identically vanishing torsion (see Exercise 24 of § 2). Moreover, the torsion $\tau$ of a $t$-parameterized space curve is given by (see Exercise 50 of § 2):
$$\tau=\frac{\dot{\mathbf{r}}\cdot(\ddot{\mathbf{r}}\times\dddot{\mathbf{r}})}{|\dot{\mathbf{r}}\times\ddot{\mathbf{r}}|^2}$$
So, according to this equation if $\dot{\mathbf{r}}\cdot(\ddot{\mathbf{r}}\times\dddot{\mathbf{r}})$ vanishes identically then $\tau$ also vanishes identically and hence the curve is plane. On the other hand, if the curve is plane, and

[47] This is based on a proven theorem that states: if all the tangent lines of a space curve have a common point of intersection then the curve is a straight line.

hence $\tau$ vanishes identically, then $\dot{\mathbf{r}} \cdot (\ddot{\mathbf{r}} \times \dddot{\mathbf{r}})$ (which is equal to $\tau \left|\dot{\mathbf{r}} \times \ddot{\mathbf{r}}\right|^2$) should also vanish identically because if $\dot{\mathbf{r}} \cdot (\ddot{\mathbf{r}} \times \dddot{\mathbf{r}}) \neq 0$ then $\tau \neq 0$ which is a contradiction. We note that for space curve with non-vanishing curvature $\left|\dot{\mathbf{r}} \times \ddot{\mathbf{r}}\right|^2$ is a positive finite real number as can be seen from the equation $\kappa = \frac{|\dot{\mathbf{r}} \times \ddot{\mathbf{r}}|}{|\dot{\mathbf{r}}|^3}$. If the space curve is straight (and hence $\kappa = 0$ and $|\dot{\mathbf{r}} \times \ddot{\mathbf{r}}| = 0$) then it is obviously a plane curve without need for a formal proof although it can be easily provided.

8. Prove that having an identically vanishing torsion is a necessary and sufficient condition for a curve to be a plane curve.
**Answer**: A full proof is given in Exercise 24 of § 2.

9. Show that two curves are plane curves if they have the same binormal lines at each pair of their corresponding points.
**Answer**: Let the curves be $C_1$ and $C_2$ and they are naturally parameterized by $s_1$ and $s_2$ (and hence their other parameters are subscripted accordingly). Since $C_1$ and $C_2$ have the same binormal lines at their corresponding points then their binormal vectors are parallel, i.e. $\mathbf{B}_1 = \pm\mathbf{B}_2$. Hence, on differentiating this relation with respect to $s_1$ we get:

$$\begin{aligned}
\frac{d\mathbf{B}_1}{ds_1} &= \pm\frac{d\mathbf{B}_2}{ds_1} \\
\frac{d\mathbf{B}_1}{ds_1} &= \pm\frac{d\mathbf{B}_2}{ds_2}\frac{ds_2}{ds_1} \\
\tau_1\mathbf{N}_1 &= \pm\tau_2\mathbf{N}_2\frac{ds_2}{ds_1}
\end{aligned}$$

where in line 2 we use the chain rule, and in line 3 we use Eq. 8. Accordingly, $\mathbf{N}_1$ and $\mathbf{N}_2$ are also parallel.
Also, because $C_1$ and $C_2$ have common binormal lines at their corresponding points then if $\mathbf{r}_1$ and $\mathbf{r}_2$ are two corresponding points then we have the following relation:

$$\mathbf{r}_2 = \mathbf{r}_1 + k\mathbf{B}_1$$

where $k$ is a real parameter. On differentiating this relation with respect to $s_1$ we get:

$$\begin{aligned}
\frac{d\mathbf{r}_2}{ds_1} &= \frac{d\mathbf{r}_1}{ds_1} + \frac{dk}{ds_1}\mathbf{B}_1 + k\frac{d\mathbf{B}_1}{ds_1} \\
\frac{d\mathbf{r}_2}{ds_2}\frac{ds_2}{ds_1} &= \frac{d\mathbf{r}_1}{ds_1} + \frac{dk}{ds_1}\mathbf{B}_1 + k\frac{d\mathbf{B}_1}{ds_1} \\
\mathbf{T}_2\frac{ds_2}{ds_1} &= \mathbf{T}_1 + \frac{dk}{ds_1}\mathbf{B}_1 - k\tau_1\mathbf{N}_1 \\
\mathbf{T}_2\frac{ds_2}{ds_1} &= \mathbf{T}_1 + \frac{dk}{ds_1}\mathbf{B}_1 \mp k\tau_1\mathbf{N}_2
\end{aligned}$$

where in line 1 we use the sum and product rules of differentiation, in line 2 we use the chain rule of differentiation, in line 3 we use Eqs. 5 and 8, and in line 4 we use the fact that $\mathbf{N}_1$ and $\mathbf{N}_2$ are parallel. Now, $\mathbf{T}_2$ is perpendicular to $\mathbf{N}_2$ and hence $\mathbf{T}_2$ has no component in the direction of $\mathbf{N}_2$.[48] Noting that $k \neq 0$ in general we should have $\tau_1 = 0$, i.e. $C_1$ is a plane curve. By reversing the role of $C_1$ and $C_2$ in the above argument we also conclude that $\tau_2 = 0$ and hence $C_2$ is also a plane curve.[49]

10. Define, rigorously, involute and evolute curves making a simple plot to outline their relation. Also explain the role of the tangent surface of the evolute in this context.
**Answer**: If $C_e$ is a space curve with a tangent surface $S_T$ and $C_i$ is a curve embedded in $S_T$ and it is orthogonal to all the tangent lines of $C_e$ at their intersection points, then $C_i$ is called an involute of $C_e$ while $C_e$ is called an evolute of $C_i$. The plot should look like Figure 21. As stated in the definition, the involute is embedded in the tangent surface of the evolute. Also, the involute is an orthogonal trajectory of the generators of the tangent surface of its evolute.

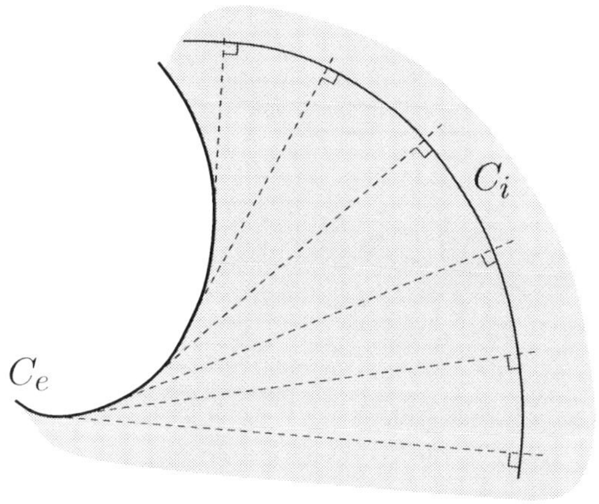


Figure 21: Evolute $C_e$, involute $C_i$, tangent lines (dashed) and tangent surface (shaded).

11. Explain all the symbols used in the following equation which is related to involute curves:

$$\mathbf{r}_i = \mathbf{r}_e + (c - s)\,\mathbf{T}_e \tag{51}$$

Make sense of this equation using your plot in the previous exercise.
**Answer**: $\mathbf{r}_i$ is an arbitrary point on the involute $C_i$, $\mathbf{r}_e$ is the point on the evolute $C_e$ corresponding to $\mathbf{r}_i$, $c$ is a given constant, $s$ is a natural parameter of $C_e$ and $\mathbf{T}_e$ is the unit tangent vector to $C_e$ at $\mathbf{r}_e$. As we see in the plot of Figure 21, any point $\mathbf{r}_i$ on $C_i$ is reached from the spatial position of the corresponding point $\mathbf{r}_e$ on $C_e$ by a scalar multiple of the unit tangent vector $\mathbf{T}_e$ of $C_e$ at $\mathbf{r}_e$. The role of $s$ in this equation will be clarified by considering the visual demonstration about generating an involute by using a taut string attached to the evolute, as explained in the next exercise.

[48] This argument also applies to the second term on the right hand side (i.e. $\frac{dk}{ds_1}\mathbf{B}_1$ which can be expressed as $\pm\frac{dk}{ds_1}\mathbf{B}_2$ since $\mathbf{B}_1$ and $\mathbf{B}_2$ are parallel) and hence $\frac{dk}{ds_1} = 0$ because $\mathbf{T}_2$ has no component in the direction of $\mathbf{B}_2$ but this is not needed in this proof.

[49] In fact, reversing the role is not needed because the labeling of $C_1$ and $C_2$ is arbitrary.

12. Outline the visual demonstration which is commonly used to explain the relation between an involute and its evolute. Use the plot mentioned in the last two questions in your explanation.
**Answer**: The generation of an involute $C_i$ of a curve $C_e$, when $(c - s)$ in Eq. 51 is positive, may be visualized by detaching a taut string attached to $C_e$ where the string is kept in the tangent direction as it is detached. A fixed point $P$ on the string, where the distance between $P$ and the point of contact of the string with $C_e$ represents a natural parameter of $C_e$, then traces an involute of $C_e$. This description clearly matches the plot of Figure 21 where the string in its different tangent states corresponds to the dashed tangent lines with $s$ being representing the length of the string between point $\mathbf{r}_i$ on the involute $C_i$ and point $\mathbf{r}_e$ on the evolute $C_e$.

13. Show that the tangent line of a curve and the principal normal line of its evolute are parallel at their corresponding points.
**Answer**: On taking the $s$-derivative of Eq. 51 we get:

$$\begin{aligned} \mathbf{r}'_i s'_i &= \mathbf{r}'_e - \mathbf{T}_e + (c - s)\,\mathbf{T}'_e \\ \mathbf{T}_i s'_i &= \mathbf{T}_e - \mathbf{T}_e + (c - s)\,\mathbf{T}'_e \\ \mathbf{T}_i s'_i &= (c - s)\,\mathbf{T}'_e \\ \mathbf{T}_i s'_i &= (c - s)\,\kappa_e \mathbf{N}_e \end{aligned}$$

where in line 1 we use the chain rule on the left (with $s_i$ being a natural parameter of the involute) and the sum and product rules on the right, in line 2 we use $\mathbf{r}' = \mathbf{T}$ (Eq. 5) and in line 4 we use $\mathbf{T}' = \kappa \mathbf{N}$ (Eq. 6) with $\kappa_e$ being the curvature of evolute. The last equation means that the unit tangent vector $\mathbf{T}_i$ of the involute and the principal normal vector $\mathbf{N}_e$ of the evolute at their corresponding points have identical orientation and hence the tangent line of a curve (i.e. involute) and the principal normal line of its evolute are parallel at their corresponding points, as required.

14. Prove that the evolutes of plane curves are helices.
**Answer**: The torsion of involute $\tau_i$ is given in terms of the torsion $\tau_e$, curvature $\kappa_e$ and the natural parameter $s$ of its evolute by the following equation:[50]

$$\tau_i = \frac{(\tau_e/\kappa_e)'}{\kappa\,(c - s)\left[1 + (\tau_e/\kappa_e)^2\right]}$$

Now, if the involute is a plane curve then $\tau_i$ should vanish identically (see Exercise 24 of § 2) and hence $(\tau_e/\kappa_e)' = 0$. On integrating the last equation we get $\tau_e/\kappa_e =$ constant and hence the evolute is a helix according to the result of Exercise 28 of § 2.

15. Prove that for a plane curve $C$ the locus of the centers of curvature of $C$ is an evolute of $C$.

[50] The proof of this equation can be found in the literature of differential geometry.

**Answer**: The position of the center of curvature $\mathbf{r}_\rho$ of $C$ at a given point $P$ on $C$ is given by:[51]

$$\mathbf{r}_\rho = \mathbf{r}_P + \frac{1}{\kappa}\mathbf{N}$$

where $\mathbf{r}_P$ is the position of $P$, $\kappa$ is the curvature of $C$ at $P$ and $\mathbf{N}$ is the principal normal vector of $C$ at $P$. Also, the position of the evolutes of a plane curve is given by:

$$\mathbf{r}_e = \mathbf{r}_i + \frac{1}{\kappa_i}\mathbf{N}_i + \frac{a}{\kappa_i}\mathbf{B}_i$$

where $\mathbf{r}_e$ and $\mathbf{r}_i$ are the positions of corresponding points on the evolute and involute, $\kappa_i$ is the curvature of involute, $\mathbf{N}_i$ and $\mathbf{B}_i$ are the normal and binormal vectors of involute, and $a$ is a constant (which varies from one evolute to another). On comparing the above equations we see that the locus of the centers of curvature of $C$ (which is a plane curve) is the plane evolute of $C$ that corresponds to $a = 0$ and lies in the plane of $C$.

16. Derive the parametric equation of the involute of a circle represented by: $\mathbf{r}(\theta) = (5\cos\theta, 5\sin\theta)$ where $0 \le \theta < 2\pi$.[52]
**Answer**: Let $s$ be a natural parameter of the circle with $s = 0$ corresponding to $\theta = 0$. Now, $s = 5\theta$ and the tangent to the circle is:

$$\mathbf{T} = \mathbf{r}' = \mathbf{r}_\theta\theta' = (-5\sin\theta, 5\cos\theta)\frac{1}{5} = (-\sin\theta, \cos\theta)$$

Hence, the parametric equation of the involute is given by (noting that $c = 0$ in this case):

$$\begin{aligned}
\mathbf{r}_i &= \mathbf{r}_e - s\mathbf{T}_e \\
\mathbf{r}_i &= (5\cos\theta, 5\sin\theta) - 5\theta(-\sin\theta, \cos\theta) \\
\mathbf{r}_i &= 5(\cos\theta + \theta\sin\theta, \sin\theta - \theta\cos\theta)
\end{aligned}$$

17. Prove that any two involutes of a plane curve are associated Bertrand curves.
**Answer**: The proof is very easy because it is no more than an application of the definitions of involute and Bertrand curves. So, let have a plane curve $C$ (i.e. the evolute). Now, for any given point $P$ on $C$ all the involutes of $C$ are by definition perpendicular to the tangent line of $C$ at $P$. Hence, any two involutes, $C_1$ and $C_2$, of $C$ will be perpendicular to the tangent line $L_t$ of $C$ at $P$ and hence $L_t$ is their common principal normal line at their corresponding points which correspond to the point $P$ on $C$. Therefore, by definition $C_1$ and $C_2$ are Bertrand curves.[53]

[51] We changed in this answer the symbol for the position of the center of curvature from $\mathbf{r}_C$ (which we use in the book and in other exercises) to $\mathbf{r}_\rho$ to avoid confusion.
[52] Because circle is a plane curve we are using only two spatial coordinates.
[53] The reader should notice that although the condition "plane" may look redundant in this proof, it is needed for having "common principal normal lines" which is the required criterion for Bertrand curves.

18. How many involutes a given curve can have? How these involutes are related to each other through the constant $c$ (see Eq. 51)?
**Answer**: Infinitely many. These involutes correspond to different values of $c$ in Eq. 51.

19. How many evolutes a given curve can have? How these evolutes are related to each other through the constant $c$ (see Eq. 51)?
**Answer**: Infinitely many. These evolutes correspond to different values of $c$ in Eq. 51.

20. Justify the fact that the involutes of a circle are congruent with a clear explanation of how these involutes are related to each other.
**Answer**: Noting the rotational symmetry of the circle and considering the visual demonstration of how to generate an involute (see Exercise 12), it should be obvious that all the involutes of a circle are identical but they differ in their point of contact with the circle and hence they can be obtained from each other by rotation around the center of the circle. We should also consider the difference in the sense of natural parameter (which corresponds to the sense of the angle $\theta$ in the parametric representation as described in Exercise 16) and hence we may have a reflection symmetry with respect to the diameter of the circle. So in brief, all the involutes of a circle are identical curves (with possible difference in rotation and reflection) and hence they can be obtained from each other by rotation around the center of the circle and reflection in the diameter of the circle.
More formally, we generalize the answer of Exercise 16 (with $R$ being the circle radius), and hence we get:

$$
\begin{aligned}
\mathbf{r}_i &= \mathbf{r}_e + [c - s]\,\mathbf{T}_e \\
\mathbf{r}_i &= (R\cos\theta, R\sin\theta) + [c - R\theta]\,(-\sin\theta, \cos\theta) \\
\mathbf{r}_i &= R\left(\cos\theta - \left[\frac{c}{R} - \theta\right]\sin\theta, \sin\theta + \left[\frac{c}{R} - \theta\right]\cos\theta\right)
\end{aligned}
$$

The last equation shows the dependence of the involutes on three factors: $R$ which is fixed for a given circle, $c$ which determines the point of contact (and hence possible rotation around the center), and $\theta$ which (assuming that it corresponds to a given point on the circle) determines the sense of rotation (and hence possible reflection in the diameter).

21. Define Bertrand curves outlining two of their main characteristic features.
**Answer**: Bertrand curves are space curves that have common principal normal lines at their corresponding points. The distance between the corresponding points of two Bertrand curves is constant, and the angle between their corresponding tangent lines is constant.

22. Show that a helix has an infinite number of Bertrand associates and identify these associates.
**Answer**: It is given in the book that if $C_1$ is a curve with non-vanishing torsion then a necessary and sufficient condition for $C_1$ to be a Bertrand curve (i.e. it possesses

an associate curve $C_2$ such that $C_1$ and $C_2$ are Bertrand curves) is that there are two constants $A$ and $B$ such that:[54]

$$\kappa = A\tau + B$$

where $\kappa$ and $\tau$ are the curvature and torsion of the curve $C_1$ and where $B$ determines the relation between $C_1$ and $C_2$ (as we will see in the end of the answer). Now, circular helix is a curve with constant non-vanishing $\tau$ and $\kappa$ (see Eqs. 9 and 10 in Exercise 28 of § 2). Hence, for any arbitrary finite real constant $B \neq 0$ we can find a constant $A$ such that $\kappa = A\tau + B$ (i.e. $A = \frac{\kappa - B}{\tau}$). Since $B$ is arbitrary then we have an infinite number of Bertrand associates $C_2$ (where each one of these $C_2$ corresponds to a given $B$ and hence a given $A$ as determined by the above condition). It can be shown that all these Bertrand curves $C_2$ that associate $C_1$ are given by:

$$\mathbf{r}_2 = \mathbf{r}_1 + \frac{1}{B}\mathbf{N}_1$$

where $\mathbf{r}_1$ and $\mathbf{r}_2$ are the spatial representation of corresponding points on $C_1$ and $C_2$ and $\mathbf{N}_1$ is the principal normal vector of $C_1$.

23. Prove that on a pair of Bertrand curves, the angle between their tangents at their corresponding points is constant.
**Answer**: Let $C_1$ and $C_2$ be two Bertrand curves that are naturally parameterized by $s_1$ and $s_2$ respectively and $\mathbf{r}_1$ is the position of an arbitrary point on $C_1$ and hence by definition $\mathbf{r}_2 = \mathbf{r}_1 + a\mathbf{N}_1$ is the position of the corresponding point on $C_2$ (with $a$ being a constant and $\mathbf{N}_1$ is the principal normal vector of $C_1$ at that point). On differentiating the dot product of the unit tangent vectors, $\mathbf{T}_1$ and $\mathbf{T}_2$, at these corresponding points with respect to $s_1$ we get:

$$\begin{aligned}
\frac{d}{ds_1}(\mathbf{T}_1 \cdot \mathbf{T}_2) &= \frac{d\mathbf{T}_1}{ds_1} \cdot \mathbf{T}_2 + \mathbf{T}_1 \cdot \frac{d\mathbf{T}_2}{ds_1} \\
&= \kappa_1 \mathbf{N}_1 \cdot \mathbf{T}_2 + \mathbf{T}_1 \cdot \frac{d\mathbf{T}_2}{ds_1} \\
&= \kappa_1 (\mathbf{N}_1 \cdot \mathbf{T}_2) + \mathbf{T}_1 \cdot \frac{d\mathbf{T}_2}{ds_2}\frac{ds_2}{ds_1} \\
&= \kappa_1 (\mathbf{N}_1 \cdot \mathbf{T}_2) + \kappa_2 \frac{ds_2}{ds_1}(\mathbf{T}_1 \cdot \mathbf{N}_2)
\end{aligned}$$

where in line 1 we use the product rule of differentiation, in line 2 we use $\mathbf{T}' = \kappa\mathbf{N}$ (Eq. 6) with $\kappa_1$ being the curvature of $C_1$, in line 3 we use the chain rule of differentiation, and in line 4 we use $\mathbf{T}' = \kappa\mathbf{N}$ (Eq. 6) with $\kappa_2$ being the curvature of $C_2$. Now, because Bertrand curves have common principal normal lines then $\mathbf{N}_1 = \pm\mathbf{N}_2$ and hence $\mathbf{N}_1 \cdot \mathbf{T}_2 = \pm\mathbf{N}_2 \cdot \mathbf{T}_2 = 0$ since $\mathbf{N}_2$ and $\mathbf{T}_2$ are perpendicular. Similarly, $\mathbf{T}_1 \cdot \mathbf{N}_2 =$

[54] The proof of this theorem can be found in the literature of differential geometry. Also, the equation in the book is $\kappa = c_1\tau + c_2$ but we changed $c_1$ and $c_2$ here to $A$ and $B$ to avoid confusion with $C_1$ and $C_2$ which are the labels of the curves.

$\mp\mathbf{T}_1 \cdot \mathbf{N}_1 = 0$. Therefore, $\frac{d}{ds_1}(\mathbf{T}_1 \cdot \mathbf{T}_2) = 0$ and hence $\mathbf{T}_1 \cdot \mathbf{T}_2 =$ constant.[55] Now, the angle $\theta$ between $\mathbf{T}_1$ and $\mathbf{T}_2$ (which are unit vectors) is given by $\cos\theta = \mathbf{T}_1 \cdot \mathbf{T}_2$ and hence this angle is constant, i.e. the angle between the tangents of $C_1$ and $C_2$ at their corresponding points is constant, as required.

24. State a sufficient and necessary condition for a curve to be a Bertrand curve by having an associate Bertrand curve.
**Answer**: It is the condition $\kappa = A\tau + B$ as explained in Exercise 22.

25. Show that a plane curve has always a Bertrand associate.
**Answer**: This should be obvious because if $C_1$ is a plane curve then any curve $C_2$ that lies in its plane such that:
$$\mathbf{r}_2 = \mathbf{r}_1 + a\mathbf{N}_1$$
is a Bertrand associate where $\mathbf{r}_1$ and $\mathbf{r}_2$ are the positions of corresponding points on $C_1$ and $C_2$, $a$ is a constant (see Exercise 27) and $\mathbf{N}_1$ is the principal normal vector of $C_1$ at $\mathbf{r}_1$. Now, since the binormal vector $\mathbf{B}$ is constant and it is common to both curves (because $\mathbf{B}$ is the unit vector perpendicular to the plane) and since $C_1$ and $C_2$ have parallel principal normal vectors at their corresponding points then their unit tangent vectors, $\mathbf{T}_1$ and $\mathbf{T}_2$, at their corresponding points should also be parallel. Hence, if $C_1$ and $C_2$ are naturally parameterized by $s_1$ and $s_2$ respectively then from Eq. 5 we should have:
$$\mathbf{T}_1 = \frac{d\mathbf{r}_1}{ds_1} \qquad \text{and} \qquad \mathbf{T}_2 = \frac{d\mathbf{r}_2}{ds_2} = \frac{d\mathbf{r}_2}{ds_1}\frac{ds_1}{ds_2}$$
So, if $C_1$ is sufficiently smooth and $s_1$ is a differentiable function of $s_2$ (which should be always the case) then the existence of $C_1$ will guarantee the existence of $C_2$.

26. Prove that the product of torsions at the corresponding points of a pair of associated Bertrand curves is constant.
**Answer**: Bertrand curves have parallel principal normal vectors at their corresponding points. Moreover, the angle between their tangents at their corresponding points $\theta$ is constant (see Exercise 23) and hence the angle between their binormal vectors at their corresponding points is also constant (which is equal to $\theta$). Now, in Exercise 27 we obtained the following result:
$$\mathbf{T}_2\frac{ds_2}{ds_1} = (1 - a\kappa_1)\,\mathbf{T}_1 + a\tau_1\mathbf{B}_1 \tag{52}$$
On dot producting the two sides with $\mathbf{B}_2$ we obtain:
$$0 = (1 - a\kappa_1)\,\mathbf{T}_1 \cdot \mathbf{B}_2 + a\tau_1\mathbf{B}_1 \cdot \mathbf{B}_2 \tag{53}$$

[55] We note that the above process can be repeated but with differentiation with respect to $s_2$ instead of $s_1$. The final result should be the same. In fact, since the labels 1 and 2 are arbitrary there is no need for any repetition.

because $\mathbf{T}_2$ and $\mathbf{B}_2$ are orthogonal. Now, from Exercise 23 we conclude that $\mathbf{B}_1 \cdot \mathbf{B}_2 = \mathbf{T}_1 \cdot \mathbf{T}_2 = \cos\theta$. Moreover, from Eq. 52 we can see that $\mathbf{T}_2$ is a linear combination of $\mathbf{T}_1$ and $\mathbf{B}_1$ (which are orthogonal) and hence if $\mathbf{B}_2$ (which is perpendicular to $\mathbf{T}_2$) makes an angle $\theta$ with $\mathbf{B}_1$ then it must make an angle $\frac{\pi}{2} - \theta$ with $\mathbf{T}_1$ and therefore $\mathbf{T}_1 \cdot \mathbf{B}_2 = \sin\theta$. So, from Eq. 53 we get:

$$(1 - a\kappa_1)\sin\theta + a\tau_1\cos\theta = 0$$

Similarly, since $\mathbf{T}_2$ is a linear combination of $\mathbf{T}_1$ and $\mathbf{B}_1$ (which are orthogonal) then from geometric considerations we can see that if $\mathbf{T}_2$ makes an angle $\theta$ with $\mathbf{T}_1$ than it should make an angle $\frac{\pi}{2} + \theta$ with $\mathbf{B}_1$ (see Figure 22). Hence, we should also have (noting that $\cos\left[\frac{\pi}{2} + \theta\right] = -\sin\theta$):

$$\mathbf{T}_2 = \mathbf{T}_1\cos\theta - \mathbf{B}_1\sin\theta \tag{54}$$

On comparing Eq. 54 with Eq. 52 we conclude:

$$\sin\theta = -a\tau_1\frac{ds_1}{ds_2} \tag{55}$$

Now, being a Bertrand curve is a symmetric relation and hence if we reverse the labeling of $C_1$ and $C_2$ then we should get the following relation:

$$\sin\theta = -a\tau_2\frac{ds_2}{ds_1} \tag{56}$$

So, if we multiply the two sides of Eq. 55 by the two sides of Eq. 56 then we get:

$$\sin^2\theta = a^2\tau_1\tau_2$$

and hence:

$$\tau_1\tau_2 = \frac{\sin^2\theta}{a^2}$$

i.e. the product $\tau_1\tau_2$ is constant because $\theta$ is constant (according to Exercise 23) and $a$ is constant (according to Exercise 27). A bonus of this solution is that we can conclude from the result that the torsions at the corresponding points have the same sign since $\frac{\sin^2\theta}{a^2}$ is definitely non-negative.

27. Prove that on a pair of Bertrand curves, the distance between their corresponding points is constant.
**Answer**: Let $C_1$ and $C_2$ be two associated Bertrand curves that are naturally parameterized by $s_1$ and $s_2$ respectively (and their other parameters are also subscripted with 1 and 2 respectively). Hence, from the definition of Bertrand curves the positions of their corresponding points, $\mathbf{r}_1$ and $\mathbf{r}_2$, are related by:

$$\mathbf{r}_2 = \mathbf{r}_1 + a\mathbf{N}_1$$

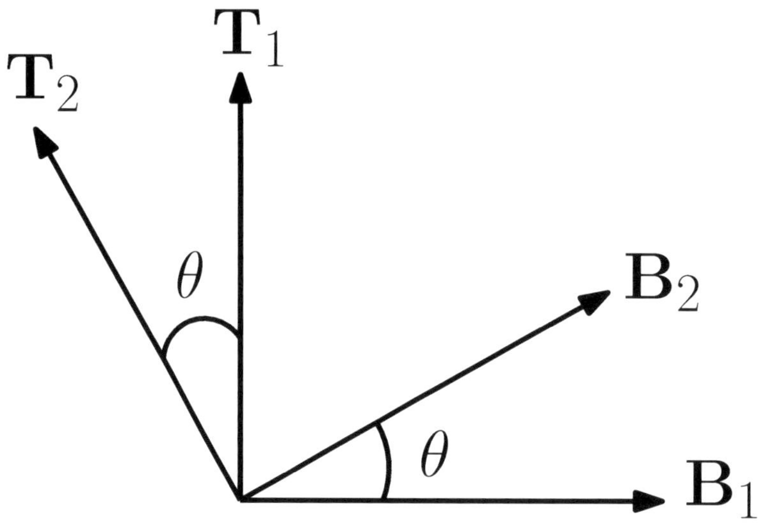


Figure 22: Demonstration of the relation between the vectors $\mathbf{T}_1, \mathbf{T}_2, \mathbf{B}_1, \mathbf{B}_2$ of associated Bertrand curves.

where $a$ is a scalar function of $s_1$. Accordingly, the distance between their corresponding points is $|\mathbf{r}_2 - \mathbf{r}_1| = |a\mathbf{N}_1| = |a|$. On differentiating the above equation with respect to $s_1$ we get:

$$\begin{aligned} \frac{d\mathbf{r}_2}{ds_1} &= \frac{d\mathbf{r}_1}{ds_1} + \frac{da}{ds_1}\mathbf{N}_1 + a\frac{d\mathbf{N}_1}{ds_1} \\ \frac{d\mathbf{r}_2}{ds_2}\frac{ds_2}{ds_1} &= \frac{d\mathbf{r}_1}{ds_1} + \frac{da}{ds_1}\mathbf{N}_1 + a\frac{d\mathbf{N}_1}{ds_1} \\ \mathbf{T}_2\frac{ds_2}{ds_1} &= \mathbf{T}_1 + \frac{da}{ds_1}\mathbf{N}_1 + a\left(\tau_1\mathbf{B}_1 - \kappa_1\mathbf{T}_1\right) \\ \mathbf{T}_2 &= \frac{ds_1}{ds_2}\left[\left(1 - a\kappa_1\right)\mathbf{T}_1 + \frac{da}{ds_1}\mathbf{N}_1 + a\tau_1\mathbf{B}_1\right] \end{aligned}$$

where in line 1 we used the sum and product rules, in line 2 we used the chain rule, and in line 3 we used Eqs. 5 and 7. Now, $\mathbf{T}_2$ is perpendicular to $\mathbf{N}_2$ and hence it is perpendicular to $\mathbf{N}_1$ (since $\mathbf{N}_1 = \pm\mathbf{N}_2$). Therefore, $\mathbf{T}_2$ should have no component along $\mathbf{N}_1$ and hence $\frac{da}{ds_1}$ must be zero (noting that $\frac{ds_1}{ds_2}$ cannot be zero). Therefore, $a =$ constant and the distance $|a|$ must also be constant.

28. Give a brief definition of spherical indicatrix with a simple sketch of the spherical normal indicatrix $\bar{C}_{\mathbf{N}}$ of a space curve to illustrate this concept.
**Answer**: The spherical indicatrix of a continuously-varying unit vector is a continuous curve $\bar{C}$ on the origin-based unit sphere generated by mapping the unit vector (e.g. $\mathbf{T}$ or $\mathbf{N}$ or $\mathbf{B}$) of a particular space curve $C$ on an equal unit vector represented by a point on the origin-based unit sphere. The sketch should look like Figure 23.

29. Prove the following equation: $\kappa_{\mathbf{T}}^2 = \frac{\kappa^2+\tau^2}{\kappa^2}$.
**Answer**: Let have a naturally parameterized space curve $C$ that is spatially represented

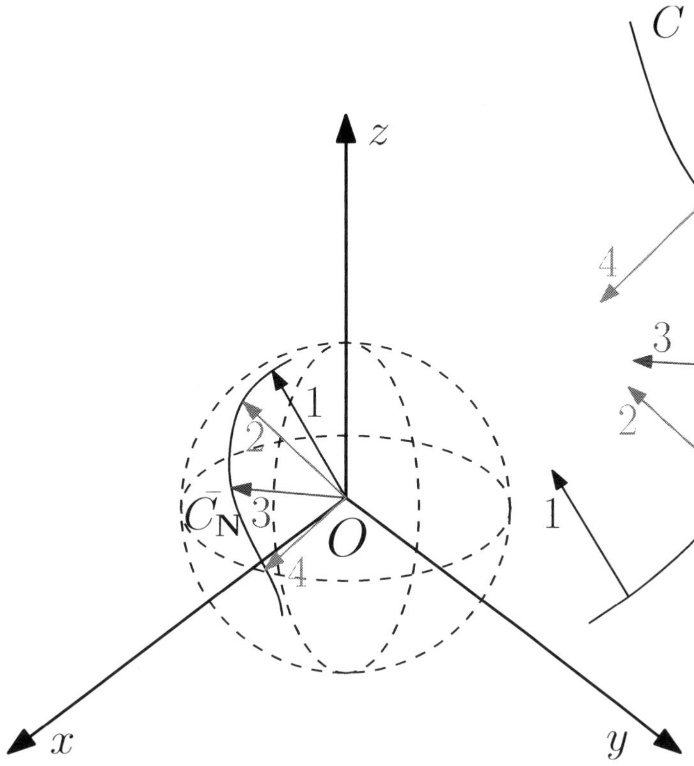


Figure 23: The spherical normal indicatrix $\bar{C}_{\mathbf{N}}$ of a space curve $C$ where the numbers indicate the correspondence between the principal normal vectors of $C$ and their map on $\bar{C}_{\mathbf{N}}$.

by $\mathbf{r}(s)$ with a spherical tangent indicatrix $\bar{C}_{\mathbf{T}}$ that is spatially represented by $\mathbf{T}(s)$. So, we have:

$$\begin{aligned}
\kappa_{\mathbf{T}} &= \frac{|\mathbf{T}' \times \mathbf{T}''|}{|\mathbf{T}'|^3} \\
&= \frac{|\mathbf{r}'' \times \mathbf{r}'''|}{|\mathbf{r}''|^3} \\
&= \frac{|\mathbf{r}'' \times \mathbf{r}'''|}{\kappa^3} \\
&= \frac{|\kappa^2\tau\mathbf{T} + \kappa^3\mathbf{B}|}{\kappa^3}
\end{aligned}$$

where $\kappa_{\mathbf{T}}$ is the curvature of $\bar{C}_{\mathbf{T}}$, $\kappa$ and $\tau$ are the curvature and torsion of $C$, $\mathbf{T}$ and $\mathbf{B}$ are the unit tangent and binormal vectors of $C$, and the prime represents derivative with respect to $s$. We note that in line 1 we use the curvature formula for a $t$-parameterized curve (see Exercise 49 of § 2) with $t = s$,[56] in line 2 we use $\mathbf{T} = \mathbf{r}'$ (Eq. 5), in line 3

[56] This is because $s$ is a natural parameter of $C$ and hence it is a $t$-parameter for $\bar{C}_{\mathbf{T}}$ which is spatially represented by $\mathbf{r}_t = \mathbf{T}(s)$ where $\mathbf{r}_t$ belongs to $\bar{C}_{\mathbf{T}}$ while $\mathbf{T}(s)$ belongs to $C$. Accordingly, the formula

we use $\kappa = |\mathbf{r}''|$, and in line 4 we use $\mathbf{r}'' \times \mathbf{r}''' = \kappa^2\tau\mathbf{T} + \kappa^3\mathbf{B}$ (see Eq. 19 in Exercise 50 of § 2). Hence:

$$\begin{aligned} \kappa_{\mathbf{T}}^2 &= \frac{|\kappa^2\tau\mathbf{T} + \kappa^3\mathbf{B}|^2}{\kappa^6} \\ &= \frac{(\kappa^2\tau\mathbf{T} + \kappa^3\mathbf{B}) \cdot (\kappa^2\tau\mathbf{T} + \kappa^3\mathbf{B})}{\kappa^6} \\ &= \frac{\kappa^4\tau^2 + \kappa^6}{\kappa^6} \\ &= \frac{\tau^2 + \kappa^2}{\kappa^2} \end{aligned}$$

as required.

30. Discuss the similarities and differences between Gauss mapping and spherical indicatrix mapping.
**Answer**: We note the following:
• Both mappings map unit vectors onto the origin-centered unit sphere.
• Gauss mapping belongs to surfaces while spherical indicatrix mapping belongs to curves.
• Gauss mapping maps the unit normal vector $\mathbf{n}$ while spherical indicatrix mapping maps the tangent, normal or binormal unit vector $\mathbf{T}, \mathbf{N}, \mathbf{B}$.

31. Justify, using a simple fact about helices, that the spherical images of $\mathbf{T}, \mathbf{N}, \mathbf{B}$ of a helix rotating around the $z$-axis are circles centered around the $z$-axis.
**Answer**: The fact is that each one of the vectors $\mathbf{T}, \mathbf{N}, \mathbf{B}$ of a helix rotating around the $z$-axis makes a constant angle with the $z$-axis (see Exercises 34 and 37 of § 2 and Exercise 16 of § 6)[57] and hence their spherical indicatrices $\bar{C_{\mathbf{T}}}, \bar{C_{\mathbf{N}}}, \bar{C_{\mathbf{B}}}$ should be circles centered around the $z$-axis. This can be seen more formally and generally from the results of Exercise 77 of § 2 where we found that for a $t$-parameterized circular helix given by $\mathbf{r}(t) = (a\cos t, a\sin t, bt)$ the vectors $\mathbf{T}, \mathbf{N}, \mathbf{B}$ are given by:

$$\begin{aligned} \mathbf{T} &= \frac{(-a\sin t, a\cos t, b)}{\sqrt{a^2 + b^2}} \\ \mathbf{N} &= (-\cos t, -\sin t, 0) \\ \mathbf{B} &= \frac{(b\sin t, -b\cos t, a)}{\sqrt{a^2 + b^2}} \end{aligned}$$

and hence $\bar{C_{\mathbf{T}}}$, $\bar{C_{\mathbf{N}}}$ and $\bar{C_{\mathbf{B}}}$ are circles centered on the $z$-axis since $a$ and $b$ are constants. Similarly, in Exercise 34 of § 2 we found that the principal normal vector of a circular

---

$\kappa = \frac{|\dot{\mathbf{r}} \times \ddot{\mathbf{r}}|}{|\dot{\mathbf{r}}|^3}$ becomes:

$$\kappa_{\mathbf{T}} = \frac{|\dot{\mathbf{r}}_t \times \ddot{\mathbf{r}}_t|}{|\dot{\mathbf{r}}_t|^3} = \frac{|\mathbf{T}' \times \mathbf{T}''|}{|\mathbf{T}'|^3}$$

[57] For example, in Exercise 37 of § 2 we see that the $z$ component of all these unit vectors is constant which means that these vectors make constant angle with the $z$-axis which is the axis of rotation.

helix rotating around the $z$-axis is parallel to the $xy$ plane and hence its spherical indicatrix should be a circle centered on the $z$-axis (in fact it is centered on the origin). Therefore, $\bar{C_\mathbf{N}}$ is a unit circle centered on the origin.
Also, the formulae that are given in the book[58] for $\kappa_\mathbf{T}$, $\tau_\mathbf{T}$, $\kappa_\mathbf{B}$ and $\tau_\mathbf{B}$ confirm this fact (partly) for $\bar{C_\mathbf{T}}$ and $\bar{C_\mathbf{B}}$ because for circular helix $\kappa$ and $\tau$ are constants (see Eqs. 9 and 10 in Exercise 28 of § 2) and hence $\kappa' = \tau' = 0$. Therefore, we get $\kappa_\mathbf{T} =$ constant, $\tau_\mathbf{T} = 0$, $\kappa_\mathbf{B} =$ constant and $\tau_\mathbf{B} = 0$, which means that $\bar{C_\mathbf{T}}$ and $\bar{C_\mathbf{B}}$ are circles because they have constant curvature and they are plane curves so they must be circles (also see Exercises 28 and 53 of § 2).

32. Justify the following statement: "The binormal indicatrix is a single point for a plane curve, and the tangent indicatrix is a single point for a straight line".
**Answer**: For plane curve, $\mathbf{B}$ is constant and hence its binormal indicatrix should be a single point. For straight line, $\mathbf{T}$ is constant and hence its tangent indicatrix should be a single point.

33. Prove, rigorously, that the tangent indicatrix of a helix is a circle.
**Answer**: From Exercise 31 we see that the tangent indicatrix of a helix is spatially represented by:

$$\mathbf{T} = \frac{(-a\sin t, a\cos t, b)}{\sqrt{a^2+b^2}} = \frac{(a\sin(-t), a\cos(-t), b)}{\sqrt{a^2+b^2}}$$

This is a parametric equation of a circle whose plane is parallel to the $xy$ plane with radius $a/\sqrt{a^2+b^2}$ and center $\left(0, 0, b/\sqrt{a^2+b^2}\right)$.

34. Prove that the tangent to the spherical indicatrix of the tangent to a given space curve $C$ and the principal normal of $C$ are parallel.
**Answer**: The spherical indicatrix $\bar{C_\mathbf{T}}$ of the tangent of $C$ is spatially represented by $\mathbf{T}$ of $C$. Hence, the tangent to $\bar{C_\mathbf{T}}$ is the $s$-derivative of $\mathbf{T}$, that is $\mathbf{T}' = \kappa\mathbf{N}$ (where $\kappa$ and $\mathbf{N}$ belong to $C$ and where Eq. 6 is used). Accordingly, the tangent to $\bar{C_\mathbf{T}}$ and the principal normal of $C$ are both along the $\mathbf{N}$ orientation and hence they are parallel.

35. Write down the mathematical formula for the torsion of the spherical binormal indicatrix $\bar{C_\mathbf{B}}$ of a space curve $C$ explaining all the symbols used in the formula.
**Answer**:

$$\tau_\mathbf{B} = \frac{\kappa'\tau - \kappa\tau'}{\tau\left(\kappa^2+\tau^2\right)}$$

where $\tau_\mathbf{B}$ is the torsion of $\bar{C_\mathbf{B}}$, $\kappa$ and $\tau$ are the curvature and torsion of $C$, and the prime represents derivative with respect to a natural parameter $s$ of $C$.

36. What "spherical curve" means? give a common example of a spherical curve.
**Answer**: It is a curve that lies completely on the surface of a sphere. The spherical tangent indicatrix $\bar{C_\mathbf{T}}$ is an example of spherical curves.

[58] These formulae are: $\kappa_\mathbf{T}^2 = \frac{\kappa^2+\tau^2}{\kappa^2}$, $\tau_\mathbf{T} = \frac{\kappa'\tau-\kappa\tau'}{\kappa(\kappa^2+\tau^2)}$, $\kappa_\mathbf{B} = \frac{\kappa^2+\tau^2}{\kappa^2}$ and $\tau_\mathbf{B} = \frac{\kappa'\tau-\kappa\tau'}{\tau(\kappa^2+\tau^2)}$.

37. State, mathematically, the sufficient and necessary condition for a curve to be a spherical curve explaining all the symbols involved.
**Answer**: A curve $C$ is spherical *iff*:

$$\frac{R_\kappa}{R_\tau} + \frac{d}{ds}\left(R_\tau \frac{dR_\kappa}{ds}\right) = 0$$

where $R_\kappa$ and $R_\tau$ are the radii of curvature and torsion and $s$ is a natural parameter of $C$.

38. Investigate if the curve represented parametrically by: $\mathbf{r}(t) = (5\cos t, 5\cos t \sin t, 5\sin^2 t)$ is a spherical curve or not.
**Answer**: The obvious criterion for a curve to be spherical is that any point on the curve should have a constant distance $d$ from a given point (because the curve is on the surface of a sphere). The last equation obviously satisfies this criterion (where the given point is the origin of coordinates) because for any $t$ we have:

$$\begin{aligned} d &= \sqrt{(5\cos t - 0)^2 + (5\cos t \sin t - 0)^2 + \left(5\sin^2 t - 0\right)^2} \\ &= 5\sqrt{\cos^2 t + \cos^2 t \sin^2 t + \sin^4 t} \\ &= 5\sqrt{\cos^2 t + \sin^2 t \left(\cos^2 t + \sin^2 t\right)} \\ &= 5\sqrt{\cos^2 t + \sin^2 t} \\ &= 5 \end{aligned}$$

Hence, the curve is spherical.

39. Show that the spherical image of a curve $C(t)$ is a closed curve when the vector that generates the spherical image is a periodic function of $t$ although $C$ may not be periodic. Discuss in this context the helix as an example.
**Answer**: Let $C_i$ be the spherical image of $C$ and $\mathbf{v}(t)$ be the vector that generates the spherical image with a period $\Pi$. Now, since $\mathbf{v}$ is a periodic function of $t$ then $\mathbf{v}(t) = \mathbf{v}(t + n\Pi)$ $(n = \ldots, -1, 0, 1, \ldots)$ which means that all these $\mathbf{v}$'s with periodic values of $t$ will map on the same point of the sphere. So, if $C$ is continuous (which should be) then the spherical image that is generated by $\mathbf{v}$ will repeat itself and hence it should be closed. This argument is valid even if $C$ is not periodic because the closeness of the spherical image depends on the periodicity of the vector that generates this image and not on the periodicity of the original curve $C$.
As discussed earlier (see Exercises 31 and 33), the spherical images (as represented by $\bar{C}_\mathbf{T}$, $\bar{C}_\mathbf{N}$ and $\bar{C}_\mathbf{B}$) of a circular helix (which is not a periodic curve) are circles (which are closed curves) because for helix the vectors $\mathbf{T}, \mathbf{N}, \mathbf{B}$ that generate $\bar{C}_\mathbf{T}$,$\bar{C}_\mathbf{N}$,$\bar{C}_\mathbf{B}$ are periodic (although the helix itself is not).[59] So, the findings of the previous exercises about helix support the theorem of this exercise.

[59] The periodicity of $\mathbf{T}, \mathbf{N}, \mathbf{B}$ can be seen from their expressions which are given in Exercise 31.

40. What is the characteristic feature of geodesic curves?
**Answer**: Their geodesic curvature $\kappa_g$ is identically zero.

41. Give a rigorous mathematical definition of geodesic curve.
**Answer**: If $S : \Omega \to \mathbb{R}^3$ is a space surface defined on a set $\Omega \subseteq \mathbb{R}^2$ and $C(t) : I \to \mathbb{R}^3$ (where $I \subseteq \mathbb{R}$) is a regular curve on $S$, then $C$ is a geodesic curve *iff* $\kappa_g(t) = 0$ on all points $t \in I$ in its domain (with $\kappa_g$ being the geodesic curvature).

42. Give examples of geodesic curves on the following surfaces: plane, sphere and cylinder.
**Answer**: Plane: straight lines. Sphere: arcs of great circles. Cylinder: generating straight lines.

43. On a surface of revolution, what type of curve is necessarily geodesic and what type is potentially geodesic?
**Answer**: The meridians are necessarily geodesic. The arcs of parallels are potentially geodesic.

44. Define geodesic curve variationally using the concepts of calculus of variations.
**Answer**: A curve is geodesic if the first variation of its length is zero.

45. Show that any helix on a circular cylinder is a geodesic curve.
**Answer**: In Exercise 114 of § 4 we found that the unit normal vector for circular cylinder is $\mathbf{n} = (\cos\phi, \sin\phi, 0)$. Also, in Exercise 31 of the present chapter we found that the principal normal vector for circular helix is $\mathbf{N} = (-\cos\phi, -\sin\phi, 0)$ (with $t \equiv \phi$ which is justified by the given parameterization in that exercise). Accordingly, $\mathbf{n}$ and $\mathbf{N}$ have the same orientation and hence the sine of the angle $\theta$ between them is identically zero. So, from Eq. 35 (with $\phi$ in Eq. 35 being replaced by $\theta$ here since $\phi$ is already in use here for another purpose) we have: $\kappa_g = \kappa \sin\theta = 0$, i.e. the geodesic curvature $\kappa_g$ of the cylindrical helix is identically zero and hence the helix is a geodesic curve according to the criterion of Exercises 40 and 41.

46. Outline the relation between the concept of geodesic curve and the concept of curve of shortest distance between two points.
**Answer**: Geodesic curve is usually the curve of shortest distance but it is not necessarily so. In brief, being a shortest path is a sufficient but not necessary condition for being a geodesic.

47. Prove that the following equation is a sufficient and necessary condition for a curve to be geodesic:

$$\frac{d^2u^\alpha}{ds^2} + \Gamma^\alpha_{\beta\gamma}\frac{du^\beta}{ds}\frac{du^\gamma}{ds} = 0 \qquad (57)$$

**Answer**: Referring to Exercises 40 and 41, a surface curve is geodesic *iff* $\kappa_g = 0$ identically. So, all we need for this proof is to show that the condition $\kappa_g = 0$ is equivalent to the condition of Eq. 57. Now, $\kappa_g$ is given by Eq. 31 which can be easily put in the following form:

$$\kappa_g = \sqrt{a}\left[-\left\{\frac{d^2u^1}{ds^2} + \Gamma^1_{11}\left(\frac{du^1}{ds}\right)^2 + 2\Gamma^1_{12}\frac{du^1}{ds}\frac{du^2}{ds} + \Gamma^1_{22}\left(\frac{du^2}{ds}\right)^2\right\}\frac{du^2}{ds} + \left\{\frac{d^2u^2}{ds^2} + \Gamma^2_{11}\left(\frac{du^1}{ds}\right)^2 + 2\Gamma^2_{12}\frac{du^1}{ds}\frac{du^2}{ds} + \Gamma^2_{22}\left(\frac{du^2}{ds}\right)^2\right\}\frac{du^1}{ds}\right] \tag{58}$$

Since $a \neq 0$, $\frac{du^2}{ds} \neq 0$ and $\frac{du^1}{ds} \neq 0$ (see the following note) then the condition $\kappa_g = 0$ is equivalent to the following two conditions:

$$\frac{d^2u^1}{ds^2} + \Gamma^1_{11}\left(\frac{du^1}{ds}\right)^2 + 2\Gamma^1_{12}\frac{du^1}{ds}\frac{du^2}{ds} + \Gamma^1_{22}\left(\frac{du^2}{ds}\right)^2 = 0 \tag{59}$$

$$\frac{d^2u^2}{ds^2} + \Gamma^2_{11}\left(\frac{du^1}{ds}\right)^2 + 2\Gamma^2_{12}\frac{du^1}{ds}\frac{du^2}{ds} + \Gamma^2_{22}\left(\frac{du^2}{ds}\right)^2 = 0 \tag{60}$$

Noting that Eq. 57 is a compact form of the last two equations (using tensor notation), we conclude that the condition $\kappa_g = 0$ (which is a sufficient and necessary condition for being geodesic according to Exercise 41) is equivalent to Eq. 57 and hence Eq. 57 is a sufficient and necessary condition for a curve to be geodesic, as required.
Note: the conditions $\frac{du^2}{ds} \neq 0$ and $\frac{du^1}{ds} \neq 0$ are based on assuming that the curve is not oriented along a coordinate curve. So, if the curve is oriented (wholly or partly) along a coordinate curve then we have only one condition (i.e. Eq. 59 if the curve is oriented along a $u^2$ coordinate curve since $\frac{du^1}{ds} = 0$ in this case, and Eq. 60 if the curve is oriented along a $u^1$ coordinate curve since $\frac{du^2}{ds} = 0$ in this case) and hence these conditions are also necessary and sufficient conditions in these cases although they apply individually.

48. Find an analytical expression representing the geodesic curves on a circular cone using one of its parametric representations.
**Answer**: Referring to Exercise 76 of § 3, we have:

$$\begin{aligned}
\mathbf{E}_1 &= \partial_\rho \mathbf{r} = (\cos\phi,\ \sin\phi,\ c) \\
\mathbf{E}_2 &= \partial_\phi \mathbf{r} = (-\rho\sin\phi,\ \rho\cos\phi,\ 0) \\
E &= \mathbf{E}_1 \cdot \mathbf{E}_1 = 1 + c^2 \\
F &= \mathbf{E}_1 \cdot \mathbf{E}_2 = 0 \\
G &= \mathbf{E}_2 \cdot \mathbf{E}_2 = \rho^2 \\
a &= EG - F^2 = \rho^2\left(1 + c^2\right) \\
\Gamma^1_{11} &= \frac{GE_\rho - 2FF_\rho + FE_\phi}{2a} = 0 \\
\Gamma^2_{11} &= \frac{2EF_\rho - EE_\phi - FE_\rho}{2a} = 0 \\
\Gamma^1_{12} &= \frac{GE_\phi - FG_\rho}{2a} = 0
\end{aligned}$$

$$\begin{aligned}
\Gamma^2_{12} &= \frac{EG_\rho - FE_\phi}{2a} = \frac{1}{\rho} \\
\Gamma^1_{22} &= \frac{2GF_\phi - GG_\rho - FG_\phi}{2a} = \frac{-\rho}{1+c^2} \\
\Gamma^2_{22} &= \frac{EG_\phi - 2FF_\phi + FG_\rho}{2a} = 0
\end{aligned}$$

Hence, Eqs. 59 and 60 become:

$$\begin{aligned}
\frac{d^2\rho}{ds^2} - \frac{\rho}{(1+c^2)}\left(\frac{d\phi}{ds}\right)^2 &= 0 \\
\frac{d^2\phi}{ds^2} + \frac{2}{\rho}\frac{d\rho}{ds}\frac{d\phi}{ds} &= 0
\end{aligned}$$

Now, if we set $\frac{d\phi}{ds} \equiv t$ then from the second equation we get:

$$\begin{aligned}
\frac{dt}{ds} + \frac{2}{\rho}\frac{d\rho}{ds}t &= 0 \\
\frac{1}{t}\frac{dt}{ds} &= -\frac{2}{\rho}\frac{d\rho}{ds} \\
\ln t &= -2\ln\rho + A \\
t &= e^{-2\ln\rho + A} \\
\frac{d\phi}{ds} &= Be^{-2\ln\rho} \\
\frac{d\phi}{ds} &= \frac{B}{\rho^2}
\end{aligned}$$

where $A$ and $B$ are constants. If we note that $s$ is a natural parameter and hence $\left|\frac{d\mathbf{r}}{ds}\right| = 1$, then we should have:

$$\begin{aligned}
\left|\frac{d\mathbf{r}}{ds}\right|^2 &= 1 \\
\frac{d\mathbf{r}}{ds}\cdot\frac{d\mathbf{r}}{ds} &= 1 \\
\left(\frac{\partial\mathbf{r}}{\partial\rho}\frac{d\rho}{ds} + \frac{\partial\mathbf{r}}{\partial\phi}\frac{d\phi}{ds}\right)\cdot\left(\frac{\partial\mathbf{r}}{\partial\rho}\frac{d\rho}{ds} + \frac{\partial\mathbf{r}}{\partial\phi}\frac{d\phi}{ds}\right) &= 1 \\
\left(\mathbf{E}_1\frac{d\rho}{ds} + \mathbf{E}_2\frac{d\phi}{ds}\right)\cdot\left(\mathbf{E}_1\frac{d\rho}{ds} + \mathbf{E}_2\frac{d\phi}{ds}\right) &= 1 \\
E\left(\frac{d\rho}{ds}\right)^2 + 2F\frac{d\rho}{ds}\frac{d\phi}{ds} + G\left(\frac{d\phi}{ds}\right)^2 &= 1 \\
(1+c^2)\left(\frac{d\rho}{ds}\right)^2 + \rho^2\left(\frac{d\phi}{ds}\right)^2 &= 1 \\
(1+c^2)\left(\frac{d\rho}{ds}\right)^2 + \rho^2\left(\frac{B}{\rho^2}\right)^2 &= 1
\end{aligned}$$

$$
\begin{aligned}
\left(1+c^2\right)\left(\frac{d\rho}{ds}\right)^2+\frac{B^2}{\rho^2} &= 1 \\
\frac{d\rho}{ds} &= \left(1+c^2\right)^{-1/2}\left(1-\frac{B^2}{\rho^2}\right)^{1/2} \\
\frac{d\rho}{ds} &= \left(1+c^2\right)^{-1/2}\frac{\sqrt{\rho^2-B^2}}{\rho}
\end{aligned}
$$

Hence, from the equation of $\frac{d\rho}{ds}$ and the equation of $\frac{d\phi}{ds}$ plus the chain rule we obtain:

$$
\begin{aligned}
\frac{d\rho}{d\phi} &= \frac{d\rho}{ds}\frac{ds}{d\phi} \\
\frac{d\rho}{d\phi} &= \left(1+c^2\right)^{-1/2}\frac{\sqrt{\rho^2-B^2}}{\rho}\frac{\rho^2}{B} \\
\frac{d\rho}{d\phi} &= \frac{\rho\sqrt{\rho^2-B^2}}{B\sqrt{1+c^2}} \\
\frac{d\rho}{\rho\sqrt{\rho^2-B^2}} &= \frac{d\phi}{B\sqrt{1+c^2}} \\
\frac{d\rho}{\rho B\sqrt{\left(\rho/B\right)^2-1}} &= \frac{d\phi}{B\sqrt{1+c^2}} \\
\frac{d\rho}{\rho\sqrt{\left(\rho/B\right)^2-1}} &= \frac{d\phi}{\sqrt{1+c^2}} \\
\frac{d(\rho/B)}{(\rho/B)\sqrt{\left(\rho/B\right)^2-1}} &= \frac{d\phi}{\sqrt{1+c^2}} \\
\operatorname{arcsec}\left(\rho/B\right) &= \frac{\phi}{\sqrt{1+c^2}}+D \\
\rho &= B\sec\left[\frac{\phi}{\sqrt{1+c^2}}+D\right]
\end{aligned}
$$

which is the required analytical expression (where $D$ is a constant and noting that $\rho \ge 0$).

49. Outline the concept of geodesic curve on a surface as perceived by a 2D inhabitant of the surface.
**Answer**: From intrinsic perspective (which is the perspective of 2D inhabitant), the geodesic curves are straight lines in the sense that a 2D inhabitant will see them straight because he cannot detect their curvature. This is due to the fact that only the geodesic part of the curvature is an intrinsic property and hence it can be detected by a 2D inhabitant. Therefore, if this part of the curvature vanished the 2D inhabitant will fail to detect any curvature to the curve which is equivalent for him to having a straight line.

50. Prove that all the geodesic curves on a plane are straight lines.
**Answer**: It should be obvious that all plane curves have no normal curvature $\kappa_n$ because their osculating plane (which is their own plane) is perpendicular to the normal vector $\mathbf{n}$ of the surface (which is the plane). Hence, if they have any curvature $\kappa$ it must be purely geodesic curvature $\kappa_g$ (i.e. $\kappa = |\kappa_g|$). Hence, if $\kappa_g = 0$ identically and hence they are geodesic (according to the criterion seen in the previous exercises) then their curvature $\kappa$ should also vanish identically and hence they must be straight (because $\kappa \equiv 0$ is a sufficient and necessary condition for a curve to be straight). On the other hand, if their curvature $\kappa$ vanished identically (and hence they are straight) then $\kappa_g \equiv 0$ and hence they are geodesic. In brief, all the geodesic curves on a plane are straight lines and vice versa, and hence being a straight line is a sufficient and necessary condition for a curve on a plane to be geodesic.

51. Does a geodesic curve necessarily exist between two given points on a space surface, and if it does exist is the geodesic curve necessarily unique? Support your answer with illustrating examples for both cases.
**Answer**: It does not necessarily exist and if it does exist it is not necessarily unique. For example, in the $xy$ plane there is no geodesic connecting two points on a straight line passing through the origin if the origin is excluded. Also, all semi-circular meridians of longitude connecting the two poles of a sphere are geodesics and hence there is an infinite number of geodesics between the two poles.

52. Discuss the following statement and its implications: "Being a shortest path is a sufficient but not necessary condition for being a geodesic curve".
**Answer**: This statement means that all shortest paths connecting two given points on a surface are geodesic curves but not all geodesic curves on a surface are necessarily shortest paths. As explained earlier, the defining property of geodesic curve is that its geodesic curvature $\kappa_g$ vanishes identically and this condition does not necessarily imply being shortest path although it is commonly so. One implication of this is that when we solve geodesic problems we should be systematic in our search by applying the rigorous condition $\kappa_g = 0$ rather than by relying on our intuition although intuition may help in the start.

53. Correct, if necessary, the following statement: "In the neighborhood of a given point $P$ on a surface, there is exactly one geodesic curve that passes through $P$".
**Answer**: We should restrict this by direction and hence the statement should be: "In the neighborhood of a given point $P$ on a surface and for any specific direction, there is exactly one geodesic curve passing through $P$ in that direction".

54. Write down, with full explanation, the differential equations which provide the necessary and sufficient conditions for a naturally parameterized curve on a surface to be geodesic.
**Answer**:

$$\frac{d^2u^1}{ds^2} + \Gamma^1_{11}\left(\frac{du^1}{ds}\right)^2 + 2\Gamma^1_{12}\frac{du^1}{ds}\frac{du^2}{ds} + \Gamma^1_{22}\left(\frac{du^2}{ds}\right)^2 = 0$$

$$\frac{d^2u^2}{ds^2} + \Gamma^2_{11}\left(\frac{du^1}{ds}\right)^2 + 2\Gamma^2_{12}\frac{du^1}{ds}\frac{du^2}{ds} + \Gamma^2_{22}\left(\frac{du^2}{ds}\right)^2 = 0$$

where $u^1$ and $u^2$ are the surface coordinates, $s$ is a natural parameter for the curve, and the indexed $\Gamma$ are the Christoffel symbols for the surface.

55. Correct the following relations which represent the geodesic differential equations for a Monge patch of the form $\mathbf{r}(u,v) = (u, v, f(u,v))$:

$$\begin{aligned} \left(1 + f_u^2 + f_v^2\right) u' + f_u f_{uu}(u')^2 + 2f_u f_{uv} u'v' + f_u f_{vv}(v')^2 &= 0 \\ \left(1 + f_u^2 + f_v^2\right) v'' + f_v f_u(u')^2 + 2f_v f_{uv} u'v' + f_v f_{vv}(v')^2 &= 0 \end{aligned}$$

**Answer**: $u'$ in the first term of the first equation should be $u''$ and $f_u$ in the second term of the second equation should be $f_{uu}$.

56. Why the normal vector $\mathbf{n}$ to the surface at any point on a geodesic curve should be contained in the osculating plane of the curve at that point? Give a clear technical justification.
**Answer**: Because for geodesic curve the geodesic component $\mathbf{K}_g$ of the curvature vector $\mathbf{K}$ vanishes identically (since $\kappa_g = 0$) and hence the curvature vector becomes $\mathbf{K} = \kappa\mathbf{N} = \kappa_n\mathbf{n} = \mathbf{K}_n$. As we see, $\kappa\mathbf{N} = \kappa_n\mathbf{n}$ means that $\mathbf{N}$ and $\mathbf{n}$ are collinear and hence the osculating plane, which by definition contains $\mathbf{N}$, should also contain $\mathbf{n}$.

57. Using Gauss-Bonnet theorem, explain why a surface with negative Gaussian curvature $K$ cannot have a geodesic that intersects itself.
**Answer**: Because if we introduce two artificial corners at two regular points on the curve then we will have a geodesic triangle whose interior angles add up to more than $\pi$ which is not possible on a surface with $K < 0$ according to the Gauss-Bonnet theorem.

58. Using Eq. 57, prove that all meridians of a surface of revolution are geodesic curves.
**Answer**: A surface of revolution can be spatially represented by:

$$\mathbf{r}(\phi, t) = (\rho\cos\phi,\ \rho\sin\phi,\ t)$$

where $\rho$ and $\phi$ are cylindrical coordinates and $\rho$ is a function of $t$, i.e. $\rho = \rho(t)$. Accordingly, the meridians of a surface of revolution are $v$ coordinate curves. Referring to Exercise 47, Eq. 57 is based on Eq. 58 which is a modified form of Eq. 31. As stated in the book, on the $v$ coordinate curves Eq. 31 simplifies to $\kappa_{gv} = -\frac{\sqrt{a}}{G^{3/2}}\Gamma^1_{22}$. Accordingly, we have:

$$\begin{aligned} \mathbf{E}_1 &= \partial_\phi\mathbf{r} = (-\rho\sin\phi,\ \rho\cos\phi,\ 0) \\ \mathbf{E}_2 &= \partial_t\mathbf{r} = (\dot\rho\cos\phi,\ \dot\rho\sin\phi,\ 1) \\ E &= \mathbf{E}_1\cdot\mathbf{E}_1 = \rho^2 \\ F &= \mathbf{E}_1\cdot\mathbf{E}_2 = 0 \\ G &= \mathbf{E}_2\cdot\mathbf{E}_2 = (\dot\rho)^2 + 1 \end{aligned}$$

$$\begin{aligned} a &= EG - F^2 = \rho^2 \left[ (\dot{\rho})^2 + 1 \right] \\ \Gamma_{22}^1 &= \frac{2GF_t - GG_\phi - FG_t}{2a} = 0 \\ \kappa_{gv} &= -\frac{\sqrt{a}}{G^{3/2}} \Gamma_{22}^1 = -\frac{\sqrt{\rho^2 \left[ (\dot{\rho})^2 + 1 \right]}}{\left[ (\dot{\rho})^2 + 1 \right]^{3/2}} \times 0 = 0 \end{aligned}$$

i.e. all meridians of a surface of revolution are geodesic curves, as required.

59. Discuss the following statement in the context of the perception of geodesic curves by a 2D inhabitant: "Geodesic curves on a developable surface become straight lines when the surface is developed into a plane".
**Answer**: Because developable surfaces are intrinsically planes and hence when they become extrinsically planes as well by unrolling them then they should keep their intrinsic attributes; one of which is their geodesic curves. Now, all geodesic curves on plane are straight lines (see Exercise 50), hence the geodesic curves of the developable surface (which map on these geodesics of plane) should become straight lines when the surface is developed into a plane. This may be demonstrated by the perception of a 2D inhabitant of the surface who will fail to observe any difference to the geodesic curve when the surface is developed into a plane and the geodesic curve necessarily becomes a straight line on the plane. In other words, on a developable surface any curve that has no geodesic curvature (and hence it is geodesic) will be perceived by a 2D inhabitant before and after unrolling in the same way and hence this curve which is extrinsically non-straight will become straight after unrolling (since geodesic on plane is necessarily straight) as it keeps its property as geodesic.

60. Give an example of a surface curve whose normal curvature $\kappa_n$ and geodesic curvature $\kappa_g$ are identically zero over the whole curve.
**Answer**: If $\kappa_n$ and $\kappa_g$ are identically zero then the curvature $\kappa$ should also be identically zero. This can be seen from Eq. 30 since we have:

$$\kappa_n^2 + \kappa_g^2 = \kappa^2 \cos^2 \phi + \kappa^2 \sin^2 \phi = \kappa^2$$

and hence if $\kappa_n = \kappa_g = 0$ then $\kappa = 0$. Now, the condition $\kappa = 0$ characterizes straight lines and hence a surface curve whose $\kappa_n$ and $\kappa_g$ are identically zero must be straight.

61. What is the relation between a line of curvature on a surface and the principal directions at the points of the curve?
**Answer**: The tangent to the line of curvature at each point on the line is collinear with one of the principal directions of the surface at that point.

62. Prove that if a plane and a surface are intersecting at a constant angle then their curve of intersection is a line of curvature.
**Answer**: Let the plane and surface be labeled as $\Pi$ and $S$ (and hence their parameters are subscripted accordingly with $\Pi$ and $S$), and the curve of intersection $C$ (which is

common to Π and $S$) be spatially represented by $\mathbf{r}(t)$. Hence, $\mathbf{n}_\Pi \cdot \mathbf{n}_S$ (which is equal to the cosine of the angle of intersection) should be constant (because the angle is constant), that is:

$$\begin{aligned}
\mathbf{n}_\Pi \cdot \mathbf{n}_S &= \text{constant} \\
\frac{d}{dt}(\mathbf{n}_\Pi \cdot \mathbf{n}_S) &= 0 \\
\frac{d\mathbf{n}_\Pi}{dt} \cdot \mathbf{n}_S + \mathbf{n}_\Pi \cdot \frac{d\mathbf{n}_S}{dt} &= 0 \\
\mathbf{n}_\Pi \cdot \frac{d\mathbf{n}_S}{dt} &= 0
\end{aligned}$$

where in line 3 we use the product rule, while in line 4 we use the fact that $\mathbf{n}_\Pi$ is constant and hence $\frac{d\mathbf{n}_\Pi}{dt} = \mathbf{0}$. The last equation (i.e. line 4) means that $\frac{d\mathbf{n}_S}{dt}$ is perpendicular to $\mathbf{n}_\Pi$. However, $\frac{d\mathbf{n}_S}{dt}$ is also perpendicular to $\mathbf{n}_S$ because $\mathbf{n}_S$ is a unit vector.[60] Accordingly, $\frac{d\mathbf{n}_S}{dt}$ is perpendicular to both $\mathbf{n}_\Pi$ and $\mathbf{n}_S$ and hence it must be along the orientation of $\frac{d\mathbf{r}}{dt}$. Therefore, $\frac{d\mathbf{n}_S}{dt}$ can be expressed in terms of $\frac{d\mathbf{r}}{dt}$ and a real number $\kappa_S$ as:

$$\frac{d\mathbf{n}_S}{dt} = -\kappa_S \frac{d\mathbf{r}}{dt}$$

Now, from the Rodrigues curvature theorem (see Exercise 63 of § 4) we conclude that $C$ should be a line of curvature of $S$, as required.[61]

63. Repeat the previous exercise replacing plane with sphere.
**Answer**: If we accept the convention that line of curvature can include umbilical points (see next Exercise), then this is no more than an application of the theorem of Exercise 68 because in this case any curve on a sphere is a line of curvature and hence it should also be a line of curvature for the surface according to the theorem of Exercise 68.[62] The reader is also referred to the footnote of Exercise 63 of § 4 about the applicability of the Rodrigues curvature theorem at umbilical points.

64. Can a line of curvature include umbilical points? Discuss this issue considering the question of allowing more than two principal directions at a point or not.
**Answer**: This issue is related to allowing (or not) of having more than two principal directions at a surface point by following (or not) the convention that any direction at umbilical point is a principal direction. In brief, it may be claimed that the definition

[60] If $\mathbf{v}$ is a unit vector then $\mathbf{v} \cdot \mathbf{v} = 1$ and hence $\frac{d(\mathbf{v}\cdot\mathbf{v})}{dt} = 2\mathbf{v} \cdot \frac{d\mathbf{v}}{dt} = 0$ which means that $\mathbf{v}$ and $\frac{d\mathbf{v}}{dt}$ are orthogonal. This also applies to any vector of constant magnitude since it can be expressed as a constant scalar times a unit vector.

[61] If we follow the convention that any direction on a plane surface is a principal direction then the curve $C$ should also be a line of curvature for the plane but this is a trivial thing that does not need a proof because in this case any curve on plane is a line of curvature.

[62] In fact, this short answer can also be used for the previous question but we preferred detailed answer for diversity and to minimize potential clash with convention.

of line of curvature is based on the existence of two distinct principal directions. If so, then umbilical points should be excluded from the definition of line of curvature due to the absence of two distinct principal directions at these points (i.e. either because there is no principal direction at all or because there is more than two principal directions). However, we think there is no contradiction in including umbilical points over the path of the line of curvature. Accordingly, any curve on plane or sphere should be a line of curvature.

65. Prove that for any sufficiently smooth surface of revolution, the parallels and meridians are lines of curvature.
**Answer**: A surface of revolution can be spatially represented by:

$$\mathbf{r}(\phi, t) = (\rho\cos\phi,\ \rho\sin\phi,\ t)$$

where $\rho$ and $\phi$ are cylindrical coordinates and $\rho$ is a function of $t$, i.e. $\rho = \rho(t)$. Accordingly, we have:

$$\begin{aligned}
\mathbf{E}_1 &= \partial_\phi \mathbf{r} = (-\rho\sin\phi,\ \rho\cos\phi,\ 0) \\
\mathbf{E}_2 &= \partial_t \mathbf{r} = (\dot\rho\cos\phi,\ \dot\rho\sin\phi,\ 1) \\
\partial_t \mathbf{E}_1 &= (-\dot\rho\sin\phi,\ \dot\rho\cos\phi,\ 0) \\
\mathbf{n} &= \frac{\mathbf{E}_1 \times \mathbf{E}_2}{|\mathbf{E}_1 \times \mathbf{E}_2|} = \frac{(\cos\phi,\ \sin\phi,\ -\dot\rho)}{\sqrt{1 + (\dot\rho)^2}} \\
F &= \mathbf{E}_1 \cdot \mathbf{E}_2 = 0 \\
f &= \mathbf{n} \cdot \partial_t \mathbf{E}_1 = 0
\end{aligned}$$

The condition that should be satisfied by a line of curvature is:

$$(eF - fE)\, du^1 du^1 + (eG - gE)\, du^1 du^2 + (fG - gF)\, du^2 du^2 = 0$$

where $(du^1, du^2)$ correspond to $(d\phi, dt)$ in our problem. Now, on parallels we have $dt = 0$ and hence the condition becomes:

$$(eF - fE)\, d\phi d\phi = 0$$

which is an identity because $f = F = 0$. Similarly, on meridians we have $d\phi = 0$ and hence the condition becomes:

$$(fG - gF)\, dt dt = 0$$

which is an identity because $f = F = 0$. So, parallels and meridians of any sufficiently smooth surface of revolution satisfy the condition for line of curvature and hence they are lines of curvature, as required.

66. Show that for a given non-umbilical point[63] $P$ on a sufficiently smooth surface there is a patch that contains $P$ where the directions of the coordinate curves at $P$ are principal directions.

**Answer**: Let have an arbitrary patch $\Pi$ containing $P$ where this patch is represented by $\mathbf{r}(u,v)$ and we label the two principal directions at $P$ with $dv_1/du_1$ and $dv_2/du_2$. This $u,v$ parameterization can be transformed to a $\theta,\phi$ parameterization by a linear transformation:

$$u = du_1\theta + du_2\phi \qquad \text{and} \qquad v = dv_1\theta + dv_2\phi$$

where the existence of such a transformation is justified by the fact that the two principal directions are distinct and hence the directions $dv_1/du_1$ and $dv_2/du_2$ are linearly independent. Accordingly, the patch $\Pi$ will be represented by $\mathbf{r}(\theta,\phi)$ where:

$$\begin{aligned}\mathbf{r}_\theta &= \mathbf{r}_u\frac{\partial u}{\partial\theta} + \mathbf{r}_v\frac{\partial v}{\partial\theta} = \mathbf{r}_u du_1 + \mathbf{r}_v dv_1 \\ \mathbf{r}_\phi &= \mathbf{r}_u\frac{\partial u}{\partial\phi} + \mathbf{r}_v\frac{\partial v}{\partial\phi} = \mathbf{r}_u du_2 + \mathbf{r}_v dv_2\end{aligned}$$

The last equations mean that the $\theta$ and $\phi$ coordinate curves at $P$ are oriented along the principal directions $dv_1/du_1$ and $dv_2/du_2$, as required.

67. Prove Hilbert lemma using the proposal that the coordinate curves on a patch can coincide with the lines of curvature in the neighborhood of a non-umbilical point.

**Answer**: Referring to Exercise 81 of § 4, we note that $P$ is not an umbilical point because $\kappa_1 \neq \kappa_2$ (also see Exercise 84 of § 4). Now, let choose coordinate curves oriented along the lines of curvature in the neighborhood of $P$ (see previous Exercise) and hence from the derivatives of $e/E$ and $g/G$ plus the Codazzi-Mainardi equations we have:

$$\frac{\partial\kappa_1}{\partial v} = \frac{E_v}{2E}(\kappa_2-\kappa_1) \qquad\qquad \frac{\partial\kappa_2}{\partial u} = \frac{G_u}{2G}(\kappa_1-\kappa_2)$$

where $\kappa_1$ and $\kappa_2$ are the principal curvatures at $P$. These relations are justified in the note in the end of this exercise. Now, since $\kappa_1 \neq \kappa_2$ and $\kappa_1$ and $\kappa_2$ are extremum at $P$ and hence $\partial\kappa_1/\partial v = \partial\kappa_2/\partial u = 0$ then these equations imply $E_v = G_u = 0$. On differentiating these equations we get:

$$\frac{\partial^2\kappa_1}{\partial v^2} = \frac{EE_{vv}-E_v^2}{2E^2}(\kappa_2-\kappa_1) + \frac{E_v}{2E}\left(\frac{\partial\kappa_2}{\partial v}-\frac{\partial\kappa_1}{\partial v}\right) = \frac{E_{vv}}{2E}(\kappa_2-\kappa_1) \tag{61}$$

$$\frac{\partial^2\kappa_2}{\partial u^2} = \frac{GG_{uu}-G_u^2}{2G^2}(\kappa_1-\kappa_2) + \frac{G_u}{2G}\left(\frac{\partial\kappa_1}{\partial u}-\frac{\partial\kappa_2}{\partial u}\right) = \frac{G_{uu}}{2G}(\kappa_1-\kappa_2) \tag{62}$$

where the last equalities are obtained by substitution from $E_v = G_u = 0$. Now, from Eq. 61 (noting that $\kappa_1 > \kappa_2$, $E > 0$ and $\kappa_1$ is a local maximum and hence $\partial^2\kappa_1/\partial v^2 \le 0$)

[63] If we accept the convention that every direction at umbilical point is a principal direction then it should be obvious that this theorem also applies to umbilical points with no need for formal proof because any direction is a principal direction and the job is done by any patch.

we get $E_{vv} \geq 0$ at $P$. Similarly, from Eq. 62 (noting that $\kappa_1 > \kappa_2$, $G > 0$ and $\kappa_2$ is a local minimum and hence $\partial^2\kappa_2/\partial u^2 \geq 0$) we get $G_{uu} \geq 0$ at $P$. So, from Eq. 46 (noting that $F = 0$ since the coordinate curves are lines of curvature), we get:

$$\begin{aligned}
K &= \frac{1}{2\sqrt{a}}\left[\partial_u\left(0 - \frac{G_u}{\sqrt{a}}\right) + \partial_v\left(0 - \frac{E_v}{\sqrt{a}} - 0\right)\right] \\
&= \frac{1}{2\sqrt{EG}}\left[-\frac{\sqrt{EG}G_{uu} - \left(\sqrt{EG}\right)_u G_u}{EG} - \frac{\sqrt{EG}E_{vv} - \left(\sqrt{EG}\right)_v E_v}{EG}\right] \\
&= \frac{1}{2\sqrt{EG}}\left[-\frac{\sqrt{EG}G_{uu} - 0}{EG} - \frac{\sqrt{EG}E_{vv} - 0}{EG}\right] \\
&= -\frac{G_{uu} + E_{vv}}{2EG}
\end{aligned}$$

where in line 2 we used the quotient rule and $a = EG$, and in line 3 we used $E_v = G_u = 0$. Now, since $E_{vv} \geq 0$ and $G_{uu} \geq 0$ (as obtained earlier) and $EG > 0$ (since $E = \mathbf{E}_1 \cdot \mathbf{E}_1 = |\mathbf{E}_1|^2 > 0$ and $G = \mathbf{E}_2 \cdot \mathbf{E}_2 = |\mathbf{E}_2|^2 > 0$ ) then $K \leq 0$, as required.
Note: The Codazzi-Mainardi equations are given by:

$$\begin{aligned}
f_u - e_v &= g\Gamma^2_{11} - f\left(\Gamma^2_{12} - \Gamma^1_{11}\right) - e\Gamma^1_{12} \\
g_u - f_v &= g\Gamma^2_{12} - f\left(\Gamma^2_{22} - \Gamma^1_{12}\right) - e\Gamma^1_{22}
\end{aligned}$$

When the coordinate curves are lines of curvature then $f = F = 0$, and hence $f_u = f_v = 0$, $\Gamma^2_{11} = \frac{-E_v}{2G}$, $\Gamma^1_{12} = \frac{E_v}{2E}$, $\Gamma^2_{12} = \frac{G_u}{2G}$ and $\Gamma^1_{22} = \frac{-G_u}{2E}$. The equations then become:

$$\begin{aligned}
-e_v &= g\Gamma^2_{11} - e\Gamma^1_{12} = g\frac{-E_v}{2G} - e\frac{E_v}{2E} = -\frac{E_v}{2}\left(\frac{g}{G} + \frac{e}{E}\right) \\
g_u &= g\Gamma^2_{12} - e\Gamma^1_{22} = g\frac{G_u}{2G} - e\frac{-G_u}{2E} = \frac{G_u}{2}\left(\frac{g}{G} + \frac{e}{E}\right)
\end{aligned}$$

that is:

$$e_v = \frac{E_v}{2}\left(\frac{e}{E} + \frac{g}{G}\right) \qquad\qquad g_u = \frac{G_u}{2}\left(\frac{e}{E} + \frac{g}{G}\right)$$

Hence:

$$\left(\frac{e}{E}\right)_v = \frac{e_vE - eE_v}{E^2} = \frac{\left[\frac{E_v}{2}\left(\frac{e}{E} + \frac{g}{G}\right)\right]E - eE_v}{E^2} = \frac{E_v}{2E}\left(\frac{g}{G} - \frac{e}{E}\right)$$

where in step 1 we used the quotient rule, in step 2 we substituted from the above equation of $e_v$, and in step 3 we simplified the expression. Similarly:

$$\left(\frac{g}{G}\right)_u = \frac{g_uG - gG_u}{G^2} = \frac{\left[\frac{G_u}{2}\left(\frac{e}{E} + \frac{g}{G}\right)\right]G - gG_u}{G^2} = \frac{G_u}{2G}\left(\frac{e}{E} - \frac{g}{G}\right)$$

Now, when the coordinate curves at a given point $P$ are aligned along the principal directions at $P$, the principal curvatures $\kappa_1$ and $\kappa_2$ at $P$ are given by (refer to the book):

$$\kappa_1 = \frac{e}{E} \qquad\qquad \kappa_2 = \frac{g}{G}$$

Hence, the above equations become:

$$\frac{\partial \kappa_1}{\partial v} = \frac{E_v}{2E}(\kappa_2 - \kappa_1) \qquad \frac{\partial \kappa_2}{\partial u} = \frac{G_u}{2G}(\kappa_1 - \kappa_2)$$

68. Prove that if the curve of intersection of two surfaces is a line of curvature for one surface then it is a line of curvature for the other surface when the two surfaces are intersecting each other at a constant angle.
**Answer**: If the two surfaces are labeled as $S_1$ and $S_2$ (and hence their parameters are subscripted accordingly with 1 and 2), and the curve of intersection $C$ (which is common to $S_1$ and $S_2$ and is assumed to be a line of curvature of $S_1$) is spatially represented by $\mathbf{r}(t)$, then $\mathbf{n}_1 \cdot \mathbf{n}_2$ (which is equal to the cosine of the angle of intersection) should be constant (because the angle is constant), that is:

$$\begin{aligned} \mathbf{n}_1 \cdot \mathbf{n}_2 &= \text{constant} \\ \frac{d}{dt}(\mathbf{n}_1 \cdot \mathbf{n}_2) &= 0 \\ \frac{d\mathbf{n}_1}{dt} \cdot \mathbf{n}_2 + \mathbf{n}_1 \cdot \frac{d\mathbf{n}_2}{dt} &= 0 \\ -\kappa_1 \frac{d\mathbf{r}}{dt} \cdot \mathbf{n}_2 + \mathbf{n}_1 \cdot \frac{d\mathbf{n}_2}{dt} &= 0 \\ \mathbf{n}_1 \cdot \frac{d\mathbf{n}_2}{dt} &= 0 \end{aligned}$$

where in line 3 we use the product rule, in line 4 we use the Rodrigues curvature formula (see Exercise 63) since $C$ is a line of curvature of $S_1$, and in line 5 we use the fact that $\frac{d\mathbf{r}}{dt}$ belongs to the tangent space of the surfaces and hence it is perpendicular to $\mathbf{n}_2$. The last equation (i.e. line 5) means that $\frac{d\mathbf{n}_2}{dt}$ is perpendicular to $\mathbf{n}_1$. However, $\frac{d\mathbf{n}_2}{dt}$ is also perpendicular to $\mathbf{n}_2$ because $\mathbf{n}_2$ is a unit vector. Accordingly, $\frac{d\mathbf{n}_2}{dt}$ is along the orientation of $\frac{d\mathbf{r}}{dt}$ and hence $\frac{d\mathbf{n}_2}{dt}$ can be expressed in terms of $\frac{d\mathbf{r}}{dt}$ and a real number $\kappa_2$ as:

$$\frac{d\mathbf{n}_2}{dt} = -\kappa_2 \frac{d\mathbf{r}}{dt}$$

Therefore, from the Rodrigues curvature theorem (see Exercise 63) $C$ is also a line of curvature of $S_2$, as required.

69. Outline the role of geodesic torsion in characterizing the line of curvature employing a mathematical formulation in this context.
**Answer**: The line of curvature is characterized by having identically vanishing geodesic torsion (i.e. $\tau_g = 0$). This can be seen from the equation $\tau_g = (\kappa_1 - \kappa_2)\sin\theta\cos\theta$ (which determines the geodesic torsion $\tau_g$ in terms of the principal curvatures $\kappa_1$ and $\kappa_2$ and the angle $\theta$ between the tangent vector $\mathbf{T}$ to the curve and the first principal direction $\mathbf{d}_1$) because on a line of curvature either $\theta = 0$ or $\theta = \pi/2$ and hence either $\sin\theta = 0$ or $\cos\theta = 0$ resulting in $\tau_g = 0$.

70. Give two examples for the line of curvature on specific types of surface discussing in each case why the described curve should be a line of curvature.
**Answer**: Meridians and parallels of sufficiently smooth surface of revolution are good examples of line of curvature. This was proved formally in Exercise 65. Another example is the generators of circular cone because the generators are straight lines along which the unit normal vector $\mathbf{n}$ to the surface is constant (see Exercise 11 of § 3) and hence the Rodrigues formula (i.e. $d\mathbf{n} = -\kappa \, d\mathbf{r}$) is satisfied identically because $\kappa = 0$ and $d\mathbf{n} = \mathbf{0}$ (see Exercise 63 of § 4). This also applies to the generators of circular cylinder. In fact, the generators of cones and cylinders may also be considered as examples of meridians of surface of revolution.

71. Give the formulae for the principal curvatures, $\kappa_1$ and $\kappa_2$, when the $u^1$ and $u^2$ coordinate curves of a surface patch are lines of curvature.
**Answer**:

$$\kappa_1 = \frac{e}{E} \qquad \qquad \kappa_2 = \frac{g}{G} \tag{63}$$

where $E, G, e, g$ are the coefficients of the first and second fundamental forms.

72. Using tensor notation, state the mathematical condition that should be met by a line of curvature on a space surface with full explanations of all the symbols involved.
**Answer**:

$$\underline{\epsilon}^{\gamma\delta} a_{\alpha\gamma} b_{\beta\delta} du^{\alpha} du^{\beta} = 0$$

where $\underline{\epsilon}^{\gamma\delta}$ is the surface absolute contravariant permutation tensor, $a_{\alpha\gamma}$ and $b_{\beta\delta}$ are the surface covariant metric and covariant curvature tensors respectively, and $u^{\alpha}$ and $u^{\beta}$ are the surface coordinates.

73. Which types of surface should be excluded from the following statement: "The lines of curvature on a surface are represented by an orthogonal net on its spherical image"?
**Answer**: Spheres and minimal surfaces.

74. Prove that on a Monge patch of the form $\mathbf{r}(u, v) = (u, v, f(u, v))$, the coordinate curves are orthogonal family *iff* $f_u f_v = 0$ identically.
**Answer**: The surface coordinate curves are orthogonal family *iff* $F = 0$ identically over the entire surface. This can be seen from the relation $F = \mathbf{E}_1 \cdot \mathbf{E}_2$ because if $F = 0$ then $\mathbf{E}_1$ and $\mathbf{E}_2$ must be orthogonal (and hence the coordinate curves must be orthogonal because $\mathbf{E}_1$ and $\mathbf{E}_2$ are the tangents to the coordinate curves) and if $\mathbf{E}_1$ and $\mathbf{E}_2$ are orthogonal then $F$ must be zero. Now, for a Monge patch of the form $\mathbf{r}(u, v) = (u, v, f(u, v))$ we have $F = f_u f_v$ (see Eq. 28). Hence, the coordinate curves are orthogonal family *iff* $F \equiv f_u f_v = 0$ identically, as required.

75. Give a mathematical condition for a direction on a surface at a given point to be asymptotic.
**Answer**: A direction is asymptotic *iff*:

$$b_{\alpha\beta} du^{\alpha} du^{\beta} = 0$$

where $b_{\alpha\beta}$ is the surface covariant curvature tensor and $u^\alpha$ and $u^\beta$ are the surface coordinates.

76. Prove that the asymptotic directions are bisected by the lines of curvature.
**Answer**: The number of asymptotic directions at elliptic, parabolic and hyperbolic points is 0, 1 and 2 respectively. Hence, this applies only at parabolic and hyperbolic points. Now, at a parabolic point the single asymptotic direction is along one of the principal directions and perpendicular to the other principal direction and hence this asymptotic direction (which can be imagined as a two oppositely-oriented asymptotic directions) is "bisected" by the lines of curvature corresponding to these principal directions in a "special sense".[64] So, in the strict sense we only need to deal with hyperbolic points where we have two distinct asymptotic directions and two distinct principal directions which correspond to two principal curvatures $\kappa_1 > 0$ and $\kappa_2 < 0$. Now, the angle $\theta$ which an asymptotic direction makes with the principal direction of $\kappa_1$ is given by (see Exercise 85):
$$\tan^2\theta = -\frac{\kappa_1}{\kappa_2}$$
So, if $\theta_1$ and $\theta_2$ correspond to the two asymptotic directions then we have:
$$\tan\theta_1 = +\sqrt{-\frac{\kappa_1}{\kappa_2}} \qquad \text{and} \qquad \tan\theta_2 = -\sqrt{-\frac{\kappa_1}{\kappa_2}}$$
i.e. the two asymptotic directions are bisected by the first line of curvature because $\theta_1$ and $\theta_2$ are equal in magnitude and opposite in sign. Now, since the two principal directions are perpendicular then the two asymptotic directions should also be bisected by the second line of curvature.

77. Give a rigorous technical definition of asymptotic line.
**Answer**: A $t$-parameterized surface curve $C(t) : I \to S$ (where $I \subseteq \mathbb{R}$ is an open interval and $S$ represents the surface) is an asymptotic line if at each point $t \in I$ the unit tangent vector $\mathbf{T}$ to $C$ is collinear with an asymptotic direction at that point. We note that asymptotic direction is characterized by having zero normal curvature, i.e. $\kappa_n = 0$.

78. Why the second fundamental form $II_S$ at a point of a surface should vanish in the asymptotic direction?
**Answer**: Because the normal curvature is given by $\kappa_n = II_S/I_S$ and hence if $\kappa_n = 0$ (which is the requirement for the asymptotic direction) then $II_S = 0$.

79. Prove Beltrami-Enneper theorem.
**Answer**: At each point $P$ on an asymptotic line $C$, the tangent plane to the surface coincides with the osculating plane of $C$ at $P$ (see the next exercise). Therefore, along

[64] This could also extend to flat points where all directions are asymptotic and lines of curvature in this "special sense".

$C$ we have $\mathbf{B} = \pm\mathbf{n}$. On taking the derivative of this equation with respect to a natural parameter $s$ of $C$ we get:

$$\mathbf{B}' = -\tau\mathbf{N} = \pm\mathbf{n}'$$

where we used Eq. 8 in the first equality. Accordingly:

$$\mathbf{n}' \cdot \mathbf{n}' = (-\tau\mathbf{N}) \cdot (-\tau\mathbf{N}) = \tau^2\mathbf{N} \cdot \mathbf{N} = \tau^2$$

where the last equality is justified by $\mathbf{N}$ being unit vector. Now, from Exercise 118 of § 3 we have $KI_S - 2H\,II_S + III_S = 0$. However, on asymptotic line we have $II_S = 0$ (see the previous exercise). Also, $III_S \equiv d\mathbf{n} \cdot d\mathbf{n}$ (see Exercise 115 of § 3) and hence:

$$III_S = \frac{d\mathbf{n}}{ds} \cdot \frac{d\mathbf{n}}{ds} (ds)^2 \equiv \mathbf{n}' \cdot \mathbf{n}' (ds)^2 = \tau^2 (ds)^2$$

We also have:

$$I_S = d\mathbf{r} \cdot d\mathbf{r} = \frac{d\mathbf{r}}{ds} \cdot \frac{d\mathbf{r}}{ds} (ds)^2 = \left|\frac{d\mathbf{r}}{ds}\right|^2 (ds)^2 = (ds)^2$$

where equality 1 is based on the definition of $I_S$ while the last equality is justified by the identity $\left|\frac{d\mathbf{r}}{ds}\right| = 1$ since $s$ is a natural parameter. Hence, the above equation (i.e. $KI_S - 2H\,II_S + III_S = 0$) becomes:

$$\begin{aligned} K(ds)^2 + \tau^2 (ds)^2 &= 0 \\ K + \tau^2 &= 0 \\ \tau^2 &= -K \end{aligned}$$

where line 2 is obtained by dividing the two sides by $(ds)^2$ since $ds \neq 0$.

80. One of the characteristic features of asymptotic line is that the tangent plane to the surface at each point of the line coincides with the osculating plane of the line at that point. Why?
**Answer**: Because along the asymptotic line the normal curvature $\kappa_n$ vanishes identically, and hence the curvature vector $\mathbf{K}$ has only a tangential component, i.e. $\mathbf{K} = \mathbf{K}_g = \kappa_g\mathbf{u}$. Therefore, the osculating plane (in which $\mathbf{K}$ lies) becomes tangent to the surface.

81. Show that on a smooth surface with orthogonal families of asymptotic lines the mean curvature is zero.
**Answer**: We simply align the coordinate curves along the asymptotic directions with a proper labeling, and hence we get: $F = 0$ because the coordinate curves are orthogonal, and $e = g = 0$ because the coordinate curves are asymptotic lines (see the upcoming note and the next exercise). Hence:

$$H = \frac{eG - 2fF + gE}{2(EG - F^2)} = \frac{0 - 0 + 0}{2(EG - 0)} = 0$$

Now, since $H$ is invariant under permissible coordinate transformations then this remains valid in other coordinate systems.[65]
Note: when the coordinate curves are orthogonal then $\mathbf{E}_1$ and $\mathbf{E}_2$ are orthogonal and hence we have $F = \mathbf{E}_1 \cdot \mathbf{E}_2 = 0$. Also, the normal curvatures in the directions of the $u$ and $v$ coordinate curves are given respectively by (see the book):

$$\kappa_{nu} = \frac{e}{E} \qquad \text{and} \qquad \kappa_{nv} = \frac{g}{G}$$

Now, since the coordinate curves are asymptotic lines (according to the above alignment) then the normal curvature should vanish identically (i.e. $\kappa_{nu} = \kappa_{nv} = 0$) and hence $e = g = 0$.

82. Justify the following statement: "The necessary and sufficient condition for the $u^1$ and $u^2$ coordinate curves to be asymptotic lines is that $e = 0$ identically on the $u^1$ coordinate curves and $g = 0$ identically on the $u^2$ coordinate curves".
**Answer**: The normal curvature is given by:

$$\kappa_n = \frac{II_S}{I_S} = \frac{e(du)^2 + 2f\,dudv + g(dv)^2}{E(du)^2 + 2F\,dudv + G(dv)^2} \tag{64}$$

Now, on the $u$ coordinate curve $dv = 0$ and hence $\kappa_{nu} = e/E$ while on the $v$ coordinate curve $du = 0$ and hence $\kappa_{nv} = g/G$. Moreover, along asymptotic lines the normal curvature vanishes identically. So, if the $u$ and $v$ coordinate curves are asymptotic lines then we should have $\kappa_{nu} = \kappa_{nv} = 0$ and hence $e = 0$ on the $u$ coordinate curves and $g = 0$ on the $v$ coordinate curves. On the other hand, if $e = 0$ on the $u$ coordinate curves and $g = 0$ on the $v$ coordinate curves then we will have $\kappa_{nu} = 0$ on the $u$ coordinate curves (since the first term in the numerator of Eq. 64 will vanish because $e = 0$ while the last two terms will vanish because $dv = 0$ on the $u$ coordinate curves) and $\kappa_{nv} = 0$ on the $v$ coordinate curves (since the last term in the numerator of Eq. 64 will vanish because $g = 0$ while the first two terms will vanish because $du = 0$ on the $v$ coordinate curves) and hence the $u$ and $v$ coordinate curves are asymptotic lines.

83. Using the equation $b_{11}(du^1)^2 + 2b_{12}\,du^1du^2 + b_{22}(du^2)^2 = 0$, prove that the generators of a circular cylinder are asymptotic lines.
**Answer**: Referring to Exercise 114 of § 4 and noting that $(b_{11}, b_{12}, b_{22}) \equiv (e, f, g)$ and $(u^1, u^2) \equiv (\phi, t)$, the above equation becomes $e\,(d\phi)^2 = 0$ which is an identity since along the generators we have $d\phi = 0$. Hence, the above equation (which sets the necessary and sufficient condition for being asymptotic) is true along the generators of a circular cylinder and therefore the generators are asymptotic lines.

84. According to the theorem of Beltrami-Enneper we have: $\tau^2 = -K$ along an asymptotic line where $\tau$ and $K$ stand for torsion and Gaussian curvature. Does this mean that $\tau$

[65] A possible difference in sign because of surface orientation is ruled out here because $H = 0$ and hence it is indifferent to change of sign.

is imaginary? Fully justify your answer.
**Answer**: No, because asymptotic directions are defined only at points for which $K \leq 0$. Therefore, the square of torsion in the equation $\tau^2 = -K$ is equal to a non-negative value (i.e. $\tau^2 = |K|$) and hence $\tau$ is real.

85. The angle $\theta$ which an asymptotic direction makes with the principal direction of $\kappa_1$ is given by: $\tan^2\theta = -\frac{\kappa_1}{\kappa_2}$ where $\kappa_1$ and $\kappa_2$ are the principal curvatures. Derive this equation.
**Answer**: According to the Euler theorem, the normal curvature $\kappa_n$ in a given direction is given in terms of the principal curvatures, $\kappa_1$ and $\kappa_2$, by $\kappa_n = \kappa_1\cos^2\theta + \kappa_2\sin^2\theta$ where $\theta$ is the angle between the principal direction of $\kappa_1$ and the given direction (see Exercise 46 of 4). Now, in an asymptotic direction $\kappa_n = 0$ and hence:

$$\begin{aligned}\kappa_1\cos^2\theta + \kappa_2\sin^2\theta &= 0\\ \frac{\sin^2\theta}{\cos^2\theta} &= -\frac{\kappa_1}{\kappa_2}\\ \tan^2\theta &= -\frac{\kappa_1}{\kappa_2}\end{aligned}$$

where $\kappa_2 \neq 0$.[66]

86. Classify the asymptotic directions at a given point on a surface as real and distinct, or real and coincident, or conjugate imaginary according to the determinant $b$ of the surface covariant curvature tensor at that point.
**Answer**: The asymptotic directions are:[67]
- Real and distinct if $b < 0$.
- Real and coincident if $b = 0$.
- Conjugate imaginary if $b > 0$.

87. Why the classification in the previous question can also be based on the Gaussian curvature $K$ of the point?
**Answer**: Because $K = b/a$ and $a$ is positive definite and hence the sign of $K$ is the same as the sign of $b$.

88. The classification in the two previous questions is related to the number of asymptotic directions at elliptic, hyperbolic and parabolic points. How?
**Answer**: As seen earlier (refer to Exercise 26 of § 4), $b > 0$ at elliptic points, $b < 0$ at hyperbolic points, and $b = 0$ at parabolic points. Hence, there should be a relation between the number of asymptotic directions (which is determined by the local shape of the surface as elliptic or hyperbolic or parabolic and is related to the sign of $b$) and the classification in the two previous questions which is based on the sign of $b$.

---

[66] Since asymptotic directions are defined only at points for which $K \leq 0$, then $\kappa_1$ and $\kappa_2$ should not have the same sign. So, if $\kappa_1 > 0$ and $\kappa_2 < 0$ then we have no problem. We also have no problem if $\kappa_1 = 0$ and $\kappa_2 < 0$. Yes, we have a problem if $\kappa_1 > 0$ and $\kappa_2 = 0$ in which case we take the reciprocal of this relation.

[67] We note that in this question and answer we consider non-flat points.

89. Justify the following statement: "A straight line contained in a surface is an asymptotic line".
**Answer**: Asymptotic line is characterized by the condition $\kappa_n = \mathbf{n} \cdot \mathbf{K} = 0$ while a straight line is characterized by $\kappa = 0$ (and hence $\mathbf{K} = \kappa \mathbf{N} = \mathbf{0}$). Therefore, the condition of asymptotic line (i.e. $\kappa_n = 0$) is satisfied identically on any straight line and hence a straight line contained in a surface is an asymptotic line.

90. State a mathematical condition for two directions at a given point on a surface to be conjugate directions explaining all the symbols used.
**Answer**: A direction $\delta v/\delta u$ at a point on a sufficiently smooth surface is described as conjugate to the direction $dv/du$ *iff* the relation $d\mathbf{r} \cdot \delta\mathbf{n} = 0$ holds true where $d\mathbf{r} = \mathbf{E}_1 du + \mathbf{E}_2 dv$ and $\delta\mathbf{n} = \partial_u \mathbf{n} \delta u + \partial_v \mathbf{n} \delta v$.

91. What "conjugate families of curves on a surface" means?
**Answer**: Two families of curves on a sufficiently smooth surface are conjugate families if the directions of their tangents at each intersection point of the curves are conjugate directions.

92. Show that at hyperbolic and elliptic points each direction has a unique conjugate direction.
**Answer**: Two directions, $\delta v/\delta u$ and $dv/du$, are conjugate *iff* the condition $e\, du\delta u + f\,(du\delta v + dv\delta u) + g\, dv\delta v = 0$ is satisfied (refer to the book), that is:

$$(e\, du + f dv)\, \delta u + (f\, du + g\, dv)\, \delta v = 0$$

This equation has a unique solution $\delta v/\delta u$ (with $(\delta u)^2 + (\delta v)^2 \neq 0$) *iff* its coefficients do not vanish simultaneously, i.e. $(e\, du + f dv)^2 + (f\, du + g\, dv)^2 \neq 0$, that is:

$$\left(e^2 + f^2\right) (du)^2 + 2\,(ef + fg)\, dudv + \left(f^2 + g^2\right) (dv)^2 \neq 0$$

Hence, now we are looking for a solution $dv/du$ (with $(du)^2 + (dv)^2 \neq 0$) for this "equation" which is a conjugate to the unique solution $\delta v/\delta u$. The condition $(du)^2 + (dv)^2 \neq 0$ requires that at least one of $du$ and $dv$ is not zero. So, we can assume (with no loss of generality since the labeling of $u$ and $v$ is rather arbitrary and hence it can be exchanged) that $du \neq 0$. Hence, if we divide the previous "equation" by $(du)^2$ we get:[68]

$$\left(e^2 + f^2\right) + 2\,(ef + fg)\, \lambda + \left(f^2 + g^2\right) \lambda^2 \neq 0$$

where $\lambda \equiv dv/du$. So, we have a quadratic "equation" in $\lambda$, and hence from the rules of polynomials if the quadratic expression on the left is not zero then its discriminant

[68] In fact, we can do this even with $dv \neq 0$ by dividing by $(dv)^2$ and hence we do not need to restrict ourselves by a specific condition (i.e. $du \neq 0$) which may compromise generality but we followed this way to keep with our conventions and symbolism.

must be negative,[69] that is:

$$\begin{aligned}
4\left(ef+fg\right)^2-4\left(f^2+g^2\right)\left(e^2+f^2\right) &< 0\\
\left(ef+fg\right)^2-\left(f^2+g^2\right)\left(e^2+f^2\right) &< 0\\
\left(e^2f^2+2ef^2g+f^2g^2\right)-\left(e^2f^2+e^2g^2+f^4+f^2g^2\right) &< 0\\
2ef^2g-e^2g^2-f^4 &< 0\\
-\left(e^2g^2-2ef^2g+f^4\right) &< 0\\
-\left(eg-f^2\right)^2 &< 0\\
-b^2 &< 0
\end{aligned}$$

and hence either $b<0$ (i.e. hyperbolic point) or $b>0$ (i.e. elliptic point). So in brief, the necessary and sufficient condition for having a unique conjugate direction to an arbitrary direction at a given point is equivalent to the condition that the point is either hyperbolic or elliptic. This implies that at hyperbolic and elliptic points each direction has a unique conjugate direction, as required.

93. Show that on a surface represented by $\mathbf{r}=\mathbf{r}_1(u)+\mathbf{r}_2(v)$, the coordinate curves are conjugate families of curves.
**Answer**: We have:

$$\begin{aligned}
\mathbf{E}_1 &= \partial_u\mathbf{r}=\partial_u\mathbf{r}_1+\mathbf{0}=\partial_u\mathbf{r}_1\\
\partial_v\mathbf{E}_1 &= \partial_v\left(\partial_u\mathbf{r}_1\right)=\mathbf{0}\\
f &= \mathbf{n}\cdot\partial_v\mathbf{E}_1=\mathbf{n}\cdot\mathbf{0}=0
\end{aligned}$$

Now, since $f=0$ identically is the necessary and sufficient condition for the $u$ and $v$ coordinate curves to be conjugate families (see next exercise) then the coordinate curves on this surface are conjugate families.

94. What is the necessary and sufficient condition for the $u^1$ and $u^2$ coordinate curves on a smooth surface to be conjugate families?
**Answer**: It is $f=0$ identically where $f$ is the coefficient of the second fundamental form.

[69] As it should be known, a quadratic expression with discriminant $\Delta$ has two zeros if $\Delta>0$, one zero if $\Delta=0$, and no zero if $\Delta<0$. So, the condition $\Delta<0$ is equivalent to having no solution to the *equation* $\left(e^2+f^2\right)+2\left(ef+fg\right)\lambda+\left(f^2+g^2\right)\lambda^2=0$ and hence having a solution to the "equation" $\left(e^2+f^2\right)+2\left(ef+fg\right)\lambda+\left(f^2+g^2\right)\lambda^2\neq 0$ which what we are looking for.

# Chapter 6
# Special Surfaces

1. State three features which are specific to plane surfaces.
   **Answer**: The surface curvature tensor vanishes identically over plane surfaces. All points on planes are flat umbilical. At any point on a plane surface, all the directions are asymptotic.

2. Why all directions are asymptotic at any point on a plane surface?
   **Answer**: Because at any point on a plane surface the normal curvature $\kappa_n$ vanishes identically in all directions. This may be concluded from the stated fact in the book that at any point on a plane surface $\kappa_1 = \kappa_2 = 0$ where $\kappa_1$ and $\kappa_2$ are the principal curvatures.

3. Show that having an identically vanishing surface curvature tensor is a necessary and sufficient condition for a surface to be plane.
   **Answer**: Let characterize the plane by its obvious geometric property that all its points are flat umbilical, which is equivalent to having $\kappa_1 = \kappa_2 = 0$ identically over the surface where $\kappa_1$ and $\kappa_2$ are the principal curvatures. Noting that the principal curvatures are normal curvature, the condition $\kappa_1 = \kappa_2 = 0$ identically (i.e. at all points) is equivalent to having $\kappa_n = 0$ identically (i.e. at all points and in all directions). Now, $\kappa_n = II_S/I_S$ and hence $\kappa_n = 0$ identically is equivalent to $II_S = 0$ identically, that is:

$$II_S \equiv b_{11}\left(du\right)^2 + 2b_{12}dudv + b_{22}\left(dv\right)^2 \equiv 0$$

   So, if the surface curvature tensor $b_{\alpha\beta}$ vanishes identically (i.e. $b_{11} = b_{12} = b_{22} = 0$) then $II_S = 0$ identically and hence $\kappa_n = 0$ identically which means that all the points on the surface are flat umbilical, i.e. the surface is plane. On the other hand, if the surface is plane and hence all its points are flat umbilical then $II_S = 0$ identically. Now, since for any direction on the surface $du$ and $dv$ cannot vanish simultaneously (i.e. $\left(du\right)^2 + \left(dv\right)^2 \neq 0$ ) then the condition $II_S = 0$ identically (i.e. at all points and in all directions) cannot be satisfied unless the surface curvature tensor $b_{\alpha\beta}$ vanishes identically (i.e. $b_{11} = b_{12} = b_{22} = 0$). Therefore, having an identically vanishing surface curvature tensor is a necessary and sufficient condition for a surface to be plane, as required.

4. Show that plane is the only connected surface of class $C^2$ whose all points are flat.
   **Answer**: We can construct a proof similar to that of Exercise 65 of § 4. However, it is more efficient to build on our proof in the previous exercise where we showed that having an identically vanishing surface curvature tensor is a necessary and sufficient condition for a surface to be plane. This is because a flat point is characterized by $e = f = g = 0$

(i.e. $b_{11} = b_{12} = b_{22} = 0$) and hence having an identically vanishing surface curvature tensor is equivalent to having flat point over the entire surface. In other words, having flat point over the entire surface is a necessary and sufficient condition for a surface to be plane and hence the plane is the only surface whose all points are flat.[70]

5. Give the tensor notation form of the equation that defines quadratic surfaces.
**Answer**:
$$A_{ij}x^i x^j + B_i x^i + C = 0$$
where the coefficients $A_{ij}$ and $B_i$ are real-valued tensors of rank-2 and rank-1 respectively, $C$ is a real scalar and $i, j = 1, 2, 3$.

6. Name three types of quadratic surface giving their canonical equation in Cartesian coordinates.
**Answer**:
• Ellipsoid: $(x^2/a^2) + (y^2/b^2) + (z^2/c^2) = 1$.
• Hyperboloid of one sheet: $(x^2/a^2) + (y^2/b^2) - (z^2/c^2) = 1$.
• Hyperboloid of two sheets: $(x^2/a^2) - (y^2/b^2) - (z^2/c^2) = 1$.
We note that $x, y, z$ are rectangular Cartesian coordinates and $a, b, c$ are real constants.

7. A surface is represented parametrically by: $\mathbf{r}(u, v) = (u + v, u - v, 2uv)$. Obtain the surface representation in canonical Cartesian form and hence determine its type.
**Answer**: We have $x = u + v$, $y = u - v$ and $z = 2uv$. Hence:
$$\begin{aligned} x + y &= 2u \\ x - y &= 2v \\ 4uv &= (x + y)(x - y) = x^2 - y^2 \\ 2uv &= \frac{x^2 - y^2}{2} \\ z &= \frac{x^2 - y^2}{2} \\ \frac{x^2}{2} - \frac{y^2}{2} - z &= 0 \end{aligned}$$
i.e. the surface is a hyperbolic paraboloid with $a = b = \sqrt{2}$.

8. Make a simple 3D plot of a hyperbolic paraboloid showing the Cartesian coordinate axes and indicating the parameters of the surface. Use a computer graphic package if convenient.
**Answer**: Noting that the hyperbolic paraboloid is given by $(x^2/a^2) - (y^2/b^2) - z = 0$, the plot is shown in Figure 24 where the parameters are $a = b = 1$.

[70] The conditions "connected" and "of class $C^2$" are mainly needed for the proof that follows the style of the proof of Exercise 65 of § 4.

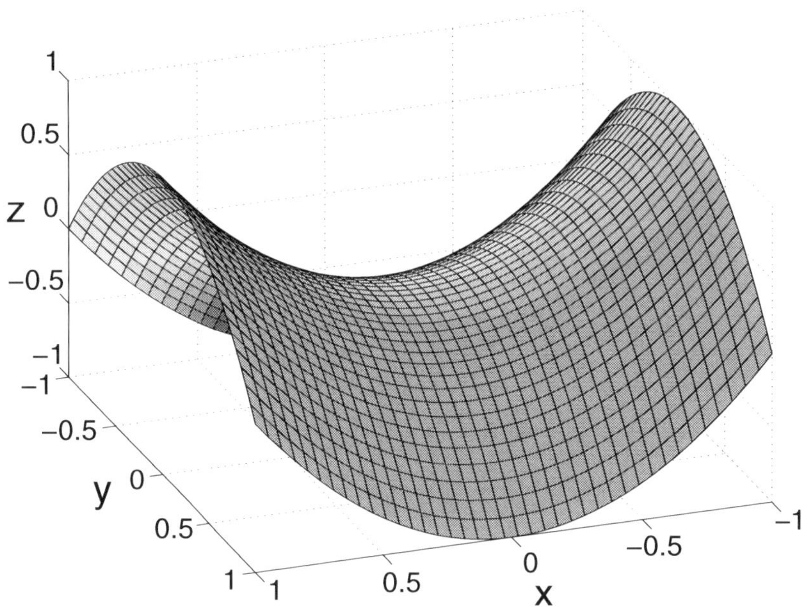


Figure 24: Hyperbolic paraboloid.

9. For which types of quadratic surface the origin of coordinates is not a valid point on the surface according to their canonical forms and why? Does this also apply to the non-canonical forms of these surfaces? Assuming a canonical form, are there any limiting conditions under which the origin can be included in these surfaces?
   **Answer**: Ellipsoid, hyperboloid of one sheet and hyperboloid of two sheets because according to their canonical forms (as given in Exercise 6) the left hand side will vanish on the origin (i.e. $x = y = z = 0$) while the right hand side will not.
   This does not necessarily apply to the non-canonical forms since this sort of mathematical restrictions is not inherent but it depends on the form.
   There are no limiting conditions under which the origin of coordinates can be included in these surfaces according to the given canonical forms.

10. Find the parametric representation of a cylindrical surface whose intersection with the $xy$ plane is given by: $4x^2 + 9y^2 = 1$ and whose central axis is the $z$-axis. What is the type of this surface?
    **Answer**: We have:
    $$4x^2 + 9y^2 = \frac{x^2}{(1/2)^2} + \frac{y^2}{(1/3)^2} = 1$$
    which is an equation of an ellipse in the $xy$ plane with center at the origin and semi-major axis $a = 1/2$ along the $x$-axis and semi-minor axis $b = 1/3$ along the $y$-axis. This ellipse can be parameterized as $(x, y) = (a \cos u, b \sin u)$ where $0 \le u < 2\pi$. Hence, the cylindrical surface can be parameterized spatially as $\mathbf{r}(u, v) = (a \cos u, b \sin u, v)$ where $-\infty < v < \infty$. About its type, it is an elliptical cylinder and a ruled surface (see next exercise) generated by a straight line parallel to the $z$-axis that follows the path of the

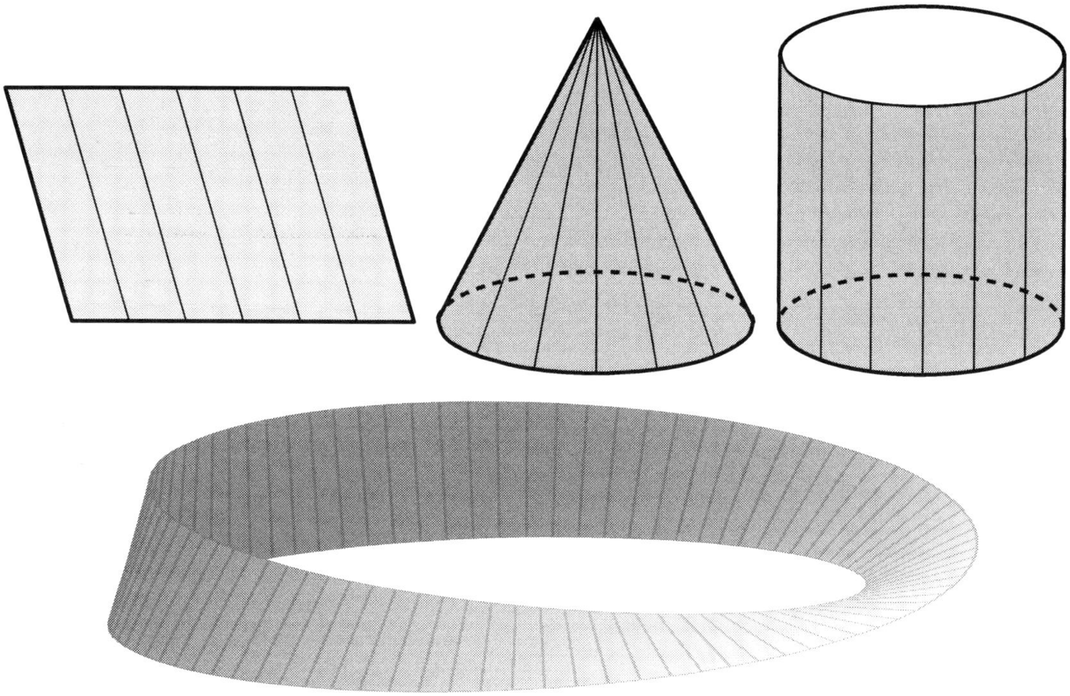

Figure 25: Plane, cone, cylinder and Mobius strip as ruled surfaces where the rulings are shown as straight lines.

aforementioned ellipse.

11. What is "ruled surface"? What is the other name given to this type of surface and why?
**Answer**: A ruled surface is a surface generated by a continuous translational-rotational motion of a straight line in space. Ruled surface is also known as scroll which is justified by the fact that the shape of this type of surface is usually associated with processes like rolling and reeling.

12. Make simple sketches for plane, cone, cylinder and Mobius strip that demonstrate their nature as ruled surfaces.
**Answer**: The sketches are shown in Figure 25.

13. What "doubly-ruled surface" means? Give an example of such a surface with a simple sketch.
**Answer**: Doubly-ruled surface is a ruled surface that can be generated by two different families of lines. Example is hyperbolic paraboloid whose sketch as a doubly-ruled surface is shown in Figure 26.

14. Prove that the tangent plane is constant along a branch of the tangent surface of a space curve.
**Answer**: The idea of the proof is to show that the unit normal vector $\mathbf{n}$ to the tangent surface is independent of the parameter that characterizes the branch (i.e. $k$ in the

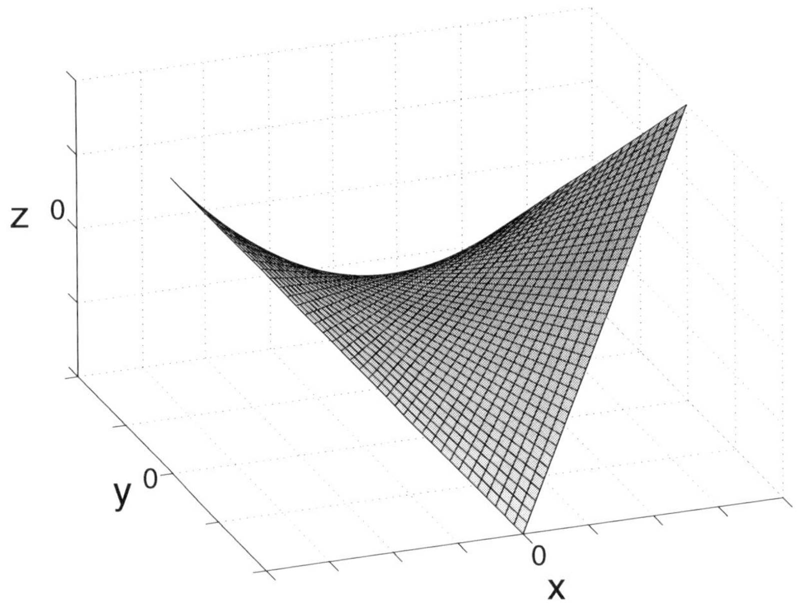


Figure 26: Hyperbolic paraboloid as a doubly-ruled surface where the grid represents the two families of rulings.

following formulations) and since a branch is defined by the variation of $k$ then the independence of $\mathbf{n}$ from $k$ means the constancy of $\mathbf{n}$ along any branch and this implies the constancy of the tangent plane along any branch of the tangent surface. As stated in the book, the equation of the tangent surface $S_T$ to a curve $C$ is given by $\mathbf{r}_T = \mathbf{r}_i + k\mathbf{T}_i$ where $\mathbf{r}_T$ is the spatial representation of an arbitrary point on the tangent surface, $\mathbf{r}_i$ is the spatial representation of a given point on the curve $C$, $k$ is a real variable ($-\infty < k < \infty$), and $\mathbf{T}_i$ is the unit vector tangent to $C$ at $\mathbf{r}_i$. Now, we have two independent parameters that characterize the tangent surface: the parameter of the curve itself (which we can consider as a natural parameter $s$) and the parameter $k$. Since these parameters are mutually independent then the partial derivative of the spatial representation of the tangent surface with respect to these parameters will produce two linearly independent vectors whose cross product defines the normal vector to the tangent surface. Noting that $\mathbf{r}_i$ and $\mathbf{T}_i$ are dependent on $s$ only (and hence they are independent of $k$), we have:

$$\begin{aligned}
\frac{\partial \mathbf{r}_T}{\partial s} &= \frac{d\mathbf{r}_i}{ds} + k\frac{d\mathbf{T}_i}{ds} = \mathbf{T}_i + k\frac{d\mathbf{T}_i}{ds} \\
\frac{\partial \mathbf{r}_T}{\partial k} &= \mathbf{T}_i \\
\frac{\partial \mathbf{r}_T}{\partial s} \times \frac{\partial \mathbf{r}_T}{\partial k} &= \left(\mathbf{T}_i + k\frac{d\mathbf{T}_i}{ds}\right) \times \mathbf{T}_i = k\frac{d\mathbf{T}_i}{ds} \times \mathbf{T}_i \\
\mathbf{n} &= \frac{\partial_s \mathbf{r}_T \times \partial_k \mathbf{r}_T}{|\partial_s \mathbf{r}_T \times \partial_k \mathbf{r}_T|} = \frac{k\,(d\mathbf{T}_i/ds) \times \mathbf{T}_i}{|k|\;|(d\mathbf{T}_i/ds) \times \mathbf{T}_i|} = \pm\frac{(d\mathbf{T}_i/ds) \times \mathbf{T}_i}{|(d\mathbf{T}_i/ds) \times \mathbf{T}_i|}
\end{aligned}$$

As we see, the unit normal vector $\mathbf{n}$ to the tangent surface is independent of $k$ and hence the tangent plane to the tangent surface is constant along any branch of the tangent surface, as required.
We should note that the tangent surface is not regular where $k = 0$ (which corresponds to the curve $C$ itself) and hence the above proof applies separately to the part with $k > 0$ and to the part with $k < 0$ (as indicated already by the $\pm$ sign).

15. Prove that the hyperbolic paraboloid is a doubly-ruled surface.
**Answer**: Hyperbolic paraboloid is given by the following canonical form: $(x^2/a^2) - (y^2/b^2) - z = 0$. However, by proper scaling it can be put in the following normalized form: $x^2 - y^2 - z = 0$.[71] So. all we need to show is that the equation $x^2 - y^2 - z = 0$ represents a doubly-ruled surface. Now, if we parameterize the surface with $u$ and $v$ where $x = u + v$, $y = u - v$ and hence $z = x^2 - y^2 = 4uv$ then the surface is spatially represented by $\mathbf{r}(u, v) = (u+v, u-v, 4uv)$. Accordingly, The $u$ and $v$ coordinate curves are given respectively by:

$$\begin{aligned} \mathbf{r}(u, v_o) &= (u + v_o, u - v_o, 4uv_o) \\ \mathbf{r}(u_o, v) &= (u_o + v, u_o - v, 4u_o v) \end{aligned}$$

where $v_o$ and $u_o$ are constants. As we see, these coordinate curves are straight lines (where the first is parameterized by $u$ while the second is parameterized by $v$) and hence one family of rulings is the family of the $u$ coordinate curves while the other family of rulings is the family of the $v$ coordinate curves, i.e. the hyperbolic paraboloid is a doubly-ruled surface, as required.[72]

16. Show that a helix embedded in a circular cylinder intersects all the generators of the cylinder with constant angle.
**Answer**: Let the axis of symmetry of the circular cylinder be the $z$-axis. Accordingly, all the generators of the cylinder are parallel to the $z$-axis. Now, the tangent vector of a helix whose axis of rotation is the $z$-axis makes a constant angle with the $z$-axis (see the following note) and hence it makes a constant angle with the generators of the cylinder. Therefore, the helix intersects all the generators of the cylinder with constant angle, as required.
Note: a circular helix whose axis of rotation is the $z$-axis is generally parameterized by $\mathbf{r}(t) = (a\cos t, a\sin t, bt)$ where $a$ and $b$ are constants. Hence, its tangent unit vector is $\mathbf{T} = \dot{\mathbf{r}}/|\dot{\mathbf{r}}| = (a^2 + b^2)^{-1/2}(-a\sin t, a\cos t, b)$. Now, the unit vector along the $z$-axis is $\mathbf{k} = (0, 0, 1)$. Therefore, the cosine of the angle between the helix and the $z$-axis is given by the dot product between $\mathbf{T}$ and $\mathbf{k}$, that is $\mathbf{T} \cdot \mathbf{k} = b(a^2 + b^2)^{-1/2}$ which is constant. Hence, the angle between the helix and its axis of rotation (i.e. the $z$-axis in our case) is constant.

[71] This scaling and normalization is not needed for establishing the proof but for making the symbolism easier, clearer and more convenient.

[72] In fact, this is demonstrated graphically in Figure 26 whose plotting technique employs this type of parameterization.

17. Show that all points on the tangent surface of a given space curve are parabolic.
**Answer**: Referring to Exercise 14, we have (see Eqs. 6 and 7 and note that $s$ and $k$ are independent parameters):

$$\begin{aligned}
\partial_s \mathbf{r}_T &= \mathbf{T}_i + k\frac{d\mathbf{T}_i}{ds} = \mathbf{T}_i + k\kappa_i \mathbf{N}_i \\
\partial_k \mathbf{r}_T &= \mathbf{T}_i \\
\mathbf{n} &= \pm\frac{(d\mathbf{T}_i/ds)\times\mathbf{T}_i}{|(d\mathbf{T}_i/ds)\times\mathbf{T}_i|} = \pm\frac{\kappa_i\mathbf{N}_i\times\mathbf{T}_i}{|\kappa_i\mathbf{N}_i\times\mathbf{T}_i|} = \mp\mathbf{B}_i \\
\partial_{ss}\mathbf{r}_T &= \frac{d\mathbf{T}_i}{ds} + k\frac{d\kappa_i}{ds}\mathbf{N}_i + k\kappa_i\frac{d\mathbf{N}_i}{ds} \\
\partial_{sk}\mathbf{r}_T &= \kappa_i\mathbf{N}_i \\
\partial_{kk}\mathbf{r}_T &= \mathbf{0} \\
e &= \mathbf{n}\cdot\partial_{ss}\mathbf{r}_T \\
&= \mp\mathbf{B}_i\cdot\left(\frac{d\mathbf{T}_i}{ds} + k\frac{d\kappa_i}{ds}\mathbf{N}_i + k\kappa_i\frac{d\mathbf{N}_i}{ds}\right) \\
&= \mp\mathbf{B}_i\cdot\left(\kappa_i\mathbf{N}_i + k\frac{d\kappa_i}{ds}\mathbf{N}_i + k\kappa_i\frac{d\mathbf{N}_i}{ds}\right) \\
&= \mp k\kappa_i\mathbf{B}_i\cdot\frac{d\mathbf{N}_i}{ds} \\
&= \mp k\kappa_i\mathbf{B}_i\cdot(\tau_i\mathbf{B}_i - \kappa_i\mathbf{T}_i) \\
&= \mp k\kappa_i\tau_i \\
f &= \mathbf{n}\cdot\partial_{sk}\mathbf{r}_T = \mp\mathbf{B}_i\cdot\kappa_i\mathbf{N}_i = 0 \\
g &= \mathbf{n}\cdot\partial_{kk}\mathbf{r}_T = \mp\mathbf{B}_i\cdot\mathbf{0} = 0
\end{aligned}$$

Hence, at all points of the tangent surface we have $b = eg - f^2 = 0$ and $e^2 + f^2 + g^2 \neq 0$ which means that all points of the tangent surface are parabolic, as required.[73]

18. Justify the following statement: "At any point of a ruled surface the Gaussian curvature is non-positive". From this perspective, discuss singly- and doubly-ruled surfaces.
**Answer**: A straight line on any surface is an asymptotic line (see Exercise 89 of § 5). Hence, at any point of a ruled surface there is an asymptotic line passing through that point. Now, the number of asymptotic directions at elliptic, parabolic and hyperbolic points is 0, 1 and 2 respectively (see Exercise 88 of § 5), while at flat points all directions are asymptotic. Accordingly, all points of ruled surface are non-elliptic. So, if we add to this fact the fact that the Gaussian curvature is positive only at elliptic points then we conclude that at any point of a ruled surface the Gaussian curvature is non-positive,

[73] We should exclude the points with $k = 0$ (which correspond to the curve itself) as well as potentially isolated points at which the curvature or torsion may vanish. However, if we have to include plane curves for example (where the torsion is identically zero), then we should extend the meaning of "parabolic point" by including even flat points and hence we should drop the condition $e^2 + f^2 + g^2 \neq 0$ and keep the condition $b = eg - f^2 = 0$ which is common to parabolic points (in the strict sense) and flat points.

as required.
Regarding the second part of the question, on a doubly-ruled surface we have two asymptotic directions at each point and hence the points of doubly-ruled surface are hyperbolic which means that the Gaussian curvature is negative (i.e. $K < 0$). On the other hand, on a singly-ruled surface we generally have only one asymptotic direction at each point and hence the points of singly-ruled surface are parabolic which means that the Gaussian curvature is zero (i.e. $K = 0$). This also applies to plane surface whose Gaussian curvature is identically zero although it has an infinite number of asymptotic directions at any point.

19. Prove that the tangent plane to a cylinder or a cone is constant along their generators.
**Answer**: This was proved for cone in Exercise 11 of § 3. Regarding cylinder, we refer to Exercise 114 of § 4 where we found that the unit normal vector for circular cylinder is $\mathbf{n} = (\cos\phi, \sin\phi, 0)$. Now, along a generator of a circular cylinder only $t$ varies. But the equation of $\mathbf{n}$ shows that $\mathbf{n}$ is independent of $t$ which means that $\mathbf{n}$ is constant along the generator. In more technical terms, along the generator $\partial_\phi \mathbf{n} = \mathbf{0}$ (since $\phi$ is constant) and $\partial_t \mathbf{n} = \mathbf{0}$ (since $\mathbf{n}$ is independent of $t$) and hence $\mathbf{n}$ is constant. Therefore, the tangent plane (which is determined by $\mathbf{n}$) along any generator of cylinder is constant, as required.

20. Prove that if $P$ is a point on a curve $C$ where $C$ has a tangent surface $S$, then the tangent plane to $S$ along the ruling that passes through $P$ coincides with the osculating plane of $C$ at $P$.
**Answer**: Referring to Exercises 14 and 17, we have $\mathbf{n} = \mp\mathbf{B}_i$ where $\mathbf{n}$ is the unit normal vector to the tangent surface along the ruling and $\mathbf{B}_i$ is the binormal unit vector of the curve at the point corresponding to the ruling. Now, $\mathbf{n}$ is perpendicular to the tangent plane to $S$ along the ruling while $\mathbf{B}_i$ is perpendicular to the osculating plane of $C$ and hence the two planes are parallel. Moreover, both planes pass through the point $P$ and hence they share this point.[74] So, the two planes are identical, i.e. the tangent plane to $S$ along the ruling that passes through $P$ coincides with the osculating plane of $C$ at $P$, as required.

21. Define "developable surface" giving several examples of this type of surface with an explanation of why they are developable surfaces.
**Answer**: Developable surface is a surface that can be flattened into a plane without local distortion. The obvious examples of developable surface are circular cone and circular cylinder since they can be developed into a plane (e.g. by cutting them along one of their generators where they can be flattened into a plane) with no local distortion. More generally, surfaces of revolution with straight generators are developable surfaces for the same reason. Also, parabolic cylinder (refer to the book) is a developable surface since it obviously can be flattened into a plane. Similarly, many other types of ruled surfaces are developable surfaces.

[74] The fact that the tangent plane to $S$ passes through $P$ is implied by the fact that "the tangent plane is constant along a branch of the tangent surface of a space curve" which we proved in Exercise 14.

22. Give an example of a ruled surface which is not developable.
**Answer**: Hyperbolic paraboloid.

23. State the condition for a ruled surface to be developable.
**Answer**: A ruled surface is developable if the tangent plane is constant along every ruling of the surface as it is the case with cones and cylinders.

24. Why any two developable surfaces are equivalent to each other by having the same metric characteristics?
**Answer**: Because any developable surface is intrinsically equivalent to plane (by having the same metric characteristics as plane) and hence it is isometric to plane. Therefore, any two developable surfaces are isometric to each other because they both are isometric to plane (see Exercise 30).

25. Show that the generators of developable surfaces are lines of curvature.
**Answer**: From the previous exercise, developable surfaces are isometric to plane and hence their Gaussian curvature $K$ is identically zero, i.e. at any point on developable surface we have $K = 0$. Now, since $K \equiv \kappa_1 \kappa_2$ then at least one of the kappas must be zero. On the other hand, the generators of developable surfaces are straight lines and hence they are asymptotic lines and hence the normal curvature $\kappa_n$ along the generators vanishes identically. Accordingly, the generators are lines of curvature corresponding to $\kappa_1 = 0$ or $\kappa_2 = 0$.[75]

26. Show that a necessary and sufficient condition for a ruled surface to be developable is that its Gaussian curvature vanishes identically.
**Answer**: If a ruled surface is developable then it is isometric to plane and hence its Gaussian curvature must vanish identically. On the other hand, if the Gaussian curvature of a ruled surface vanishes identically then it is isometric to plane and hence it is developable.[76]

27. What is "isometric mapping"? Give an example of such a mapping between two common types of surface.
**Answer**: Isometric mapping from one surface to another is an injective mapping that preserves distances. An example of isometric mapping is the mapping of a circular cylinder onto a plane.

28. How are two isometric surfaces seen by a 2D inhabitant? Can he distinguish between the two and why?
**Answer**: They are seen identical because the two surfaces are intrinsically identical

---

[75] We note that the principal curvatures, $\kappa_1$ and $\kappa_2$, at a given point on a surface are the maximum and minimum of the normal curvature $\kappa_n$ at that point.

[76] We note that in the second part of the proof we are using the following theorem (which is given in the book): any surface is isometric to the plane *iff* the Gaussian curvature vanishes identically on the surface. In fact, we can create a more technical proof by building on the results of the previous exercises, but we think this proof is enough.

and the 2D inhabitant has access only to the intrinsic properties; therefore he will not be able to distinguish between the two.

29. Prove that isometry is a symmetric relation.
**Answer**: Noting that isometry is based on the property of preserving distances (or lengths), the symmetry of isometry is implied by the symmetry of equality. To be more clear, if an arbitrary curve $C_1$ on a surface $S_1$ is mapped onto a curve $C_2$ on a surface $S_2$ such that the length of $C_1$ and $C_2$ are equal $\left[\text{i.e. } L(C_1) = L(C_2)\right]$ then the reverse mapping that will map $C_2$ onto $C_1$ must also preserve length because if $L(C_1) = L(C_2)$ then $L(C_2) = L(C_1)$. In brief, if $S_1$ is isometric to $S_2$ then $S_2$ must be isometric to $S_1$ and hence isometry is a symmetric relation, as required.

30. Demonstrate symbolically that isometric mapping is an equivalence relation.
**Answer**: If the symbol $\sim$ stands for isometric relation then we have:
• $S_1 \sim S_1$ (i.e. reflective).
• $S_1 \sim S_2 \quad \Longleftrightarrow \quad S_2 \sim S_1$ (i.e. symmetric).
• If $S_1 \sim S_2$ and $S_2 \sim S_3$ then $S_1 \sim S_3$ (i.e. transitive).
Hence, isometric mapping is an equivalence relation.

31. How are two isometric surfaces characterized in terms of their first and second fundamental forms? Provide detailed explanations.
**Answer**: Referring to Exercise 28, isometric surfaces are intrinsically identical and hence any difference between them can only be perceived by an external observer residing in the enveloping space. Accordingly, two isometric surfaces possess identical first fundamental forms and therefore any potential difference between them (which should be based on their extrinsic properties) should be based on the difference between their second fundamental forms.

32. Show that catenoid and helicoid are locally isometric.
**Answer**: The catenoid was parameterized in the book as:

$$\mathbf{r}_c(\theta,\xi) = \left(a\cosh\left(\frac{\xi}{a}\right)\cos\theta,\ a\cosh\left(\frac{\xi}{a}\right)\sin\theta,\ \xi\right)$$

which can be re-parameterized with $(u,v) \equiv (\theta, \xi/a)$ as:

$$\mathbf{r}_c(u,v) = (a\cosh v\,\cos u,\ a\cosh v\,\sin u,\ av)$$

Accordingly, we have for catenoid:

$$\begin{aligned}
\mathbf{E}_1 &= \partial_u \mathbf{r}_c = (-a\cosh v\,\sin u,\ a\cosh v\,\cos u,\ 0) \\
\mathbf{E}_2 &= \partial_v \mathbf{r}_c = (a\sinh v\,\cos u,\ a\sinh v\,\sin u,\ a) \\
E &= \mathbf{E}_1\cdot\mathbf{E}_1 = a^2\cosh^2 v \\
F &= \mathbf{E}_1\cdot\mathbf{E}_2 = 0 \\
G &= \mathbf{E}_2\cdot\mathbf{E}_2 = a^2\cosh^2 v
\end{aligned}$$

Similarly, the helicoid was parameterized in the book as:

$$\mathbf{r}_h(\theta, \xi) = (a\xi \cos\theta,\ a\xi \sin\theta,\ b\theta)$$

which can be re-scaled (such that $b = a$) and re-parameterized with $(U, V) \equiv (\theta, a\xi)$ as:

$$\mathbf{r}_h(U, V) = (V \cos U,\ V \sin U,\ aU)$$

Now, let re-parameterize the helicoid such that $(U, V) \equiv (u, a \sinh v)$.[77] Accordingly, the helicoid will be finally represented by:

$$\mathbf{r}_h(u, v) = (a \sinh v \cos u,\ a \sinh v \sin u,\ au)$$

Accordingly, we have for helicoid:

$$\begin{aligned}
\mathbf{E}_1 &= \partial_u \mathbf{r}_h = (-a \sinh v \sin u,\ a \sinh v \cos u,\ a) \\
\mathbf{E}_2 &= \partial_v \mathbf{r}_h = (a \cosh v \cos u,\ a \cosh v \sin u,\ 0) \\
E &= \mathbf{E}_1 \cdot \mathbf{E}_1 = a^2 \cosh^2 v \\
F &= \mathbf{E}_1 \cdot \mathbf{E}_2 = 0 \\
G &= \mathbf{E}_2 \cdot \mathbf{E}_2 = a^2 \cosh^2 v
\end{aligned}$$

On comparing the coefficients of the first fundamental form of the two surfaces we see that they are identical and hence catenoid and helicoid are locally isometric, as required.

33. Why a surface of revolution is isometric to itself in infinitely many ways? Demonstrate your answer by an example.
**Answer**: Due to the rotational symmetry, the start and end of mapping can be determined in infinitely many ways where each one of these ways corresponds to a rotation of the surface through a given angle around its axis of symmetry. For example, a cone can be mapped isometrically onto itself in a position of $\pi/4$ angle of rotation or $\pi/3$ angle of rotation or $\pi/2$ angle of rotation, as well as infinitely many other possibilities.

34. Define "tangent surface" of a space curve descriptively and mathematically.
**Answer**: The tangent surface of a space curve is a surface generated by the assembly of all the tangent lines to the curve. Hence, the equation of a tangent surface $S_T$ to a curve $C$ is given by:

$$\mathbf{r}_T = \mathbf{r}_i + k\mathbf{T}_i$$

where $\mathbf{r}_T$ represents an arbitrary point on the tangent surface, $\mathbf{r}_i$ represents a given point on the curve $C$, $k$ is a real variable ($-\infty < k < \infty$), and $\mathbf{T}_i$ is the unit tangent vector to $C$ at $\mathbf{r}_i$. The tangent surface is generated by varying $i$ along $C$ and $k$ along the tangent line.

[77] In fact, this parameterization is what is needed to enable the mapping between the catenoid and helicoid. It can be shown that the mapping that is based on this parameterization is smooth and bijective and hence it is legitimate.

35. What is the difference between the "tangent surface" of a curve and the "tangent plane" of a surface? Make detailed comparisons between the two.
**Answer**: The answer is the same as the answer of Exercise 34 of § 3.

36. Derive the equation representing the tangent surface of a space curve represented by: $\mathbf{r}(t) = (t^2, t-2, t^3+5)$.
**Answer**: We have:

$$\begin{aligned} \mathbf{r}_i &= (t^2, t-2, t^3+5) \\ \mathbf{T}_i &= \frac{(2t, 1, 3t^2)}{\sqrt{4t^2+1+9t^4}} \end{aligned}$$

Hence, the equation representing the tangent surface is:

$$\mathbf{r}_T = (t^2, t-2, t^3+5) + \frac{(2kt, k, 3kt^2)}{\sqrt{4t^2+1+9t^4}}$$

37. Why is the tangent surface of a space curve made of two sections? How these sections meet on the curve?
**Answer**: Because there is one section corresponding to $k > 0$ and another section corresponding to $k < 0$. These two sections meet at the curve where the curve makes a border line between the two sections. Hence, the two sections are tangent to each other along the curve which forms a sharp edge between the two.

38. Make a simple 3D sketch of an arbitrary twisted space curve and its tangent surface showing parts of its two sections.
**Answer**: The sketch should be similar to Figure 27.

39. Prove that the curve made by the intersection of the normal plane of a curve $C$ at a given point $P$ with the tangent surface of $C$ has a cusp at $P$.
**Answer**: If $P$ is an arbitrary point on a naturally parameterized curve $C$ then it can be shown (see the upcoming note) that the curve can be represented canonically as:

$$x = s - \frac{\kappa^2}{6}s^3 + \cdots \qquad y = \frac{\kappa}{2}s^2 + \frac{\kappa'}{6}s^3 + \cdots \qquad z = \frac{\kappa\tau}{6}s^3 + \cdots$$

where $\kappa$ and $\tau$ are the curvature and torsion of $C$ at $P$, $s$ is a natural parameter of $C$ and the prime represents derivative with respect to $s$. Now, if we use the leading terms of this canonical form to represent the curve then we have:

$$x = s \qquad y = \frac{\kappa}{2}s^2 \qquad z = \frac{\kappa\tau}{6}s^3$$

and hence the equation of the tangent surface (i.e. $\mathbf{r}_T = \mathbf{r}_i + k\mathbf{T}_i$) becomes:

$$x_T = x_i + kx_i' = s + k$$

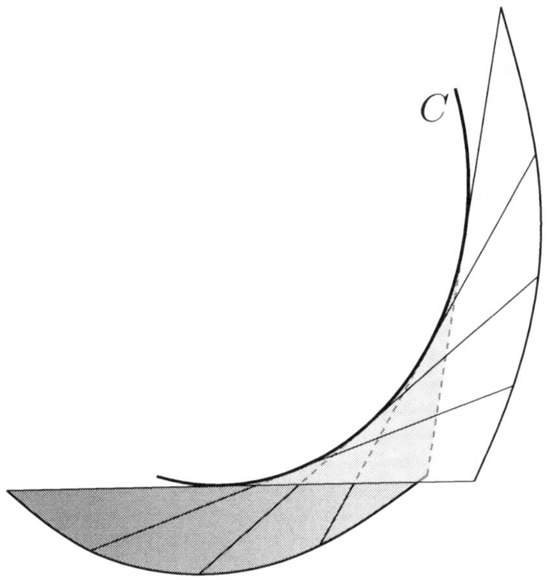


Figure 27: Space curve $C$ and the two sections of its tangent surface which are shaded differently for clarity.

$$\begin{aligned} y_T &= y_i + k y_i' = \frac{\kappa}{2} s^2 + k\kappa s \\ z_T &= z_i + k z_i' = \frac{\kappa\tau}{6} s^3 + k \frac{\kappa\tau}{2} s^2 \end{aligned}$$

where the second term in each of these equations is $k$ multiplied by the derivative of the first term (which represents $x_i, y_i, z_i$) which is justified by the relation $\mathbf{T} = \mathbf{r}'$ (see Eq. 5). Now, the normal plane of $C$ at $P$ is the plane represented by $x_T = 0$ (see the upcoming note). Hence, the intersection of the normal plane with the tangent surface should satisfy the condition: $s + k = 0$, i.e. $k = -s$. So, on substituting from this condition into the above equations of $y_T$ and $z_T$, we see that in the neighborhood of $P$ the intersection can be represented approximately by:

$$\begin{aligned} y_T &= \frac{\kappa}{2} s^2 - \kappa s^2 = -\frac{\kappa}{2} s^2 \\ z_T &= \frac{\kappa\tau}{6} s^3 - \frac{\kappa\tau}{2} s^3 = -\frac{\kappa\tau}{3} s^3 \end{aligned}$$

On combining the last equations (by substituting for $s$ from the second equation into the first), we get:

$$y_T = -\frac{\kappa}{2} \left( -\frac{3 z_T}{\kappa\tau} \right)^{2/3} = -\frac{1}{2} \kappa^{1/3} \left( \frac{3}{\tau} \right)^{2/3} z_T^{2/3}$$

which is an equation of a cusp in the neighborhood of $P$, as required.

Note: let $P$ be a point on an $s$-parameterized sufficiently-smooth curve such that: $s = 0$ at $P$, $P$ is at the origin of coordinates, and the curve is oriented such that $(\mathbf{T}, \mathbf{N}, \mathbf{B})$

vectors of $C$ at $P$ coincide with the coordinate unit basis vectors $(\mathbf{e}_1, \mathbf{e}_2, \mathbf{e}_3)$. So, if $\mathbf{r}(s)$ is the spatial representation of $C$ then we have: $\mathbf{r}(0) = \mathbf{0}$, $\mathbf{r}'(0) = \mathbf{T}(0) = \mathbf{e}_1$ and $\mathbf{r}''(0) = \mathbf{T}'(0) = \kappa\mathbf{N}(0) = \kappa\mathbf{e}_2$ where $\kappa$ is the curvature of $C$ at $P$ and the prime represents derivative with respect to the natural parameter $s$. Moreover, we have:

$$\mathbf{r}''' = \kappa'\mathbf{N} + \kappa\mathbf{N}' = \kappa'\mathbf{N} + \kappa(\tau\mathbf{B} - \kappa\mathbf{T}) = \kappa'\mathbf{N} + \kappa\tau\mathbf{B} - \kappa^2\mathbf{T}$$

where we used the product rule in the first equality and Eq. 7 in the second equality. So, at $P$ we have:

$$\mathbf{r}'''(0) = \kappa'\mathbf{e}_2 + \kappa\tau\mathbf{e}_3 - \kappa^2\mathbf{e}_1$$

Now, on using Taylor series expansion at the origin we obtain:

$$\begin{aligned}
\mathbf{r}(s) &= \mathbf{r}(0) + \mathbf{r}'(0)\,s + \mathbf{r}''(0)\frac{s^2}{2!} + \mathbf{r}'''(0)\frac{s^3}{3!} + \cdots \\
&= \mathbf{0} + \mathbf{e}_1 s + \kappa\mathbf{e}_2\frac{s^2}{2!} + \left(\kappa'\mathbf{e}_2 + \kappa\tau\mathbf{e}_3 - \kappa^2\mathbf{e}_1\right)\frac{s^3}{3!} + \cdots \\
&= \left(s - \frac{\kappa^2}{6}s^3 + \cdots\right)\mathbf{e}_1 + \left(\frac{\kappa}{2}s^2 + \frac{\kappa'}{6}s^3 + \cdots\right)\mathbf{e}_2 + \left(\frac{\kappa\tau}{6}s^3 + \cdots\right)\mathbf{e}_3
\end{aligned}$$

that is:

$$x = s - \frac{\kappa^2}{6}s^3 + \cdots \qquad y = \frac{\kappa}{2}s^2 + \frac{\kappa'}{6}s^3 + \cdots \qquad z = \frac{\kappa\tau}{6}s^3 + \cdots$$

which are the equations of the aforementioned canonical representation of the curve $C$ at point $P$.

40. Write down the equation representing the tangent surface of a space curve explaining all the symbols used in the equation.
**Answer**:

$$\mathbf{r}_T = \mathbf{r}_i + k\mathbf{T}_i$$

where $\mathbf{r}_T$ is an arbitrary point on the tangent surface, $\mathbf{r}_i$ is a given point on the curve, $k$ is a real variable ($-\infty < k < \infty$), and $\mathbf{T}_i$ is the unit tangent vector to the curve at $\mathbf{r}_i$.

41. In the context of tangent surface of a curve, what "branch", "generator", and "edge of regression" mean? Make an attempt to justify these names.
**Answer**: "Branch" and "generator" mean the tangent lines to the curve, while "edge of regression" means the curve itself. The justification is that the tangent lines to the curve are branches of the tangent surface which is made of the union of all these tangent lines. Similarly, the tangent lines to the curve generate the tangent surface since it is made of the collection of all these tangents and hence these tangent lines are aptly described as generators. The curve is described as the edge of regression because it is the border that separates the two sections of the tangent surface (which correspond to $k < 0$ and $k > 0$) and hence the curve is the edge where the tangent surface reverts and retreats to divide itself into these two sections.

42. What is the meaning of "minimal surface"? Give geometric and physical examples of this type of surface.
**Answer**: "Minimal surface" means a surface whose area is minimum compared to the area of any other surface sharing the same boundary. A geometric example of minimal surface shape is a plane surface. A physical example of minimal surface shape is a soap film formed between two coaxial circular rings where surface tension minimizes the area of the film.

43. Check if the surface represented parametrically by: $\mathbf{r}(\theta, \phi) = (\cosh\theta\cos\phi, \cosh\theta\sin\phi, \theta)$ is minimal or not.
**Answer**: This is a catenoid with $a = 1$, $\xi \equiv \theta$ and $\theta \equiv \phi$ (according to the parameterization given in the book). Hence, it is a minimal surface. To verify this, we have:

$$\begin{aligned}
\mathbf{E}_1 &= \partial_\theta \mathbf{r} = (\sinh\theta\cos\phi, \sinh\theta\sin\phi, 1) \\
\mathbf{E}_2 &= \partial_\phi \mathbf{r} = (-\cosh\theta\sin\phi, \cosh\theta\cos\phi, 0) \\
\mathbf{n} &= \frac{\mathbf{E}_1 \times \mathbf{E}_2}{|\mathbf{E}_1 \times \mathbf{E}_2|} = \frac{(-\cos\phi, -\sin\phi, \sinh\theta)}{\cosh\theta} \\
\partial_\theta \mathbf{E}_1 &= (\cosh\theta\cos\phi, \cosh\theta\sin\phi, 0) \\
\partial_\phi \mathbf{E}_1 &= (-\sinh\theta\sin\phi, \sinh\theta\cos\phi, 0) \\
\partial_\phi \mathbf{E}_2 &= (-\cosh\theta\cos\phi, -\cosh\theta\sin\phi, 0) \\
E &= \mathbf{E}_1 \cdot \mathbf{E}_1 = \cosh^2\theta \\
F &= \mathbf{E}_1 \cdot \mathbf{E}_2 = 0 \\
G &= \mathbf{E}_2 \cdot \mathbf{E}_2 = \cosh^2\theta \\
e &= \mathbf{n} \cdot \partial_\theta \mathbf{E}_1 = -1 \\
f &= \mathbf{n} \cdot \partial_\phi \mathbf{E}_1 = 0 \\
g &= \mathbf{n} \cdot \partial_\phi \mathbf{E}_2 = +1 \\
H &= \frac{eG - 2fF + gE}{2(EG - F^2)} = \frac{-\cosh^2\theta - 0 + \cosh^2\theta}{2(EG - F^2)} = 0
\end{aligned}$$

Since a minimal surface is characterized by having an identically vanishing mean curvature (i.e. $H = 0$), then this surface is a minimal surface.

44. Why minimal surfaces are characterized by having identically vanishing mean curvature?
**Answer**: Because the mean curvature $H$ of a surface at a given point $P$ is a measure of the rate of change of area of the surface elements in the neighborhood of $P$.

45. What is the implication of having vanishing mean curvature $H$ on the principal curvatures of the surface at the points where $H$ vanishes?
**Answer**: Since $H = \frac{\kappa_1 + \kappa_2}{2}$ then having zero mean curvature at a surface point means that the principal curvatures, $\kappa_1$ and $\kappa_2$, of the surface at that point have the same magnitude and opposite signs, i.e. $\kappa_2 = -\kappa_1$.

46. Should a surface with orthogonal families of asymptotic lines be a minimal surface? If so, why?
**Answer**: Yes, because it was shown in Exercise 81 of § 5 that such a surface has identically vanishing mean curvature (i.e. $H = 0$) and hence it is a minimal surface according to the criterion that we stated in the book and in the previous exercises.

# Chapter 7
# Tensor Differentiation over Surfaces

1. Summarize the main rules that govern the differentiation of tensor fields over surfaces and compare these rules to those of $n$D spaces ($n > 2$).
**Answer**: The main rules are:
• The sum and product rules of differentiation apply to covariant and absolute differentiation as usual.
• The covariant and absolute derivatives of tensors are tensors.
• The covariant and absolute derivatives of scalars and tensors with invariant basis of higher ranks are the same as the ordinary derivatives.
• The covariant and absolute derivative operators commute with the contraction of indices.
• The covariant and absolute derivatives of the metric, Kronecker and permutation tensors vanish identically in any coordinate system.
As seen above, tensor differentiation over a 2D surface generally follows similar rules to those of tensor differentiation over general $n$D curved spaces. However, there are some exceptions. For example, the covariant derivative of the space basis vectors is identically zero, but this is not the case with the surface basis vectors whose covariant derivatives do not vanish identically (see next exercise).

2. Is there any rule of tensor differentiation that applies to $n$D spaces ($n > 2$) but not to surfaces? If so, which and why? State your answer with a full formal explanation.
**Answer**: As indicated in the previous exercise, the covariant derivative of the space basis vectors, $\mathbf{E}_i$ and $\mathbf{E}^i$, is identically zero, that is:

$$\mathbf{E}_{i;j} = \partial_j \mathbf{E}_i - \Gamma^k_{ij} \mathbf{E}_k = +\Gamma^k_{ij} \mathbf{E}_k - \Gamma^k_{ij} \mathbf{E}_k = \mathbf{0} \tag{65}$$

$$\mathbf{E}^i_{;j} = \partial_j \mathbf{E}^i + \Gamma^i_{kj} \mathbf{E}^k = -\Gamma^i_{kj} \mathbf{E}^k + \Gamma^i_{kj} \mathbf{E}^k = \mathbf{0} \tag{66}$$

but this is not the case with the surface basis vectors, $\mathbf{E}_\alpha$ and $\mathbf{E}^\alpha$, whose covariant derivative do not vanish identically. The reason is that, due to the curvature of the surface in the embedding space, the partial derivatives of the surface basis vectors generally do not lie in the tangent plane and hence the relations $\partial_j \mathbf{E}_i = \Gamma^k_{ij} \mathbf{E}_k$ and $\partial_j \mathbf{E}^i = -\Gamma^i_{kj} \mathbf{E}^k$ which are valid in the enveloping space and are used in Eqs. 65 and 66 (and hence they contributed to setting the covariant derivative to zero), are not valid on the surface.

3. At the points of a smooth surface with geodesic surface coordinates and Cartesian spatial coordinates of a flat embedding 3D space, what happens to the covariant and absolute derivatives of tensor fields?

**Answer**: The covariant and absolute derivatives of tensor fields reduce respectively to the partial and total derivatives.

4. Derive the following identity: $\mathbf{E}_{\alpha;\beta} = \mathbf{E}_{\beta;\alpha}$.
**Answer**:

$$\begin{aligned}
\mathbf{E}_{\alpha;\beta} &= \partial_\beta \mathbf{E}_\alpha - \Gamma^\gamma_{\alpha\beta} \mathbf{E}_\gamma \\
&= \partial_\alpha \mathbf{E}_\beta - \Gamma^\gamma_{\beta\alpha} \mathbf{E}_\gamma \\
&= \mathbf{E}_{\beta;\alpha}
\end{aligned}$$

where line 1 is based on the definition of covariant derivative, line 2 is based on the commutativity of the partial differential operators (i.e. $\partial_\alpha \mathbf{E}_\beta = \partial_\alpha \partial_\beta \mathbf{r} = \partial_\beta \partial_\alpha \mathbf{r} = \partial_\beta \mathbf{E}_\alpha$) and the symmetry of the Christoffel symbols in their paired indices (i.e. $\Gamma^\gamma_{\alpha\beta} = \Gamma^\gamma_{\beta\alpha}$), and line 3 is based on the definition of covariant derivative.

5. Express the identity in the previous exercise in full tensor notation.
**Answer**:

$$x^i_{\alpha;\beta} = x^i_{\beta;\alpha}$$

where $i$ ranges over 1,2,3 while $\alpha$ and $\beta$ range over 1,2.

6. Is $\mathbf{E}_{\alpha;\beta}$ a surface vector or a space vector? Discuss the possible different meanings of these attributes.
**Answer**: $\mathbf{E}_{\alpha;\beta}$ is normal to the surface with no tangential component and hence it is a space vector in this sense, i.e. it exists in the enveloping space but not in the surface (as represented by the tangent space). However, $\mathbf{E}_{\alpha;\beta}$ is also a surface vector since it represents the covariant derivative of the surface basis vector and hence it is an attribute of the surface. So in brief, it is a space vector from the attribute of its existence (as represented by its components) and it is a surface vector from the attribute of belonging and relation to the surface.

7. Explain, in detail, the following equation related to the covariant derivative of space tensors with respect to surface coordinates: $A^i_{\ ;\alpha} = A^i_{\ ;k} x^k_\alpha$.
**Answer**: It means: the covariant derivative $A^i_{\ ;\alpha}$ of a space tensor $A^i$ with respect to a surface coordinate $u^\alpha$ is formed by the inner product of the covariant derivative $A^i_{\ ;k}$ of the tensor $A^i$ with respect to the space coordinates $x^k$ by the tensor $x^k_\alpha$ which is the tensor form of the surface basis vector.

8. Write down the mathematical expression for the covariant derivative of the tensor $A^{\delta j}_{kn\beta}$ with respect to the surface coordinate $u^\gamma$ where the Latin and Greek indices represent space and surface general coordinates.
**Answer**:

$$A^{\delta j}_{kn\beta;\gamma} = \frac{\partial A^{\delta j}_{kn\beta}}{\partial u^\gamma} + \Gamma^\delta_{\omega\gamma} A^{\omega j}_{kn\beta} + \Gamma^j_{bc} A^{\delta b}_{kn\beta} \frac{\partial x^c}{\partial u^\gamma} - \Gamma^a_{kc} A^{\delta j}_{an\beta} \frac{\partial x^c}{\partial u^\gamma} - \Gamma^a_{nc} A^{\delta j}_{ka\beta} \frac{\partial x^c}{\partial u^\gamma} - \Gamma^\omega_{\beta\gamma} A^{\delta j}_{kn\omega}$$

9. Write down the tensor equation for the covariant derivative of the surface basis vector $x_\gamma^m$ with respect to the index $\beta$.
**Answer**:
$$x^m_{\gamma;\beta} = \frac{\partial^2 x^m}{\partial u^\beta \partial u^\gamma} + \Gamma^m_{jk} x^j_\gamma x^k_\beta - \Gamma^\delta_{\gamma\beta} x^m_\delta$$

10. Complete the following equation which involves the tensor $\mathbf{B}$ where the indices represent surface coordinates:
$$B^\alpha_{;\gamma\delta} - B^\alpha_{;\delta\gamma} = ?$$
**Answer**:
$$B^\alpha_{;\gamma\delta} - B^\alpha_{;\delta\gamma} = R^\alpha_{\omega\gamma\delta} B^\omega$$
where $R^\alpha_{\omega\gamma\delta}$ is the Riemann-Christoffel curvature tensor of the second kind for the surface.

11. Give a brief descriptive definition of the absolute differentiation of a tensor field along a curve.
**Answer**: Absolute differentiation of a tensor field along a $t$-parameterized curve in an $n$D space with respect to the parameter $t$ is the inner product of the covariant derivative of the tensor and the tangent vector to the curve.

12. What is the other name given to the absolute differentiation?
**Answer**: It is also known as intrinsic differentiation.

13. Explain the mathematical pattern of absolute differentiation of tensor fields along surface curves illustrating this by an example.
**Answer**: The pattern is that: the covariant derivative of the tensor field is inner multiplied with the tangent vector $du^\beta/dt$ to the curve where the index $\beta$ is contracted with the differentiation index of the covariant derivative. So, if the covariant derivative of a tensor field $A_\alpha$ is $A_{\alpha;\beta}$ then the absolute derivative of this tensor field along a $t$-parameterized surface curve is:
$$\frac{\delta A_\alpha}{\delta t} = A_{\alpha;\beta}\frac{du^\beta}{dt} = \left(\frac{\partial A_\alpha}{\partial u^\beta} - \Gamma^\gamma_{\alpha\beta} A_\gamma\right)\frac{du^\beta}{dt}$$

14. Is the pattern of absolute differentiation of surface tensor fields along surface curves identical to the pattern of absolute differentiation of space tensor fields along space curves? If there is any difference, identify and explain.
**Answer**: The pattern is essentially the same. However, the indices differ since they represent the coordinates of the corresponding spaces (i.e. 2D for surface and 3D for space) and hence they range differently as indicated by the use of Greek and Latin letters.

15. Write, in expanded form, the mathematical equation of the following intrinsic derivative: $\frac{\delta B^k_\gamma}{\delta t}$ where $\mathbf{B}$ is a tensor and $k$ and $\gamma$ are space and surface indices.

**Answer**:

$$\frac{\delta B_{\gamma}^{k}}{\delta t}=\left(\frac{\partial B_{\gamma}^{k}}{\partial u^{\beta}}+\Gamma_{bc}^{k}B_{\gamma}^{b}\frac{\partial x^{c}}{\partial u^{\beta}}-\Gamma_{\gamma\beta}^{\omega}B_{\omega}^{k}\right)\frac{du^{\beta}}{dt}$$

16. What are the covariant and absolute derivatives of space and surface metric, permutation and Kronecker tensors in their covariant, contravariant and mixed forms?
**Answer**: They are zero.

17. Do the operators of covariant and absolute differentiation of space and surface fields commute with the metric tensor involved in an inner or outer product with another tensor? Explain why.
**Answer**: Yes, because the metric is like a constant with respect to covariant and absolute differentiation.

18. Explain the following identity giving detailed definitions of all the symbols and notations involved: $\underline{\epsilon}_{ijk|\gamma}=0$.
**Answer**: The covariant and absolute derivative of the absolute permutation tensor of 3D space with respect to the surface coordinate $u^{\gamma}$ is identically zero. In this identity, $\underline{\epsilon}_{ijk}$ is the absolute permutation tensor of 3D space, the symbol $|$ represents covariant or absolute differentiation, and $\gamma$ is a surface index representing the surface coordinate $u^{\gamma}$.

19. Do the nabla based differential operations apply to the surface tensor fields as to the tensor fields in curved spaces of higher dimensionality?
**Answer**: Yes.

20. What is the Laplacian of a differentiable coordinate-dependent surface scalar field $h$ (i.e $\nabla^2 h$)? Write in your answer the mathematical equation for this operation defining all the symbols used.
**Answer**:

$$\nabla^{2}h=\frac{1}{\sqrt{a}}\partial_{\alpha}\left(\sqrt{a}a^{\alpha\beta}\partial_{\beta}h\right)$$

where $\nabla^2$ symbolizes the Laplacian operator, $a$ is the determinant of the surface covariant metric tensor, $\partial_\alpha$ and $\partial_\beta$ are partial derivative operators with respect to the $\alpha^{th}$ and $\beta^{th}$ surface coordinates, $a^{\alpha\beta}$ is the surface contravariant metric tensor, and $\alpha$ and $\beta$ range over 1,2.

21. Compare the equation in the previous question with the equation of the Laplacian of a scalar field defined over a general $n$D space.
**Answer**: They are identical in form although they usually differ in the symbols used to represent the metric tensor and its determinant (where $g^{ij}$ and $g$ may be used to represent the contravariant metric tensor and the determinant of its covariant form instead of $a^{\alpha\beta}$ and $a$) as well as in the range of the indices (which may be indicated by using Latin indices instead of Greek indices).

# Index

Made in the USA
San Bernardino, CA
30 July 2020

76243876R00131